A FIRST COURSE IN DISCRETE MATHEMATICS

A FIRST COURSE IN

DISCRETE MATHEMATICS

John C. Molluzzo
PACE UNIVERSITY

Fred Buckley
BARUCH COLLEGE, CITY UNIVERSITY OF NEW YORK

WAVELAND
PRESS, INC.
Prospect Heights, Illinois

To Maria, the guys,
and Mui

For information about this book, write or call:
Waveland Press, Inc.
P.O. Box 400
Prospect Heights, Illinois 60070
847/634-0081

Preface

A First Course in Discrete Mathematics was written to fill the need for a text in elementary discrete mathematics which emphasizes computer applications. Our audience is the first or second year computer science or computer information systems student. Most recent books in computer or discrete mathematics were written for mathematically mature students—students who would otherwise major in mathematics. Our book, is intended for non-mathematically oriented students, those who several years ago would have majored in business, education, or the social sciences. We therefore:

1. Introduce definitions and concepts through examples.
2. Provide many examples with solutions.
3. Include Pascal programming examples and exercises.
4. Discuss many applications to computer science and computer information systems.
5. Stress the use of theorems rather than their proofs.

The major goal of this book is to provide the student with a core of mathematical terminology and concepts. The book is not intended to develop mathematical maturity or to teach the reader how to construct mathematical proofs, although it will still require the reader to think.

The mathematical background we assume is high school mathematics through intermediate algebra. We include an appendix on Pascal to support the programming examples and exercises in the text. These examples and exercises enable the student to implement many mathematical concepts on the computer.

Most of the material in this book can be covered in a one-semester, three credit-hour course. The amount of time an instructor spends on any one topic will depend to a large extent on how many exercises are assigned, especially the programming exercises. In a class of well-prepared students, the instructor can include proofs of some of the theorems.

INDEPENDENT STUDY PROJECTS

The case studies at the end of each chapter can be covered in class or assigned as independent study projects. Many texts include case studies that are brief, cryptic, cursory treatments of some highly technical area. Our case studies, on the other hand, are carefully chosen applications developed with the same detail and attention used for all the other sections in the book. The case studies also provide more challenging and lengthy programming problems for the better students.

DEPTH OF COVERAGE

Chapter 1 discusses the binary, octal, and hexadecimal number systems. It also covers the basic terminology of number theory and the algebraic properties of the rational and real numbers. In the case studies these topics are applied to fixed-point and floating-point arithmetic, elementary error analysis, and memory addressing and data encoding.

Chapter 2 deals with the usual topics in elementary set theory and logic, including logical equivalence, valid and fallacious arguments, and the construction of decision tables. The case study discusses sorting and searching as operations on sets, the sequential and binary searches, and the method of successive minima.

Chapter 3 presents the basic ideas of combinatorics—the fundamental counting principle, permutations, combinations, the binomial coefficients, and partitions. The frequently neglected topic of mathematical induction is given careful treatment early in the chapter, and is used in Case Study 3A to show how simple algorithms can be proven correct. In Case Study 3B we analyze the efficiency of several algorithms.

Chapter 4 covers finite probability spaces, including conditional probability, independence, compound experiments, random variables, and expected value. In the case study we apply probability and number theory to develop a pseudorandom number generator and use it to program several simple simulations.

Chapter 5 includes a discussion of relations, functions, and graphs of functions. We also discuss frequently used functions (polynomials, rational functions, exponential and logarithmic functions) and their important properties. Case Study 5A discusses built-in and user-defined functions as well as recursively defined functions. Case Study 5B introduces "big O" and "little o" notation in its discussion of orders of magnitude and the relative efficiency of algorithms. Finally, Case Study 5C discusses encoding functions, in particular modular encodings.

Chapter 6 defines vectors and matrices, and the more useful matrix operations and functions, including inverses and Gauss–Jordan reduction. In Case Study 6A, we discuss the definition and use of stacks, queues, and deques. Case Study 6B discusses the implementation and comparison of the insertion, bubble, and quicksort sorting algorithms.

Chapter 7 includes the definition and basic properties of Boolean algebra and functions. Care is taken to emphasize the relation among sets, logic, and Boolean algebra. Karnaugh maps are treated in detail and used in the minimization of

Boolean functions and expressions. The case study applies the concepts and techniques of the chapter to the design of simple logic networks.

Chapter 8 introduces basic graph and digraph theory. Various classes of graphs are covered including trees, bipartite, and planar graphs. We define several matrices associated with graphs and discuss their uses. We also include discussions of connectivity, and of Eulerian and Hamiltonian graphs. Continuing our discussion of sorting, Case Study 8A introduces and analyzes the heap sort algorithm. Case Study 8B discusses the Critical Path Method.

ELEMENTARY APPLICATIONS TO COMPUTING

The unique feature of this book is the inclusion of many elementary applications to computing. These applications serve several purposes:

1. They motivate students, many of whom believe mathematics to be a necessary evil—like the annual dental checkup.
2. They use mathematics in a meaningful way. We emphasize, through example not preaching, that mathematics is a useful and necessary tool for the practicing professional.
3. They unify much of the mathematical material. Sorting algorithms, for example, appear in one form or another in Chapters 2, 3, 5, 6 and 8.

All applications are self-contained and do not require previous knowledge or experience on the part of the student.

We wish to thank Rich Jones, managing editor, for his encouragement and enthusiasm for this project, and his guidance in its initial stages. We would also like to thank Ray Coleman, our original contact. We are grateful to our editor Heather Bennett and the entire production staff for the wonderful job of keeping the project going and seeing it to a successful conclusion. In addition, we wish to thank the following academic reviewers for their helpful suggestions: David G. Cantor, University of California, Los Angeles; Paul L. Emerick, DeAnza College; Henry A. Etlinger, Rochester Institute of Technology; Ladnor Geissinger, University of North Carolina; Allison Girod, Erie Community College; Robert W. Hayden, Winona State University; Gordon W. Hoagland, Ricks College; Louie C. Huffman, Midwestern State University; Edward C. Polhamus Jr., Danville Community College; John Watson, Arkansas Technical University. Finally, a special thanks to Maria Molluzzo for her careful and critical reading of the original version of the manuscript.

John C. Molluzzo

Fred Buckley

Contents

A FIRST COURSE IN DISCRETE MATHEMATICS

CHAPTER 1

Some computer languages, like Pascal, can store integers in binary (or integer) form. What is binary form? Why can the value of a Pascal integer variable be no larger than MAXINT?

Virtually all large computers can store numbers in floating-point (or real) form. What is floating-point form? If you multiply two large numbers on a scientific calculator, why does the answer appear in E format? What does this format mean?

If you perform a calculation on a calculator or a computer, are you *sure* you will get the correct answer? Try $1 \div 3 \times 3 - 1 =$ on a calculator. What answer do you get? What answer *should* you get?

How does a computer store your name and address?

In this chapter you will learn some of the mathematics needed to understand how character and numeric information are stored and manipulated in a computer. This will help you understand the limitations on both the range of numbers stored and the accuracy of numeric calculations performed on a computer.

1.1 THE BINARY SYSTEM

We look first at something familiar—the decimal, or base 10, number system. What does 3572 mean? The decimal system is a positional system; that is, each digit represents a quantity by virtue of its position in the number. The right-most digit, 2 in our example, represents the number of units in the number. The next digit to the left, 7, gives the number of tens. The next digit gives the number of hundreds,

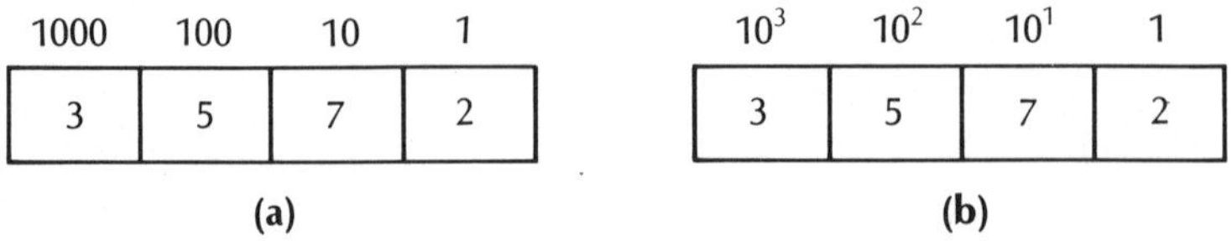

1000	100	10	1
3	5	7	2

(a)

10^3	10^2	10^1	1
3	5	7	2

(b)

Figure 1.1

NUMBER SYSTEMS

the next digit the number of thousands, and so on. Schematically, 3572 may be represented as in Figure 1.1. In so-called **expanded notation,** we have

$$3572 = 3(10)^3 + 5(10)^2 + 7(10)^1 + 2(1)$$

Because numbers are written as sums of powers of ten, our number system is called the **decimal,** or **base 10,** system (the prefix **deci** means "ten").

Each digit position corresponds to a power of ten. The powers $1 = 10^0$, $10 = 10^1$, $100 = 10^2$, $1000 = 10^3$, and so on are called the **place values** of the digit positions. The digits in each position may be any of the usual ten decimal digits, 0, 1, 2, 3, 4, 5, 6, 7, 8, 9. Only ten digits are required in the decimal system. The number thirteen, for example, is not written as thirteen units but as $1(10) + 3$. As a general rule, Note 1.1 applies.

Note 1.1

If the base of a number system is b, then b digits are used to write numbers in that system.

In the binary (or base 2) number system, numbers are written as sums of powers of 2. Thus, in the binary system, the place values for the positions are, from right to left, $1, 2, 2^2, 2^3, 2^4$, and so on. Binary place values are represented schematically as in Figure 1.2.

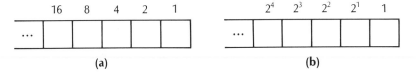

(a) (b)

Figure 1.2

Figure 1.3

By Note 1.1 only two digits are allowed in representing numbers in the binary system. These are the binary digits (or **bits**), 0 and 1. A 1 in a position indicates the presence of the corresponding power of 2, while a 0 indicates its absence. For example, the binary number 110101 is depicted in Figure 1.3. In expanded notation the number is

$$110101 = 1(2)^5 + 1(2)^4 + 0(2)^3 + 1(2)^2 + 0(2)^1 + 1$$
$$= 32 + 16 + 0 + 4 + 0 + 1$$
$$= 53$$

To gain more familiarity with the binary system, let us count in binary. One is simply written as 1. Two is written as 10; that is, $10 = 1(2)^1 + 0$. But wait! How do we distinguish the binary number two, 10, from the decimal number ten, also written 10? To avoid confusion among numbers written in different bases, we shall follow the notational convention described in Note 1.2.

Note 1.2

We indicate the base of a number, *when the base is not 10*, as a subscript to the lower right of that number. If the number is a base 10 number, no subscript is written.

Thus, 10_2 is a binary number, but 10 is a decimal number since no base is indicated.

Continuing, three is written as 11_2, that is, $2 + 1$. Four is $100_2 = 4 + 0 + 0$; five is $101_2 = 4 + 0 + 1$; six is $110_2 = 4 + 2 + 0$; and so on. The numbers from 0 to 16 in both the decimal and binary systems are given in Table 1.1.

EXAMPLE 1.1 _____

Convert 1101101_2 to decimal.

SOLUTION

$$1101101_2 = 1(2)^6 + 1(2)^5 + 0(2)^4 + 1(2)^3 + 1(2)^2 + 0(2)^1 + 1$$
$$= 64 + 32 + 0 + 8 + 4 + 0 + 1$$
$$= 109$$

Decimal	Binary	Decimal	Binary
0	0_2	9	1001_2
1	1_2	10	1010_2
2	10_2	11	1011_2
3	11_2	12	1100_2
4	100_2	13	1101_2
5	101_2	14	1110_2
6	110_2	15	1111_2
7	111_2	16	10000_2
8	1000_2		

Table 1.1

If you understand positional notation, you will find it easy to convert a binary number to its equivalent decimal number. But, how can we convert from a decimal number to its equivalent binary number? For small decimal numbers, it is easy to count in binary up to the required number, as we did earlier. But what if we want to convert 77 to binary? We can do it by counting up to 77 using expanded binary notation, but that will take a long time. Instead we use another, more efficient method: the method of successive short division by 2, described in Note 1.3.

Note 1.3

To convert a decimal number to binary:

1. Divide the number by 2 and save the quotient for the next step. Write the remainder on the side.
2. If the quotient in step 1 is 0, stop.
3. If the quotient in step 1 is not 0, repeat step 1, using the quotient as the new number.

   Obtain the answer by reading the remainders in the inverse order in which they are produced.

Example 1.2 illustrates step by step this procedure of converting a decimal number to binary.

EXAMPLE 1.2

Convert 25 to binary.

SOLUTION

Remainder

Step 1   2 | 25
              ‾‾‾‾
              12   1

Step 2   2 | 25
             2 | 12   1
                ‾‾‾‾
                 6   0

Step 3   2 | 25
             2 | 12   1
             2 | 6   0
                ‾‾‾
                 3   0

Step 4   2 | 25
             2 | 12   1
             2 | 6   0
             2 | 3   0
                ‾‾‾
                 1   1

Step 5   2 | 25
             2 | 12   1 ⎫
             2 | 6   0 ⎪
             2 | 3   0 ⎬ Read up
             2 | 1   1 ⎪
                ‾‾‾
                 0   1 ⎭

Answer: $25 = 11001_2$

In practice, the decimal can be converted to binary more compactly, as follows:

             2 | 25
             2 | 12   1 ⎫
             2 | 6   0 ⎪
             2 | 3   0 ⎬ Read up
             2 | 1   1 ⎪
                ‾‾‾
                 0   1 ⎭

EXAMPLE 1.3

Write a Pascal program to convert a decimal integer into binary.

SOLUTION

```
program decimaltobinary (input, output);
var number, remainder, quotient: integer;
begin
    writeln ('Input a positive integer');
    readln (number);
    while (number <> 0) do
        begin
            quotient := number div 2;
            remainder := number mod 2;
            writeln (remainder);
            number := quotient;
        end;
end.
```

The Pascal DIV operator performs integer division. The value of 7 DIV 3 is 2. The MOD operator gives the integer remainder when one integer is divided by another. The value of 7 MOD 3 is 1.

In the following sections, we examine how the elementary arithmetic operations can be done on binary numbers. Although we operate on numbers in a base other than 10, the methods learned in grade school for adding, subtracting, multiplying, and dividing still work for binary numbers.

Addition

When we learned to add decimal, or base 10, numbers, we memorized an addition table that described how to add any pair of numbers from 0 through 9. This made adding such numbers easier. Likewise, to help us add binary numbers we should know the binary addition table. However, this table, illustrated in Figure 1.4, has only two rows and two columns.

$+$	0_2	1_2
0_2	0_2	1_2
1_2	1_2	10_2

Figure 1.4

The only entry in the table that we must remember in particular is $1_2 + 1_2 = 10_2$; that is, 1 plus 1 is 2. When adding $1_2 + 1_2$ we write the 0 and carry the 1. It is also convenient to remember that $1_2 + 1_2 + 1_2 = 11_2$. Here, we write the 1 and carry a 1.

In the following examples, carries are indicated by circled digits above the appropriate column.

EXAMPLE 1.4

Add: $1010_2 + 1011_2$ (that is, $10 + 11$).

SOLUTION

Reason

Step 1   1010
              1011
           ——————
                 1         $1_2 + 0_2 = 1_2$

                 ①
Step 2   1010
              1011
           ——————
                01         $1_2 + 1_2 = 10_2$
                           Hence, 0 with a carry of 1.

①
Step 3   1010
1011
‾‾‾‾‾
101      $1_2 + 0_2 + 0_2 = 1_2$

① ①
Step 4   1010
1011 $1_2 + 1_2 = 10_2$
‾‾‾‾‾
0101   Hence, 0 with a carry of 1.

① ①
Step 5   1010
1011
‾‾‾‾‾
10101 $1_2 + 0_2 = 1_2$

Answer: $10101_2 \, (= 21)$

When adding more than two binaries, arrange the addition in column form. Then add the bits in a given column in decimal, convert to and write the answer—including the carries—in binary. Then move to the next column, continuing in the same way.

EXAMPLE 1.5 _____

Add: $110101_2 + 100110_2 + 101101_2 + 101110_2$ (that is, $53 + 38 + 45 + 46$).

SOLUTION

Reason

①
Step 1   110101 $0 + 1 + 0 + 1 = 2$. In binary, 2 is 10_2. Write the 0
100110   and carry the 1.
101101
101110
‾‾‾‾‾‾
0

① ①
Step 2   1 1 0 1 0 1 $1 + 0 + 1 + 0 + 1 = 3$. In binary, 3 is 11_2. Write
1 0 0 1 1 0   the 1 and carry the 1.
1 0 1 1 0 1
1 0 1 1 1 0
‾‾‾‾‾‾‾‾
1 0

① ① ①
Step 3   1 1 0 1 0 1   Now $1 + 1 + 1 + 1 + 1 = 5$. In binary, 5 is 101_2.
1 0 0 1 1 0   Write the right-most 1 and carry the middle and
1 0 1 1 0 1   left bits—0 and 1, respectively—to the next two
1 0 1 1 1 0   columns.
‾‾‾‾‾‾‾‾
1 1 0

①
①⒪①①
Step 4   1 1 0 1 0 1
   1 0 0 1 1 0
   1 0 1 1 0 1
   1 0 1 1 1 0
     0 1 1 0

Since $1_2 + 1_2 = 10_2$, the second column from the left now has two carry bits.

①①
①⒪①①
Step 5   1 1 0 1 0 1
   1 0 0 1 1 0
   1 0 1 1 0 1
   1 0 1 1 1 0
    1 0 1 1 0

Now add the two carries and the top bit to get 11_2, and carry a 1 to the left-most column.

  ①①
 ①⒪①①
Step 6 1 1 0 1 0 1
    1 0 0 1 1 0
    1 0 1 1 0 1
    1 0 1 1 1 0
  1 0 1 1 0 1 1 0

Finally, add the left-most column and convert the 5 to 101_2.

Answer: $10110110_2\ (= 182)$

Subtraction

The binary subtraction algorithm uses the same method of borrowing as does the decimal algorithm for subtraction. Consider the decimal subtraction $57 - 39$.

First, a 1 is borrowed from the tens position, reducing the 5 to 4 and making the 7 a 17.

$$
\begin{array}{r}
4 \\
1 \\
\cancel{5}7 \\
-39 \\
\hline
\end{array}
$$

Now 9 is subtracted from 17 and 3 from 4 to obtain

$$
\begin{array}{r}
4 \\
1 \\
\cancel{5}7 \\
-39 \\
\hline
18
\end{array}
$$

EXAMPLE 1.6

Subtract: $1010_2 - 100_2$ (that is, $10 - 4$).

SOLUTION

Reason

Step 1   1010 $0_2 - 0_2 = 0_2$
          100
          ———
            0

Step 2   1010 $1_2 - 0_2 = 1_2$
          100
          ———
         10

         0
         1
Step 3 $\cancel{1}$010   Borrow 1, reducing the left-most 1 to 0. The 0 in the third posi-
         100   tion from the right becomes 10_2 (that is, 2), and $10_2 - 1_2 = 1_2$.
         ———
        110

Answer: 110_2 (= 6)

As in decimal subtraction, binary subtraction is complicated by the need for borrowing from a zero digit. Consider first a decimal example.

EXAMPLE 1.7 _____

Subtract: $5007 - 3948$.

SOLUTION

Reason

        499
         1
Step 1 $\cancel{5}\cancel{0}\cancel{0}7$ You cannot borrow a 1 from a 0, so you must borrow
       3948   from the first nonzero digit to the left, which is 5. The in-
      ————   tervening 0s become 9s, and the 7 becomes 17. Subtract 8
         9   from 17 to get 9.

          499
           1
Steps 2, 3, 4 $\cancel{5}\cancel{0}\cancel{0}7$ Subtract the remaining digits.
           3948
           ————
           1059

EXAMPLE 1.8 _____

Subtract: $101001_2 - 10111_2$ (that is, $41 - 23$).

SOLUTION

Reason

Step 1   101001 $1_2 - 1_2 = 0_2$
         10111
         ————
           0

Step 2 $\overset{\overset{\displaystyle 0\,1}{\scriptstyle 1}}{10\!\!\not1\!\!\not001}$

          10111
          ─────
             10

You must borrow a 1 from the fourth position from the right. The intervening 0 becomes a 1. Subtract $10_2 - 1_2 = 1_2$.

Steps 3, 4 $\overset{\overset{\displaystyle 0\,1}{\scriptstyle 1}}{10\!\!\not1\!\!\not001}$

             10111
             ─────
             0010

$1_2 - 1_2 = 0_2$
$0_2 - 0_2 = 0_2$

Step 5 $\overset{\overset{\displaystyle 0\;\;0\,1}{\scriptstyle 1\;\;\;1}}{\not1\!0\!\!\not1\!\!\not001}$

          10111
          ─────
          10010

Borrow 1, and subtract $10_2 - 1_2 = 1_2$.

Answer: $10010_2 \; (= 18)$

Multiplication

As with subtraction, the algorithm used for decimal multiplication works for binary multiplication as well. The binary multiplication table, illustrated in Figure 1.5, is simple.

$\times$	0_2	1_2
0_2	0_2	0_2
1_2	0_2	1_2

Figure 1.5

EXAMPLE 1.9 _____

Multiply: $1011_2 \times 101_2$ (that is, 11×5).

SOLUTION

                              Reason
Step 1   1011   Multiply the top number by the right-most bit, 1, in the bot-
            101   tom number.
            ────
            1011

(Multiplying a binary number by 1 leaves the number unchanged.)

Step 2 1011   Multiply the top number by the next bit in the bottom number,
     101   0. Shift to the left one position when writing the answer.
    1011
  0000

(Multiplying a binary number by 0 yields 0.)

Step 3   1011   Multiply the top number by the left-most bit in the bottom
     101   number, 1. Shift left one position when writing the answer.
   1011
  0000
 1011

Step 4   1011   Add column by column.
     101
 ①1011
 0000
 1011
 110111

Answer: 110111_2 ($= 55$)

Division

Binary division proceeds in the same manner as decimal division. We consider only those divisions that result in integer quotients.

EXAMPLE 1.10

Divide: $10010_2 \div 11_2$ (that is, $18 \div 3$).

SOLUTION

Reason

Step 1 $11 \overline{)\, 10010}$ 11_2 will not divide into 10_2. Therefore, divide 11_2 into
       $\underline{11}$ 100_2. It divides once.

Step 2 $11 \overline{)\, 10010}$ Subtract 11_2 from 100_2 to get 1_2.
       $\underline{11}$
       1

Step 3 $11 \overline{)\, 10010}$ Bring down the next bit in the dividend.
       $\underline{11}$
      11

$$\text{Step 4} \quad 11 \overline{)\begin{array}{l} 11 \\ 10010 \\ \underline{11} \\ 11 \\ \underline{11} \end{array}}$$

Divide 11_2 into 11_2.

$$\text{Steps 5, 6} \quad 11 \overline{)\begin{array}{l} 11 \\ 10010 \\ \underline{11} \\ 11 \\ \underline{11} \\ 00 \end{array}}$$

Subtract and bring down the next bit in the dividend.

$$\text{Step 7} \quad 11 \overline{)\begin{array}{l} 110 \\ 10010 \\ \underline{11} \\ 11 \\ \underline{11} \\ 00 \\ \underline{00} \\ 0 \end{array}}$$

Divide 11_2 into 0_2.

Answer: $110_2 \ (= 6)$

EXERCISES 1.1

1. Continue the counting started in Table 1.1; that is, count in binary from 17 to 32.
2. Write the following decimal numbers in expanded notation:
   a. 142   **b.** 7045
3. Convert the following decimal numbers to binary:
   a. 38   **b.** 71   **c.** 142   **d.** 279
4. Write the following binary numbers in expanded notation:
   a. 1001101_2 **b.** 1010001_2
5. Convert the following binary numbers to decimal:
   a. 101101_2 **b.** 1010110_2 **c.** 111001101_2
6. Add the following binary numbers:
   a. $1011010_2 + 101110_2$ **b.** $10110_2 + 10100_2 + 11100_2$
   c. $11111_2 + 10001_2$ **d.** $1010101_2 + 1100111_2$
7. Subtract the following binary numbers:
   a. $10110_2 - 101_2$ **b.** $1011011_2 - 110111_2$
   c. $11100_2 - 10011_2$ **d.** $111111_2 - 10101_2$
8. Multiply the following binary numbers:
   a. $1101_2 \times 110_2$ **b.** $1010_2 \times 1010_2$
   c. $11011_2 \times 1101_2$ **d.** $11001_2 \times 1101_2$

9. Divide the following binary numbers:

   a. $1001101_2 \div 111_2$ **b.** $1010100_2 \div 1110_2$
   c. $1111001_2 \div 1011_2$ **d.** $10101010_2 \div 10001_2$

Problems and Projects

10. When the program of Example 1.3 converts a decimal integer to binary, it outputs the bits in the same order as generated by the hand-written solution. Write a program that converts a decimal integer to binary and outputs the binary number in correct form; that is, if 77 is input, the output should be 1001101, written from left to right.

11. It is possible to write binary number fractions. A decimal number fraction, say 23.25, means, in expanded notation, $2(10) + 3 + 2/10 + 5/(10)^2$. Thus, the place values to the right of the decimal point are, from left to right, $1/10$, $1/10^2$, $1/10^3$, and so on. In the same way, binary number fractions are interpreted using binary expanded notation. The binary 101.1101_2 means

$$1(2)^2 + 0(2) + 1 + \frac{1}{2} + \frac{1}{(2)^2} + \frac{0}{(2)^3} + \frac{1}{(2)^4}$$

$$= 4 + 0 + 1 + 0.5 + 0.25 + 0 + 0.0625$$

$$= 5.8125$$

Note that in the binary number 101.1101_2, the dot is called the "binary point."

a. Convert the following binaries to decimal:

   i. 0.001011_2 **ii.** 1101.1_2 **iii.** 1.10101_2

To convert a decimal fraction, say 13.6875, to binary, first convert the integer part to binary ($13 = 1101_2$). To convert the fraction part to binary, invert the procedure that converts decimal integers to binary, as follows: multiply by 2; separate and write alongside the integer part of the answer; multiply the fraction part of the answer by 2; separate and write alongside the integer part of the answer; and so on. The binary equivalent of the decimal fraction is obtained by reading the separated integer parts from top to bottom. To illustrate

$$
\begin{array}{ll}
 & \textit{Integer part} \\
0.6875 \times 2 = 1.3750 & 1 \\
0.375 \times 2 = 0.750 & 0 \\
0.75 \times 2 = 1.50 & 1 \quad \text{Read down} \\
0.5 \times 2 = 1.00 & 1
\end{array}
$$

If the fraction part equals zero, the process stops. If the fraction part never equals zero, then it is a nonterminating binary fraction. In the illustration, $0.6875 = 0.1011_2$. Combining the integer and fractional parts gives $13.6875 = 1101.1011_2$.

b. Convert the following decimals to binary:

   i. 39.40625   **ii.** 3.625   **iii.** 0.8

   Think about the last one!

12. Write a program that converts a binary fraction to decimal.

13. Write a program that converts a decimal fraction to binary.

1.2   THE OCTAL AND HEXADECIMAL SYSTEMS

The Octal System

In the **octal** (or **base 8**) system, numbers are represented positionally as sums of powers of 8. Thus, the place values are, from right to left, 1, 8, 8^2, 8^3, and so on, as illustrated in Figure 1.6.

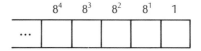

Figure 1.6

Because the base of the octal system is 8, we allow only eight digits—0, 1, 2, 3, 4, 5, 6, 7—as multipliers of the powers of 8. To distinguish these digits from the decimal digits 0 through 7, we shall refer to them as **octits.** Also, we write octal numbers with a subscript of 8.

EXAMPLE 1.11 _____

Count from 0 to 16 in octal.

SOLUTION

See Table 1.2.

Decimal	Octal	Decimal	Octal
0	0_8	8	10_8
1	1_8	9	11_8
2	2_8	10	12_8
3	3_8	11	13_8
4	4_8	12	14_8
5	5_8	13	15_8
6	6_8	14	16_8
7	7_8	15	17_8
		16	20_8

Table 1.2

We convert from octal to decimal by rewriting the octal number using expanded notation in base 8.

EXAMPLE 1.12 —————————————————————————————————————

Convert 726_8 to decimal.

SOLUTION

$$726_8 = 7(8)^2 + 2(8) + 6$$
$$= 7(64) + 2(8) + 6$$
$$= 448 + 16 + 6$$
$$= 470$$

—————————————————————————————————————

We convert from decimal to octal by a procedure similar to the one we used for converting from decimal to binary (see Note 1.3). Instead of successively dividing by 2, however, we successively divide by 8.

EXAMPLE 1.13 —————————————————————————————————————

Convert 795 to octal.

SOLUTION

$$
\begin{array}{r|r}
 & \text{\textit{Remainders}} \\
8 \mid 795 & \\
8 \underline{\mid 99} & 3 \\
8 \underline{\mid 12} & 3 \\
8 \underline{\mid 1} & 4 \\
\underline{0} & 1
\end{array}
\quad \text{Read up}
$$

Answer: $795 = 1433_8$

—————————————————————————————————————

The Hexadecimal System

In the **hexadecimal** (or **base 16**) number system, numbers are represented as sums of powers of 16. Thus, the place values are, from right to left, 1, 16, 16^2, 16^3, and so on, as shown in Figure 1.7.

Because the base of the hexadecimal system is 16, we use the usual decimal digits 0, 1, 2, 3, 4, 5, 6, 7, 8, 9 as the first ten hexadecimal digits and the letters A, B, C, D, E, F as the next six hexadecimal digits, corresponding to the integers 10 through 15. To distinguish hexadecimal digits from decimal digits, we shall call the hexadecimal digits **hexits**. Also, we write hexadecimal numbers with the subscript 16.

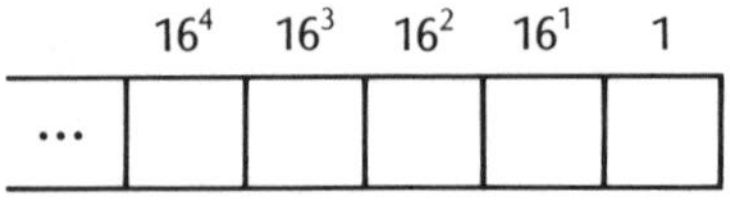

$$16^4 \quad 16^3 \quad 16^2 \quad 16^1 \quad 1$$

Figure 1.7

EXAMPLE 1.14

Count from 0 to 26 in hexadecimal.

SOLUTION

See Table 1.3.

Decimal	Hexadecimal	Decimal	Hexadecimal
0	0_{16}	13	D_{16}
1	1_{16}	14	E_{16}
2	2_{16}	15	F_{16}
3	3_{16}	16	10_{16}
4	4_{16}	17	11_{16}
5	5_{16}	18	12_{16}
6	6_{16}	19	13_{16}
7	7_{16}	20	14_{16}
8	8_{16}	21	15_{16}
9	9_{16}	22	16_{16}
10	A_{16}	23	17_{16}
11	B_{16}	24	18_{16}
12	C_{16}	25	19_{16}
		26	$1A_{16}$

Table 1.3

To convert from hexadecimal to decimal, rewrite the number using expanded notation in base 16.

EXAMPLE 1.15

Convert $90FC_{16}$ to decimal.

SOLUTION

$$
\begin{aligned}
90FC_{16} &= 9(16)^3 + 0(16)^2 + F(16) + C \\
&= 9(16)^3 + 0(16)^2 + 15(16) + 12 \\
&= 9(4096) + 0(256) + 15(16) + 12 \\
&= 36864 + 0 + 240 + 12 \\
&= 37116
\end{aligned}
$$

To convert a decimal number to the equivalent hexadecimal number, we successively divide by 16 (see Note 1.3). Care must be taken to write the remainders correctly. If a remainder is 10, 11, 12, 13, 14, or 15, it must be written as A, B, C, D, E, or F, respectively.

EXAMPLE 1.16 ───

Convert 429 to hexadecimal.

SOLUTION

$$
\begin{array}{r|r}
 & \textit{Remainders} \\
16 & 429 \\
16 & 26 \\
16 & 1 \\
\hline
 & 0
\end{array}
\quad
\begin{array}{l}
\text{D} \\
\text{A} \\
1
\end{array}
\Bigg\} \text{Read up}
$$

Answer: $429 = 1AD_{16}$

Octal and Hexadecimal Arithmetic

As with binary arithmetic, both octal arithmetic and hexadecimal arithmetic follow the same procedures as those for decimal arithmetic. To facilitate hand calculations, carry out the intermediate results mentally in decimal, and then convert to the appropriate base. We illustrate with several examples.

EXAMPLE 1.17 ───

Add: $526_8 + 345_8$ (that is, $342 + 229$).

SOLUTION

Reason

Step 1
$$
\begin{array}{r}
① \\
526 \\
345 \\
\hline
3
\end{array}
$$
In decimal, $6 + 5 = 11$. Converted to octal, $11 = 13_8$. Write the 3 and carry 1.

Step 2
$$
\begin{array}{r}
① \\
526 \\
345 \\
\hline
73
\end{array}
$$
$1 + 2 + 4 = 7$

Step 3
$$
\begin{array}{r}
① \\
526 \\
345 \\
\hline
1073
\end{array}
$$
In decimal, $5 + 3 = 8$. Converted to octal, $8 = 10_8$

Answer: $1073_8 \ (= 571)$

EXAMPLE 1.18

Multiply: $423_8 \times 215_8$ (that is, 275×141).

SOLUTION

Reason

Step 1

①
a. 423
    215
    ‾‾‾
      7

$5 \times 3 = 15$ in decimal. Converted to octal, $15 = 17_8$. Write the 7 and carry 1.

①①
b. 4 23
    2 15
    ‾‾‾‾
      37

$5 \times 2 = 10$, plus the 1 carry, is 11 in decimal and $11 = 13_8$. Write the 3 and carry 1.

① ①
c. 4   23
    2   15
    ‾‾‾‾‾
    25 37

$5 \times 4 = 20$, plus the 1 carry, is 21 in decimal and $21 = 25_8$. Write the 25.

Step 2   423
             215
             ‾‾‾‾
            2537
             423

Similar reasoning as in step 1.

Step 3   423
               215
               ‾‾‾‾
              2537
               423
              1046

Similar reasoning as in step 1.

Step 4   423
               215
               ‾‾‾‾
              2537
               423
              1046
              ‾‾‾‾‾‾
             113567

Add column by column.

Answer: $113567_8 \; (= 38775)$

EXAMPLE 1.19

Add: $A37_{16} + CAB_{16}$ (that is, $2615 + 3243$).

SOLUTION

Reason

Step 1 ①
A37 $7_{16} + B_{16} = 7 + 11 = 18 = 12_{16}$. Write the 2 and carry 1.
CAB
‾‾‾‾
2

Step 2 ①
A37 $1_{16} + 3_{16} + A_{16} = 1 + 3 + 10 = 14 = E_{16}$.
CAB
‾‾‾‾
E2

Step 3 ①
A37 $A_{16} + C_{16} = 10 + 12 = 22 = 16_{16}$.
CAB
‾‾‾‾
16E2

Answer: $16E2_{16}$ (= 5858)

EXAMPLE 1.20

Multiply $D3_{16} \times 8A_{16}$ (that is, 211 × 138).

SOLUTION

Reason

Step 1
①
a. D3 $A_{16} \times 3_{16} = 10 \times 3 = 30 = 1E_{16}$. Write E and carry 1.
8A
‾‾‾
E

①
b. D3 $A_{16} \times D_{16} = 10 \times 13 = 130$, plus the 1 carry is 131 and 131 =
8A 83_{16}.
‾‾‾
83E

Step 2
①
a. D3 $8_{16} \times 3_{16} = 8 \times 3 = 24 = 18_{16}$. Write 8 and carry 1.
8A
‾‾‾
83E
8

①
b. D3 $8_{16} \times D_{16} = 8 \times 13 = 104$, plus the 1 carry is 105, and 105 =
8A 69_{16}.
‾‾‾
83E
698

Step 3   D 3   Add column by column.
   8 A
  ‾‾‾‾‾
  8 3 E
  6 9 8
  ‾‾‾‾‾
  7 1 B E

Answer: $71BE_{16}$ ($= 29118$)

The Conversion Graph

We now have at our disposal four number systems: decimal, binary, octal, and hexadecimal. It is possible to convert a number from any of the four systems to its equivalent form in any of the other systems.

The easy conversions from one base to another are depicted in the conversion graph of Figure 1.8. In the graph, D represents the decimal system, B the binary system, O the octal system, and H the hexadecimal system. An arrow from one system to another means that numbers can be converted directly from the first base to the second.

We have already discussed how to convert from decimal to binary and conversely (D ⇄ B); how to convert from decimal to octal and conversely (D ⇄ O); how to convert from decimal to hexadecimal and conversely (D ⇄ H). Note that the conversion graph also indicates direct conversion from binary to both octal and hexadecimal, and conversely. We now discuss how this is done.

To facilitate octal to binary conversion, described in Note 1.4 (and its converse, described in Note 1.5), we should be familiar with the 3-bit binary representation of each of the eight octits. See Figure 1.9(a).

Note 1.4

To convert an octal number to binary, expand each octit to its equivalent 3-bit binary representation.

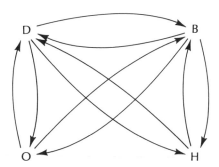

Figure 1.8

Octit	3-Bit Binary
0	000
1	001
2	010
3	011
4	100
5	101
6	110
7	111

(a)

Hexit	4-Bit Binary
0	0000
1	0001
2	0010
3	0011
4	0100
5	0101
6	0110
7	0111
8	1000
9	1001
A	1010
B	1011
C	1100
D	1101
E	1110
F	1111

(b)

Figure 1.9

EXAMPLE 1.21

Convert 2307_8 to binary.

SOLUTION

$$\begin{array}{cccc} 2 & 3 & 0 & 7 \\ \overbrace{010} & \overbrace{011} & \overbrace{000} & \overbrace{111} \end{array}$$

Note that 0_8 expands to 000_2, and that the leading 0 of the binary answer may be dropped.

Answer: $2307_8 = 10011000111_2$

Note 1.5

To convert a binary number to octal, reverse the procedure of Note 1.4. That is, group the bits into sets of three, beginning at the right of the binary number, and convert each resulting 3-bit binary to its equivalent octit. If extra bits are needed on the left to make a set of three bits, fill in with 0s.

EXAMPLE 1.22

Convert 1010110_2 to octal.

SOLUTION

Fill with 0s to make a set of three.

001010110

1 2 6

Answer: $1010110_2 = 126_8$

Analogously, hexadecimal to binary conversion, described in Note 1.6 (and its converse, described in Note 1.7), is facilitated by knowing the 4-bit binary representation of each of the 16 hexits. See Figure 1.9(b).

Note 1.6

To convert a hexadecimal number to binary, expand each hexit to its equivalent 4-bit binary representation.

EXAMPLE 1.23

Convert $7B0_{16}$ to binary.

SOLUTION

7   B   0

011110110000

The leading 0 in the answer may be dropped.

Answer: $7B0_{16} = 11110110000_2$

Note 1.7

To convert from binary to hexadecimal, reverse the procedure of Note 1.6. Group the bits into sets of four, beginning at the right, and convert each set of four bits into its equivalent hexit. If additional bits are needed on the left, fill in with 0s.

EXAMPLE 1.24

Convert 1011100101_2 to hexadecimal.

SOLUTION

Fill with 0s.
$\downarrow$

$\overbrace{001}\underbrace{011}\overbrace{100}\underbrace{101}$

2   E   5

Answer: $1011100101_2 = 2E5_{16}$

Remark   It is possible to convert directly from binary to octal or hexadecimal, and conversely, because the bases of the octal and hexadecimal systems are powers of 2, the base of the binary system. The rule for converting an octal number to binary is a shortcut for the method illustrated in the following calculation for converting 624_8 to binary. First note the expanded binary equivalent of the coefficients: $6 = 2^2 + 2 + 0, 2 = 0 \cdot 2^2 + 2 + 0, 4 = 2^2 + 0 \cdot 2 + 0$. Then

$$624_8 = 6(8)^2 + 2(8) + 4 = 6(2^3)^2 + 2(2^3) + 4$$
$$= 6 \cdot 2^6 + 2 \cdot 2^3 + 4$$
$$= (2^2 + 2 + 0) \cdot 2^6 + (0 \cdot 2^2 + 2 + 0) \cdot 2^3 + (2^2 + 0 \cdot 2 + 0)$$
$$= 2^8 + 2^7 + 0 \cdot 2^6 + 0 \cdot 2^5 + 2^4 + 0 \cdot 2^3 + 2^2 + 0 \cdot 2 + 0$$

which is the expanded binary equivalent of 110010100_2.

The final type of conversion, octal to hexadecimal and conversely, cannot be conveniently done directly. It is, however, easily accomplished by an indirect route through the conversion graph, as described in Note 1.8.

Note 1.8

To convert from octal to hexadecimal, first convert from octal to binary, then from binary to hexadecimal (O → B → H).
    To convert from hexadecimal to octal, reverse the steps (H → B → O).

EXAMPLE 1.25

Convert 506_8 to hexadecimal.

SOLUTION

5   0   6   Octal

$\overbrace{0001}\underbrace{01000}\overbrace{110}$ Binary

1   4   6   Hexadecimal

Answer: $506_8 = 146_{16}$

EXAMPLE 1.26

Convert $9F3_{16}$ to octal.

SOLUTION

$$
\begin{array}{llll}
9 & F & 3 & \text{Hexadecimal}
\end{array}
$$

$\overbrace{100111110011}$ Binary

$$
\begin{array}{llll}
4 & 7 & 6 & 3 \quad \text{Octal}
\end{array}
$$

Answer: $9F3_{16} = 4763_8$

EXERCISES 1.2

1. Convert the following decimal numbers *directly* to *both* octal and hexadecimal.
   a. 129   **b.** 342   **c.** 417
2. Convert the following from the base indicated directly to decimal:
   a. 61_8 **b.** 341_8 **c.** 617_8 **d.** $C1A_{16}$ **e.** 341_{16} **f.** FED_{16}
3. Convert the following binaries to octal:
   a. 101101101_2 **b.** 1101110111_2
4. Convert the following binaries to hexadecimal:
   a. 10001110011_2 **b.** 110111011110_2
5. Convert the following from the base indicated directly to binary:
   a. 61_8 **b.** 341_8 **c.** 617_8 **d.** $C1A_{16}$ **e.** 341_{16} **f.** FED_{16}
6. Convert the following octal numbers to hexadecimal:
   a. 721_8 **b.** 350_8
7. Convert the following hexadecimal numbers to octal:
   a. CAB_{16} **b.** $83D_{16}$
8. Add in the indicated bases:
   a. $341_8 + 61_8$ **b.** $617_8 + 531_8$
   c. $C1A_{16} + FED_{16}$ **d.** $341_{16} + DEC_{16}$
9. Multiply in the indicated bases:
   a. $35_8 \times 62_8$ **b.** $71_8 \times 17_8$ **c.** $ABC_{16} \times 16_{16}$ **d.** $81_{16} \times B5_{16}$

Subtraction and division may be carried out in the octal and hexadecimal systems in essentially the same way as in binary and decimal.

10. Subtract in the indicated bases:
    a. $3A7_{16} - 2DB_{16}$ **b.** $726_8 - 354_8$
11. Divide in the indicated bases:
    a. $4142_8 \div 35_8$ **b.** $13C4_{16} \div 5C_{16}$
12. Write the answers to the following additions in the indicated bases:
    a. $3B_{16} + 74_8$ in binary   **b.** $35 + 11011_2$ in hexadecimal
13. Before attempting this exercise, refer to exercise 11 of Exercises 1.1, which discusses fractional numbers in binary.
    a. Convert the following to decimal:
        i. 7.36_8 **ii.** $94.A3_{16}$
    b. Convert the following decimal numbers to both octal and hexadecimal:
        i. 35.5625   **ii.** 0.87890625

Problems and Projects

14. Write a program that inputs a decimal number and outputs the equivalent number in octal.
15. Write a program that accepts a decimal number as input, and outputs the equivalent hexadecimal number. (Note that this exercise is a bit—no pun intended—more difficult than decimal-to-binary and decimal-to-octal conversion because of the need for the hexits A, B, C, D, E, and F.)
16. Write a program that inputs a binary number and outputs the equivalent octal number.
17. Write a program that inputs a binary number and outputs the equivalent hexadecimal number.
18. Write a program that inputs an octal number and outputs the equivalent hexadecimal number.
19. Write a program that inputs a hexadecimal number and outputs the equivalent octal number.
20. Write a program that inputs a number written in base b_1 $(2 \le b_1 \le 16)$, and outputs the number in base b_2 $(2 \le b_2 \le 16)$. The input to the program should consist of b_1, b_2, and the number to be converted, n.

1.3   THE INTEGERS

In the first two sections of this chapter, we discussed several ways of representing integers in different number bases. Independent of their representation, however, the integers possess many algebraic properties. In this section we discuss some of these properties.

Algebraic Properties of the Integers

If two integers are added or multiplied, the resulting sum or product is also an integer. This property is called closure.

Closure Laws

For any two integers a and b, the sum $a + b$ and the product ab are integers.

The integers are not closed under all operations. If we divide one integer by another, the result may not be an integer. For example, 7 divided by 2 is 3.5, which is not an integer.

One of the first arithmetic laws learned in grade school is that the order in which integers are added or multiplied is immaterial. This property is called the commutative law.

> **Commutative Laws**
>
> For any two integers a and b,
>
> i. $a + b = b + a$ ii. $ab = ba$

Addition and multiplication of integers are binary operations; that is, they operate on two numbers at a time. If you are told to add 3, 4, and 5, you can first add 3 and 4, and then add the sum 7 to 5 to obtain the answer 12. Alternately, you can first add 4 and 5, and then add 3 to the sum 9 to obtain the answer 12. The associative law for addition states that these two ways of adding three integers are equivalent. Similar calculations can be performed to multiply 3, 4, and 5.

> **Associative Laws**
>
> For any integers a, b, and c,
>
> i. $(a + b) + c = a + (b + c)$ ii. $(ab)c = a(bc)$

Two integers, 0 and 1, are special with regard to addition and multiplication.

> **Identity Laws**
>
> For any integer a,
>
> i. $a + 0 = a$ ii. $a \cdot 1 = a$

The integer 0 is called the **additive identity** because of identity law (i), and the integer 1 is called the **multiplicative identity** because of identity law (ii).

Each integer has an **additive inverse** (or opposite, or negative). For example, the additive inverse of -2 is $+2$. They are additive inverses because their sum is the additive identity, 0. Put another way, the additive inverse of an integer is that integer which must be added to it to give 0.

> **Additive Inverse Law**
>
> For each integer a, there exists an integer $-a$ such that $a + (-a) = 0$.

What about the existence of a multiplicative inverse, that is, an integer by which another integer must be multiplied to give 1 (the multiplicative identity)? The only integers that have multiplicative inverses that are also integers are $+1$

and -1. We shall see in the next section that all rational numbers (except 0) and all real numbers (except 0) have multiplicative inverses.

The additive inverse law is the basis of the definition of the inverse operation to addition, namely subtraction. The **difference** between a and b is defined by $a - b = a + (-b)$.

Finally, the distributive law concerns the interaction of addition and multiplication.

Distributive Law

For any integers a, b, and c,

$$a(b + c) = ab + ac$$

Thus, $6(7 + 8)$ can be evaluated as:

$$6(7 + 8) = 6 \cdot 15 = 90 \quad \text{or} \quad 6(7 + 8) = 6 \cdot 7 + 6 \cdot 8 = 42 + 48 = 90$$

The distributive law is the basis of nearly all the factoring techniques of high school algebra and is used extensively to simplify algebraic expressions.

We shall see later in this chapter, in Chapter 2 on set theory and logic, and in Chapter 7 on Boolean algebra that some of these laws have counterparts for other sets of numbers, sets in general, logical propositions, and electric circuits.

Divisibility

When two integers, a and b, are multiplied to get a third integer, c, then c is said to be divisible by both a and b. Just as subtraction is the inverse operation to addition, so division is the inverse operation to multiplication. We now discuss division and its properties.

The integer a **divides** b if there is an integer k such that $b = ka$. If a divides b, we write $a \mid b$. Thus, $3 \mid 18$ because $18 = 6 \times 3$. If a does not divide b, we write $a \nmid b$. Thus, $5 \nmid 24$. If a divides b, then a is said to be a **factor** of b, or a **divisor** of b.

EXAMPLE 1.27

List all the positive divisors of 24.

SOLUTION

1, 2, 3, 4, 6, 8, 12, 24

Note that every number has 1 and itself as factors. For some special numbers these are the only factors. A **prime number**, or simply a **prime**, is an integer greater than 1 that has only 1 and itself as factors. The smallest prime, and the only even prime, is 2. (Why is 2 the only even prime?) The next few primes are 3, 5, 7, 11, 13, 17, 19, ...

Divisibility has three properties with which you should be familiar.

Theorem 1.1

  i. If $d|a$ and $d|b$, then $d|(a + b)$.
 ii. If $d|a$ and $d|b$, then $d|(a - b)$.
iii. If $d|a$, then for any integer c, $d|ac$.

Proof: We prove parts (i) and (iii) of Theorem 1.1 and leave the proof of part (ii) as an exercise.

  i. By the definition of divisibility, $d|a$ and $d|b$ mean, respectively, that $a = kd$ and $b = hd$ for some integers k and h. We must prove that d divides $a + b$, so consider $a + b = kd + hd = (k + h)d$. This last equation states that $a + b$ equals the integer $h + k$ times d. Thus, $d|(a + b)$.

iii. Since $d|a$, $a = kd$ for some integer k. Multiplying both sides by c, we obtain $ac = (kd)c = (kc)d$. Thus, ac is the integer kc times d. So, $d|ac$.

The concept of divisibility makes precise the notion of "one integer dividing another evenly," as taught in grade school. The case of "one integer not dividing another evenly," that is, division with the possibility of a remainder, must also be made precise. In grade school most students learn that the answer to a problem such as $17 \div 5$ is a quotient of 3 with a remainder of 2. This is sometimes written

$$17 \div 5 = 3R2$$

It means that

$$17 = (5 \times 3) + 2$$

To consider another case, $29 \div 6 = 4R5$ means $29 = (6 \times 4) + 5$, or the quotient is 4 and the remainder is 5. We now state, without proof, the so-called division algorithm.

Theorem 1.2   The Division Algorithm

For given positive integers a and b, $b \neq 0$, there exist unique integers q (the quotient) and r (the remainder), with $0 \leq r < b$, such that $a = bq + r$.

The division algorithm makes precise another fact taught in grade school, namely, that the remainder upon dividing one integer by another is greater than or equal to 0 and less than the number you divide by (the divisor). Note also that if $r = 0$, then $a = bq$, which states that $b|a$.

Remark   We can use the division algorithm to prove the procedure that uses repeated division to convert from decimal to another base. When converting from decimal to binary, for example, each time we divide by 2, the remainder is either 0 or 1. Consider the following conversion of 19 to binary:

$$\begin{array}{r|r} 2 & 19 \\ \hline & 9 \quad 1 \end{array}$$

$$19 = (9 \times 2) + 1$$

$$\begin{array}{r|r} 2 & 19 \\ \hline 2 & 9 \quad 1 \\ \hline & 4 \quad 1 \end{array}$$

$$9 = (4 \times 2) + 1$$
$$19 = ((4 \times 2) + 1) \times 2 + 1$$
$$= (4 \times 2^2) + (1 \times 2) + 1$$

$$\begin{array}{r|r} 2 & 19 \\ \hline 2 & 9 \quad 1 \\ \hline 2 & 4 \quad 1 \\ \hline & 2 \quad 0 \end{array}$$

$$4 = (2 \times 2) + 0$$
$$19 = ((2 \times 2) + 0) \times 2^2 + (1 \times 2) + 1$$
$$= (2 \times 2^3) + (0 \times 2^2) + (1 \times 2) + 1$$

$$\begin{array}{r|r} 2 & 19 \\ \hline 2 & 9 \quad 1 \\ \hline 2 & 4 \quad 1 \\ \hline 2 & 2 \quad 0 \\ \hline & 1 \quad 0 \end{array}$$

$$2 = (1 \times 2) + 0$$
$$19 = ((1 \times 2) + 0) \times 2^3 + (0 \times 2^2) + (1 \times 2) + 1$$
$$= (1 \times 2^4) + (0 \times 2^3) + (0 \times 2^2) + (1 \times 2^1) + 1$$

The last equation on the right is the expanded binary form of 10011_2.

Congruences

One of the most useful concepts in number theory is that of congruent integers. We give three equivalent formulations of congruence in Note 1.9.

Note 1.9

i. a is congruent to b modulo m, written $a \equiv b \bmod m$, if $m | (a - b)$. Thus, $9 \equiv 4 \bmod 5$ because $5 | (9 - 4)$. Also $72 \equiv 51 \bmod 7$ because $7 | (72 - 51)$. Note that the integer m is called the **modulus.**

ii. $a \equiv b \bmod m$ if a and b have the same remainder upon division by m. Thus, $87 \equiv 15 \bmod 6$ because both 87 and 15 leave a remainder of 3 upon division by 6.

iii. $a \equiv b \bmod m$ if $a = b + km$ for some integer k. Thus, $32 \equiv 17 \bmod 3$ because $32 = 17 + (5 \times 3)$.

To show the equivalence of (i), (ii), and (iii) in Note 1.9, we show that (ii) implies (i), (i) implies (iii), and (iii) implies (ii). This cycle of implications shows that any of (i), (ii), or (iii) implies the other two.

(ii) implies (i):
If a and b have the same remainder upon division by m, the division algorithm gives $a = mq_1 + r$ and $b = mq_2 + r$. Thus, $a - b = mq_1 + r - (mq_2 + r) = m(q_1 - q_2)$. Hence, $m|(a - b)$.

(i) implies (iii):
Suppose $m|(a - b)$. Then, $a - b = km$ so $a = b + km$.

(iii) implies (ii):
Suppose $a = b + km$. By the division algorithm, there exist unique integers q and r, $0 \leq r < m$, such that $a = qm + r$. Thus, $b = a - km = qm + r - km = (q - k)m + r$. Thus, a and b have the same remainder upon division by m.

From (ii) we have Note 1.10.

Note 1.10

Each number is congruent modulo m to its remainder upon division by m. Note also, $a \equiv 0 \bmod m$ means that $m|a$.

Thus,
$$32 \equiv 2 \bmod 3 \quad \text{and} \quad 87 \equiv 3 \bmod 6$$

By the division algorithm, a is congruent modulo m to its remainder upon division by m because $a = mq + r$, $0 \leq r < m$. Hence, $a \equiv r \bmod m$. Since the remainder r is unique, no nonnegative integer smaller than r can be congruent to a modulo m. For this reason, the remainder upon division by m is frequently called the **least residue** of a modulo m. Thus, the least residues modulo m must be 0, 1, 2, ..., $m - 1$, that is, the possible remainders upon division by m.

EXAMPLE 1.28

Find the least residues of 53 modulo 4 and of 107 modulo 11.

SOLUTION

The remainder upon dividing 53 by 4 is 1. Thus, the least residue of 53 modulo 4 is 1, that is, $53 \equiv 1 \bmod 4$.

The remainder upon dividing 107 by 11 is 8. Thus, the least residue of 107 modulo 11 is 8, that is, $107 \equiv 8 \bmod 11$.

The last paragraph is summarized by Theorem 1.3.

Theorem 1.3

Every integer is congruent modulo m to exactly one of 0, 1, ..., $m - 1$.

Congruences have many of the elementary properties of equality. We state these properties in Theorems 1.4 through 1.7.

Theorem 1.4

   i. (The Reflexive Property) For any a, $a \equiv a$ mod m.

  ii. (The Symmetric Property) For any a and b, if $a \equiv b$ mod m, then $b \equiv a$ mod m.

 iii. (The Transitive Property) For any a, b, and c, if $a \equiv b$ mod m and $b \equiv c$ mod m, then $a \equiv c$ mod m.

Proof: We prove part (iii) of Theorem 1.4 and leave the proofs of the first two parts as exercises. If $a \equiv b$ mod m and $b \equiv c$ mod m, then $a = b + mk_1$ and $b = c + mk_2$ by part (iii) of Note 1.9. Thus,

$$a = b + mk_1 = (c + mk_2) + mk_1 = c + m(k_1 + k_2)$$

Hence, by part (iii) of Note 1.9, $a \equiv c$ mod m.

A typical use of this theorem is the following: To show $794 \equiv 563$ mod 7, simply determine the least residues modulo 7 of 794 and 563, which is 3 in both cases. By (ii) and (iii) of Theorem 1.4, it follows that $794 \equiv 563$ mod 7.

Theorem 1.5

   i. If $a \equiv b$ mod m, then for any c, $a + c \equiv (b + c)$ mod m.

  ii. If $a \equiv b$ mod m, then for any c, $a - c \equiv (b - c)$ mod m.

Proof: We prove part (i) of Theorem 1.5 and leave the proof of part (ii) as an exercise. If $a \equiv b$ mod m, then $a = b + mk$ by part (iii) of Note 1.9. Thus, $a + c = b + c + mk$ for any integer c. Thus, $a + c \equiv (b + c)$ mod m by part (iii) of Note 1.9.

To illustrate Theorem 1.5, we know that $19 \equiv 4$ mod 5. Thus, $19 + 3 \equiv 4 + 3$ mod 5; that is, $22 \equiv 7$ mod 5. Sometimes application of part (ii) of Theorem 1.5 yields negative answers. For example, $13 \equiv 1$ mod 4. Thus, $13 - 2 \equiv (1 - 2)$ mod 4, or $11 \equiv -1$ mod 4. The following theorem generalizes Theorem 1.5.

Theorem 1.6

If $a \equiv b$ mod m and $c \equiv d$ mod m, then

   i. $a + c \equiv (b + d)$ mod m

  ii. $a - c \equiv (b - d)$ mod m

Proof: We prove part (ii) of Theorem 1.6 and leave the proof of part (i) as an exercise. Since $a \equiv b$ mod m, $a - b = km$, and since $c \equiv d$ mod m, $c - d = hm$. We must show that $a - c \equiv (b - d)$ mod m or that $(a - c) - (b - d)$ is a multiple of m:

$$(a - c) - (b - d) = a - c - b + d = a - b - c + d$$
$$= a - b - (c - d)$$
$$= km - hm = (k - h)m$$

It is instructive at this point to discuss the least residues modulo m of negative integers. What, for example, are the least residues mod 4 of -1, -2, -3, and so on? From the first formulation of the definition of congruence it follows that $a \equiv a + m$ mod m for any integer a. This is so because $m | (a + m) - a$. Thus, if we add the modulus to a number, we obtain a number congruent to the original number. For example, $3 \equiv (3 + 5)$ mod 5 and $7 \equiv (7 + 23)$ mod 23. So, $-1 \equiv (-1 + 4)$ mod 4, or $-1 \equiv 3$ mod 4. Likewise, $-2 \equiv (-2 + 4)$ mod 4, or $-2 \equiv 2$ mod 4.

EXAMPLE 1.29 _____

What are the least residues modulo 7 of -1, -2, -3, -4, -5, -6?

SOLUTION

Adding 7 to each, we obtain, respectively, 6, 5, 4, 3, 2, 1 as the least residues.

_____

The technique of adding the modulus to a negative number to obtain its least residue may be extended, as given in Note 1.11, to negative numbers with large absolute values.

Note 1.11

To obtain the least residue of a negative number, add to the negative number the smallest multiple of the modulus that is greater than or equal to the absolute value of the negative number.

EXAMPLE 1.30 _____

What is the least residue modulo 7 of -94?

SOLUTION

Find the first multiple of the modulus, 7, that is greater than 94, namely 98. Now add 98 to -94. Thus, $-94 \equiv (-94 + 98)$ mod 7, or $-94 \equiv 4$ mod 7, and the least residue modulo 7 of -94 is 4.

_____

Theorem 1.7

If $a \equiv b \bmod m$, then for any c, $ac \equiv bc \bmod m$.

To illustrate Theorem 1.7, $7 \equiv 4 \bmod 3$. Thus, $7 \times 5 \equiv (4 \times 5) \bmod 3$, that is, $35 \equiv 20 \bmod 3$.

Note 1.12 summarizes Theorems 1.5–1.7.

Note 1.12

We can add or subtract a number from both sides of a congruence, or multiply both sides of a congruence by a number, and in each case the results are congruent.

The last elementary arithmetic operation we consider is division. Can you divide both sides of a congruence by the same number and be left with a valid congruence? Sometimes yes and sometimes no. Consider, for example, $12 \equiv 8 \bmod 4$. If we divide both sides by 2, we obtain the incorrect result $6 \equiv 4 \bmod 4$. However, if we divide both sides of $20 \equiv 14 \bmod 3$ by 2, we obtain the correct result $10 \equiv 7 \bmod 3$.

To state the theorem that governs the division of congruences we need to introduce the concept of the greatest common divisor of two integers. We say that d is the **greatest common divisor** of a and b, written $\gcd(a, b)$, if (1) d is a common divisor of a and b, that is, $d|a$ and $d|b$; and (2) d is the largest such common divisor—that is, if c is also a common divisor of a and b, then $c \leq d$.

EXAMPLE 1.31 _____

Find $\gcd(8, 12)$, $\gcd(4, 14)$, $\gcd(5, 15)$ and $\gcd(18, 45)$.

SOLUTION

$\gcd(8, 12) = 4$; $\gcd(4, 14) = 2$; $\gcd(5, 15) = 5$; $\gcd(18, 45) = 9$.

EXAMPLE 1.32 _____

For any integer a ($\neq 0$), find $\gcd(a, 0)$ and $\gcd(a, 1)$.

SOLUTION

Since any nonzero integer divides 0, $\gcd(a, 0) = a$. Since the only divisor of 1 is 1, $\gcd(a, 1) = 1$.

For small numbers, such as those in Example 1.31, it is relatively easy to find the greatest common divisor. However, how can we find gcd(2415, 3289)? The numbers are clearly too large to find the greatest common divisor by inspection. Any greatest common divisor can be calculated by using a method called the Euclidean Algorithm.

Note 1.13   The Euclidean Algorithm

Denote the larger of the two numbers by L and the smaller by S. Divide S into L. Let the remainder be denoted by R. Now, replace L by S and S by R. Repeat the procedure until $R = 0$, in which case the greatest common divisor is the last value of S.

EXAMPLE 1.33 _____

Find gcd(2415, 3289) by the Euclidean Algorithm.

SOLUTION

Let $L = 3289$, $S = 2415$.

Remainder

Step 1   Divide:
2415 | 3289
          1   874

Step 2   874 | 2415
               2   667

Step 3   667 | 874
               1   207

Step 4   207 | 667
               3   46

Step 5   46 | 207
              4   23

Step 6   23 | 46
              2   0

Answer: gcd(2415, 3289) = 23

We can now state the following.

Theorem 1.8

If $ac \equiv bc \bmod m$ and $\gcd(m, c) = d$, then $a \equiv b \bmod (m/d)$.

In other words, you can divide both sides of a congruence by a number c, if you also divide the modulus by $\gcd(m, c)$.

EXAMPLE 1.34

Divide both sides of $12 \equiv 8 \bmod 4$ by 2.

SOLUTION

We can divide both sides by 2, if we also divide the modulus 4 by $\gcd(4, 2) = 2$. Thus, we obtain $6 \equiv 4 \bmod 2$.

Two integers a and b are **relatively prime** if $\gcd(a, b) = 1$, that is, if they have no factors in common other than 1. Thus, for example, 8 and 21 are relatively prime because $\gcd(8, 21) = 1$. A specialized case of Theorem 1.8 is formalized in Corollary 1.1.

Corollary 1.1

If $ac \equiv bc \bmod m$ and $\gcd(m, c) = 1$, then $a \equiv b \bmod m$.

Thus, if the divisor is relatively prime to the modulus, we may simply divide both sides of the congruence by the number and do nothing to the modulus. We can divide both sides of $39 \equiv 15 \bmod 8$ by 3 and obtain the congruence $13 \equiv 5 \bmod 8$.

A useful consequence of the Euclidean Algorithm is Theorem 1.9, which we state without proof.

Theorem 1.9

If p is prime and $p \mid ab$, then either $p \mid a$ or $p \mid b$.

If we know, for example, that $2 \mid 3x$, we can conclude from Theorem 1.9 that $2 \mid x$.

Remark   Obtaining the quotient Q and remainder R upon dividing the integer A by the integer B is easy in Pascal:

$$Q := A \text{ DIV } B$$

$$R := A \text{ MOD } B$$

It is, therefore, a simple matter to test for congruence and to implement the Euclidean Algorithm. (See exercises 16, 17, 18, and 19 in Exercises 1.3.)

Solving Linear Congruences

We can use Theorems 1.4 to 1.8 to solve linear congruences in much the same way as we solve linear equations. We illustrate the techniques with Example 1.35.

EXAMPLE 1.35 ───────────────────────────────────

Solve the following congruences:

a. $4x - 3 \equiv 13 \bmod 7$ b. $4x \equiv 16 \bmod 12$ c. $5x \equiv 2 \bmod 6$

d. $10x \equiv 25 \bmod 8$ e. $10x \equiv 25 \bmod 9$

SOLUTION

a. $4x - 3 \equiv 13 \bmod 7$

$\qquad 4x \equiv 16 \bmod 7$ $\qquad\qquad\qquad\qquad$ [Theorem 1.5(i)]

$\qquad x \equiv 4 \bmod 7$ $\qquad\qquad\qquad\qquad$ (Corollary 1.1)

This means that any number congruent to 4 modulo 7, such as 4, 11, ..., solves the original congruence.

b. $4x \equiv 16 \bmod 12$

$\qquad x \equiv 4 \bmod 3$ $\qquad\qquad\qquad\qquad$ (Theorem 1.8 with gcd = 4)

c. $5x \equiv 2 \bmod 6$

5 does not divide 2, so replace 2 by a number congruent to 2 modulo 6 that is a multiple of 5. You can do this by repeatedly adding 6 to the right side of the congruence until you obtain a multiple of 5.

$$5x \equiv 2 \bmod 6 \equiv 8 \bmod 6 \equiv 14 \bmod 6 \equiv 20 \bmod 6$$

Dividing by 5, we obtain

$$x \equiv 4 \bmod 6 \qquad\qquad \text{(Corollary 1.1)}$$

as the solution.

d. $10x \equiv 25 \bmod 8$

$\qquad 2x \equiv 5 \bmod 8$ $\qquad\qquad\qquad\qquad$ (Corollary 1.1)

In this case, no matter what multiple of 8 we add to 5, the resulting sum is odd and so not divisible by 2. Since we cannot divide both sides of the congruence by 2, the congruence has no solution.

e. $10x \equiv 25 \bmod 9$

$\qquad 2x \equiv 5 \bmod 9$ $\qquad$ Divide by 5. $\qquad\qquad$ (Corollary 1.1)

$\qquad 2x \equiv 14 \bmod 9$ $\qquad$ Add the modulus, 9, to the right side.

$\qquad x \equiv 7 \bmod 9$ $\qquad$ Divide by 2. $\qquad\qquad$ (Corollary 1.1)

EXERCISES 1.3

1. List all the factors of 36; of 124.
2. List the first 20 primes.
3. Find the quotient and remainder when 51 is divided by 6; when 109 is divided by 22; when 747 is divided by 41.
4. Which integers divide 0?
5. Prove that if $a|b$ and $b|c$, then $a|c$.
6. What is true if $a|b$ and $b|a$? Can you prove your answer?
7. Which of the following congruences are true?
   a. $217 \equiv 5 \bmod 9$ b. $174 \equiv 0 \bmod 6$
   c. $342 \equiv 30 \bmod 8$ d. $371 \equiv 37 \bmod 13$
8. Find the least residue of the following:
   a. $92 \bmod 7$ b. $81 \bmod 9$ c. $-15 \bmod 6$
   d. $216 \bmod 5$ e. $356 \bmod 11$ f. $-107 \bmod 13$
9. Find the following by the Euclidean Algorithm:
   a. $\gcd(168, 308)$ b. $\gcd(231, 630)$
   c. $\gcd(75, 2205)$ d. $\gcd(325, 4851)$
10. Solve the following congruences:
    a. $3x \equiv 12 \bmod 5$ b. $4x \equiv 8 \bmod 20$
    c. $3x \equiv 4 \bmod 7$ d. $12x \equiv 15 \bmod 29$

Problems and Projects

11. Prove Theorem 1.1(ii)
12. Prove Theorem 1.4(i) and (ii).
13. Prove Theorem 1.5(ii).
14. Prove Theorem 1.6(i).
15. Prove Theorem 1.7.
16. Write a program that inputs an integer N and outputs YES if N is prime and NO if N is not prime.
17. Write a program that inputs two integers N and M, and outputs the least residue of N modulo M.
18. Write a program that inputs three integers A, B, and M, and decides if $A \equiv B \bmod M$.
19. Write a program for the Euclidean Algorithm. It should input two integers and output their greatest common divisor.

1.4   THE REAL NUMBERS

What is a real number? Intuitively, any number that can represent the length of a line segment, or the negative of such a number, is a **real number.** We also include 0 as a real number, although it may be difficult to envision a line segment of length zero (if you can, you'll get the point!). Thus, all integers are real numbers, as are all rational numbers.

Rational Numbers

A **rational number** is a ratio of two integers—what we commonly call a fraction, such as 1/2, −7/5, and so on. The integers, which were discussed in the last section, are rational numbers. For example, the integer 3 can be expressed as any of the following: 3/1, 6/2, 9/3, and so on. Thus, the set of rational numbers includes the set of integers.

The algebraic properties that were established for the integers in Section 1.3 also hold for rational numbers. However, each rational number, except 0, does have a multiplicative inverse. For example, the multiplicative inverse of 2/3 is 3/2. The **multiplicative inverse** (or **reciprocal**) of a rational number ($\neq 0$) is that number by which it must be multiplied to give 1.

Multiplicative Inverse Law

For each rational number a ($\neq 0$), there is a rational number $1/a$ such that $a(1/a) = 1$.

The multiplicative inverse law is the basis for the definition of **division,** which is the inverse operation to multiplication. The **quotient** of a divided by b ($b \neq 0$), is defined by $a/b = a(1/b)$.

Remark   Be sure to remember that 0 does not have a multiplicative inverse. Another way of stating this restriction is to note that you can *never* divide by 0. You can, however, always multiply by 0. What is $a \cdot 0$?

We summarize the algebraic properties of the rational numbers in Note 1.14.

Note 1.14

Let a, b, and c be rational numbers.

1. **Closure Laws:**
   i. $a + b$ is a rational number.
   ii. ab is a rational number.
2. **Commutative Laws:**
   i. $a + b = b + a$ ii. $ab = ba$
3. **Associative Laws:**
   i. $(a + b) + c = a + (b + c)$ ii. $(ab)c = a(bc)$
4. **Identity Laws:**
   i. $a + 0 = a$ ii. $a \cdot 1 = a$
5. **Inverse Laws:**
   i. For any rational a, there exists a rational $-a$ such that $a + (-a) = 0$.
   ii. For any rational $a \neq 0$, there exists a rational $1/a$ such that $a(1/a) = 1$.
6. **Distributive Law:** $a(b + c) = ab + ac$

Irrational Numbers

For thousands of years people believed all real numbers were rational. The ancient Greeks, however, discovered that there are real numbers that are not rational. An **irrational number** is a number that is not rational; that is, it *cannot* be expressed as the ratio of two integers.

Perhaps the simplest example of an irrational number is $\sqrt{2}$. We can prove $\sqrt{2}$ is irrational by the following proof by contradiction: Assume $\sqrt{2}$ is rational, that is, $\sqrt{2} = a/b$, where a and b are integers and a and b are relatively prime (a and b have no common factors). Squaring both sides of $\sqrt{2} = a/b$ we obtain $2 = a^2/b^2$ or

$$a^2 = 2b^2 \qquad (1)$$

Since $2|2b^2$, it follows that $2|a^2$. Because 2 is prime, $2|a$ (Theorem 1.9). Therefore, $a = 2c$, or $a^2 = 4c^2$. Substituting in (1) we obtain $4c^2 = 2b^2$ or

$$2c^2 = b^2 \qquad (2)$$

Since $2|2c^2$, $2|b^2$. Thus, $2|b$. We have, therefore, shown that $2|a$ and $2|b$ contradict our assumption that a and b are relatively prime. Since assuming $\sqrt{2}$ is rational leads to a contradiction, it must be true that $\sqrt{2}$ is irrational.

Since we think of real numbers as those numbers that can represent the lengths of line segments, how can we find a line segment whose length is $\sqrt{2}$? First, construct a square all of whose sides are one unit in length. From the Pythagorean Theorem, the diagonal of this square is $\sqrt{2}$ units. See Figure 1.10.

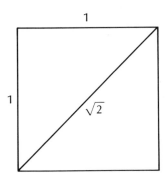

Figure 1.10

Other Representations of Real Numbers

Real numbers can also be described as numbers that have decimal expansions. Here, however, we must be careful. Integers and certain rational numbers have **terminating decimal expansions.** For example, $4 = 4.0$, $1/2 = 0.5$, $3/4 = 0.75$. In a terminating decimal expansion it is understood that all digits to the right of the right-most digit are 0s. Therefore, we do not write these 0s. Certain other rational numbers have **repeating decimal expansions.** For example, $1/3 = 0.3333\ldots$. The

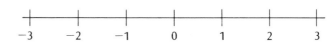

digit 3 continues ad infinitum. For 1/3, the repeating part of the decimal expansion consists of one digit, namely 3, and we write $1/3 = 0.\overline{3}$. The bar over the 3 indicates that the 3 repeats in the expansion infinitely often. To consider another example, $4/33 = 0.\overline{12}$. This indicates that the pair of digits 12 repeats infinitely often $(4/33 = 0.121212121212\ldots)$.

Irrational numbers have **nonrepeating** (and, hence, nonterminating) **decimal expansions.** The number π (the number of times the diameter of a circle is contained in its circumference, *not* 22/7) is an irrational number. Correct to ten decimal places $\pi \approx 3.1415926536\ldots$. (The $\approx$ means approximately equals). We include the $\ldots$ at the right of the number to indicate that the expansion continues and that there is no easy way to indicate the value of the next digit in the decimal expansion. Usually, when doing arithmetic with irrational numbers, we use rational approximations for them. A common approximation for π, for example, is 22/7 or $3.\overline{142857}$. This is correct to only two decimal places, but suffices for many simple calculations.

Still another and very fruitful way of representing real numbers is geometrically on the so-called **real number line.** See Figure 1.11. To construct the real number line, take a straight line (remember the line extends infinitely in both directions), and choose a point on the line. This point represents the number 0. Now choose another point (usually to the right of 0). This point represents the number 1.

Using the distance between 0 and 1 as unit distance, we can now locate the points that correspond to 2, 3, and so on by moving the appropriate number of unit distances to the right from 0. Negative integers are located on the line by moving an appropriate number of unit distances to the left from 0.

How can we locate the point corresponding to $-6/5$? First, we divide the unit distances into five equal parts. Each of these parts corresponds to 1/5. Now count off six of these parts to the left of 0. See Figure 1.12.

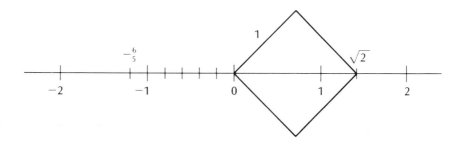

An irrational number can be represented on the line by first finding a line segment whose length equals the irrational number and then placing the segment on the number line with one end at zero. In Figure 1.12 we show how this can be done for $\sqrt{2}$.

A fundamental property of the real number line is that it contains no gaps, as stated in Note 1.15.

Note 1.15

Every real number corresponds to a point on the number line, and conversely—that is, every point on the number line corresponds to a real number.

Since we may always divide a line segment in half, Note 1.16 applies.

Note 1.16

Between any pair of real numbers there lies another real number.

To find such a number is quite easy. Simply take the average of the two numbers in question. For example, to find a number between 3.8762 and 3.9464, we do the following: $(3.8762 + 3.9464)/2 = 3.9113$. There are, of course, other numbers between 3.8762 and 3.9464. In fact, there are infinitely many such numbers, and we can continue finding numbers between numbers we have already found, as shown in Figure 1.13.

It can also be shown that between any pair of real numbers we can find both a rational and an irrational number. We summarize these points in Note 1.17.

Note 1.17

i. Between any pair of real numbers there lie infinitely many real numbers.

ii. Between any pair of real numbers there lie infinitely many rational numbers.

iii. Between any pair of real numbers there lie infinitely many irrational numbers.

Remark   If we construct a rational number line, that is, a line all of whose points represent rational numbers, we will have a line with holes in it. Notice, for example, from Figure 1.12 that there would be no point on our rational number line corresponding to $\sqrt{2}$ units to the right of 0 because $\sqrt{2}$ is an irrational number. However, Note

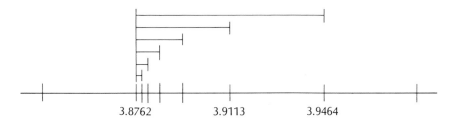

Figure 1.13

1.17(ii) implies that, in the following sense, the holes are not important: Rational numbers can be used to approximate any real number to any degree of accuracy. For example, the decimal numbers 3, 3.1, 3.14, 3.141, 3.1415, 3.14159, 3.141592 are more and more accurate rational number approximations for π.

Algebraic Properties of the Real Numbers

The real numbers obey the same set of algebraic laws as the rational numbers.

Note 1.18

Let a, b, and c be real numbers.

1. **Closure Laws:**
   i. $a + b$ is a real number.
   ii. ab is a real number.

2. **Commutative Laws:**
   i. $a + b = b + a$ ii. $ab = ba$

3. **Associative Laws:**
   i. $(a + b) + c = a + (b + c)$ ii. $(ab)c = a(bc)$

4. **Identity Laws:**
   i. $a + 0 = a$ ii. $a \cdot 1 = a$

5. **Inverse Laws:**
   i. For any real number a, there exists a real number $-a$ such that $a + (-a) = 0$.
   ii. For any real number a ($\neq 0$), there exists a real number $1/a$ such that $a(1/a) = 1$.

6. **Distributive Law:** $a(b + c) = ab + ac$

Remark   The laws in Note 1.18 are sometimes called the **field axioms.** Any set F with two binary operations ($+$ and $\cdot$) that obeys the field axioms is called a **field.** Thus, the rational numbers and the real numbers, with addition and multiplication as the binary operations, are examples of fields.

EXERCISES 1.4

1. Find the decimal expansions for the following fractions. Indicate whether they are terminating or repeating decimals. Use the bar notation where appropriate to indicate the digits that repeat.

   a. $\dfrac{3}{8}$ b. $\dfrac{2}{7}$ c. $\dfrac{2}{3}$ d. $\dfrac{23}{12}$ e. $\dfrac{7}{20}$

2. Find four numbers between 10.762 and 10.768. How many numbers are there between 10.762 and 10.768?

3. What is the next real number after 2 (*not* the next integer, which is 3, but the next real number)?

4. State the laws used in each of the following:
   a. $(2 \cdot \frac{1}{2}) \cdot 3 = 1 \cdot 3$
   b. $8 + (-8) = 0$
   c. $4(3 + 9) = (3 + 9) \cdot 4$
   d. $(4 \cdot 1) = 4$
   e. $(-3) + (3 + 9) = ((-3) + 3) + 9 = 0 + 9 = 9$

5. a. Give the additive inverse of: 4, -7, 2, -5, $-6/7$.
   b. Give the multiplicative inverse of: 8, -2, 4/5, $-4/9$.

6. a. What number (or numbers) is its own additive inverse?
   b. What number (or numbers) is its own multiplicative inverse?

7. Does the associative law hold for subtraction? That is, is it true for all real numbers a, b, c that

$$(a - b) - c = a - (b - c)$$

8. The binary relation greater than ($>$) can be defined as follows: $a > b$ if and only if $a - b$ is positive (that is, if $a - b > 0$). Prove the following properties of $>$:
   i. If $a > b$, then for any c, $a + c > b + c$.
   ii. If $a > b$ and $b > c$, then $a > c$.
   iii. If $a > b$ and $c > 0$, then $ac > bc$.

CASE STUDY 1A   COMPUTER ARITHMETIC: BINARY NUMBERS

In this chapter we have presented the conceptual and algebraic foundations of the integers, rational numbers, and real numbers. We have also discussed several ways of representing the integers in different number bases. It is important to remember that a given number, say five, has many representations. The word *five*, the arabic numeral 5, the roman numeral V, and the 3-bit binary 101 are all different symbols for the same number. Conversely, one symbol, say 11000001, can represent many different things. If interpreted as a decimal number, it represents eleven million one. If interpreted as an 8-bit binary integer, it represents one hundred ninety-three. If interpreted as an 8-bit 2's-complement binary, it represents -63. In Case Study 1D, we shall see how to interpret 11000001 as an 8-bit EBCDIC character representing the letter A. The symbol

11000001 could even be interpreted as a floating-point number, as we will see in Case Study 1B.

A computer by itself does not give meaning to a sequence of bits. The programmer, by specifying the type of variable to be stored in a particular location, determines how the computer will interpret the sequence of bits stored in that location. Once this specification is made, the computer will use rules appropriate to the type of data stored in that location to manipulate the data.

Most modern computers can perform several kinds of arithmetic. Among these are binary, floating-point, and decimal arithmetic. How these arithmetics are implemented depends on the language and computer being used. The most basic type of arithmetic, common to all computers, is binary arithmetic.

In computers, binary arithmetic takes place in special locations called **registers.** A register holds a certain fixed number of bits depending on the particular computer. For example, the Z-80 microprocessor has 8-bit registers, the PDP-11 minicomputer family has 16-bit registers, and the IBM-370 family of computers has 32-bit registers. In this section we discuss a hypothetical computer whose one and only register R holds five bits. If the number 7 is placed (that is, **loaded**) into R, the binary equivalent of 7 is placed in R. We denote R and its contents as follows:

$$R: \quad 00111$$

Note two things about this representation. First, we do not place the subscript 2 on the register contents since it is understood that R contains a binary number. Second, all bit positions are shown, even the left-most, or leading, positions when they are zero.

How many numbers can be represented in R? That is, how many numbers can be represented using a 5-bit binary representation? The answer is given in Note 1.19, which is proved in Chapter 3.

Note 1.19

Using an n-bit binary representation, we can represent 2^n numbers.

In R, therefore, we can represent $2^5 = 32$ numbers. In a PDP-11 minicomputer, $2^{16} = 65,536$ numbers may be represented in a register. In an IBM-370 computer, we can represent $2^{32} = 4,294,967,296$ different numbers in a register.

Now that we know *how many* numbers we can represent in R, we must decide *which* numbers we can represent in R. To solve this problem, we must keep in mind that we want to do arithmetic on our hypothetical computer. After all, what is a computer for?

First, when we add two 5-bit numbers, the sum may be six bits long. Thus, we must consider the size of the numbers representable in R. Suppose we load 23 into R, giving

$$R: \quad 10111$$

If we add 17 (that is, 10001_2) to the contents of R, we obtain the 6-bit integer 101000. Since this number is too large to store in R, the left-most bit of the answer is lost. The resulting contents of R will be

$$\text{R:}\quad 01000$$

The condition where the left-most (or **most significant**) bit of the answer to an arithmetic operation is lost is called **overflow.** As you see it results in an incorrect answer in R. Many modern computer systems have built-in hardware that detects overflow.

The second point we must take into account, because we want to do arithmetic in R, is how to represent negative numbers. We need negative numbers because we want to subtract numbers in R.

In algebra we usually represent negative numbers by preceding the number with a minus sign. Thus, the negative of 6 is represented by -6. Since all information is stored in a computer by using sequences of bits, we must devise a way of coding the minus sign. We also want to encode the sign so it is computationally efficient.

The most frequently used method of representing signed numbers in a computer is the **2's-complement method.** In 2's-complement, the left-most bit of the register denotes the sign of the number. If the left-most bit is 0, the number is positive. If the left-most bit is 1, the number is negative. If the number is positive, the rest of the bit positions carry the binary equivalent of the number. The largest positive number that can be stored in R is, therefore, 01111, which represents $+15$. The number $+6$ is represented by 00110.

To obtain the n-bit 2's-complement representation of a negative number, for example -9, we proceed as described in Note 1.20.

Note 1.20

To find the n-bit 2's-complement representation of a negative number:

1. Find the n-bit 2's-complement representation of the corresponding positive number. (The 2's-complement representation of $+9$ is 01001.)

2. Find the complement of this number; that is, change each 0 bit to 1 and each 1 bit to 0. (Thus, we obtain 10110.)

3. Add 1 to the resulting number. If there is overflow, the overflow bit is disregarded. (Thus, the 2's-complement representation of -9 is $10110 + 1 = 10111$.)

EXAMPLE 1.36 ——————————————————————————————————

a. Find the 8-bit 2's-complement representation of -55.

b. 10100 is the 5-bit 2's-complement of what number?

SOLUTION

a. The 8-bit 2's-complement representation of $+55$ is 00110111. Its complement is 11001000. Adding 1,

$$
\begin{array}{r}
11001000 \\
+\,1 \\
\hline
11001001
\end{array}
$$

Thus, the 2's-complement representation of -55 is 11001001.

b. To find which number 10100 represents, we reverse the steps of the procedure of Note 1.20—that is, we subtract 1 and complement the result:

$$
\begin{array}{r}
10100 \\
-\,1 \\
\hline
10011
\end{array}
$$

Complementing, we obtain 01100, which represents $+12$. Thus, 10100 is the 2's-complement representation of -12.

An equivalent procedure (we leave the justification to the reader) is to find the 2's-complement of 10100—that is, complement the bits, 01011, and add 1:

$$
\begin{array}{r}
01011 \\
+\,1 \\
\hline
01100
\end{array}
$$

Since 01100 represents $+12$, 10100 is the 2's-complement representation of -12.

Table 1.4 lists the numbers that can be stored in register R, together with their 2's-complement representations.

The smallest negative number that can be represented in R in 2's-complement form is -16.

Recalling the definition of subtraction given in Section 1.3, we can subtract in register R by adding the negative of the subtrahend. Example 1.37 illustrates this point.

EXAMPLE 1.37

Show how to subtract $13 - 5$ in R.

SOLUTION

We treat the subtraction as an addition: $13 - 5 = 13 + (-5)$. Now we use binary addition to add the representation of 13 (01101) to the representation of -5 (11011):

$$
\begin{array}{r}
01101 \\
+\,11011 \\
\hline
101000
\end{array}
$$

The overflow bit is disregarded. The answer is, therefore, 01000, which is the 2's-complement representation of $+8$.

Number	R	Number	R
0	00000		
+1	00001	−1	11111
+2	00010	−2	11110
+3	00011	−3	11101
+4	00100	−4	11100
+5	00101	−5	11011
+6	00110	−6	11010
+7	00111	−7	11001
+8	01000	−8	11000
+9	01001	−9	10111
+10	01010	−10	10110
+11	01011	−11	10101
+12	01100	−12	10100
+13	01101	−13	10011
+14	01110	−14	10010
+15	01111	−15	10001
		−16	10000

Table 1.4

If the overflow bit is disregarded in 2's-complement arithmetic, how is overflow detected? Let us see what happens if we add 11 and 13 in R. First, load 11 into R, obtaining R: 01011. Now add 13, 01101, to the contents of R. The sum, which is stored in R, is R: 11000; this is the 2's-complement representation of −8. We have obtained a negative answer when adding two positive numbers! Since this is impossible, we now have a simple way for the computer to detect arithmetic overflow—by comparing the sign of the answer with the sign of the numbers being added. See also exercise 7 at the end of this section.

EXERCISES: CASE STUDY 1A

1. How many binary numbers can be represented in a register of 6 bits? 12 bits? 36 bits?
2. Use 5-bit 2's-complement arithmetic to perform the following, if possible: $9 + 8$, $-5 - 2$, $-10 - 7$, $10 - 16$.
3. Verify all the 2's-complement entries in Table 1.4.
4. Microcomputers using the Z-80 microprocessor, the PDP-11 computers, and the IBM-370 computers all use 2's-complement binary number representation. (Recall, they use 8-, 16-, and 32-bit registers, respectively.) What are the smallest and largest numbers that can be represented in the registers of each of these computers? How would 100 and −46 be represented in each of these computers?

5. A familiar property of the decimal numbers is that the sum of a number and its negative is zero. Convince yourself (and a friend) that this property also holds for 2's-complement arithmetic in the 5-bit register R.

6. **a.** Convince yourself that the 2's-complement of the 2's-complement of a number gives the number that you started with. To what general property of the decimal numbers is this similar?

   b. There is one exception to what you might have thought was a general rule in part a. Find the 2's-complement representation of -16.

7. When adding numbers of opposite sign, why is it impossible for the register R to overflow?

8. **a.** What is the value of MAXINT in your implementation of Pascal? Is this value 1 less than a power of 2? If so, what might this tell you about the size of the registers of your computer?

   b. What happens in your implementation of Pascal if you add 1 to MAXINT? If you add -1 to $-$MAXINT?

CASE STUDY 1B   COMPUTER ARITHMETIC: FLOATING-POINT NUMBERS

In Case Study 1A we saw how to represent integers in a computer. In this case study, we describe how to represent real numbers in a computer using the floating-point system, and how the floating-point system differs from the real number system.

Real numbers are represented in a computer in **floating-point notation.** We write a number in floating-point by expressing it as a number between 0.1 and 1, times a power of 10. For example, we write 87.6 in floating-point as 0.876×10^2. We move the decimal point to the left of the first significant digit, and then multiply by the appropriate power of 10. The power of 10 is the number of places we moved the decimal point to the *left*. Moving the decimal point to the right results in a negative exponent.

EXAMPLE 1.38 _____

Express 265.0 and 0.00000107 in floating-point.

SOLUTION

$$265.0 = 0.265 \times 10^3$$
$$0.00000107 = 0.107 \times 10^{-5}$$

The decimal part of a number in floating-point form is called the **mantissa,** and the power of 10 is called the **exponent.** The **precision** of a floating-point number is the number of significant digits in its mantissa. Thus, the precision of 0.265×10^3 is 3 and the precision of 0.26072×10^{-1} is 5.

Floating-point numbers are usually displayed on a computer or calculator either in decimal form (for example, 12.23) or in E, or exponent, form (for example, 0.1223 E2).

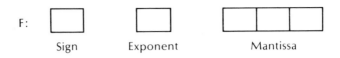

Figure 1.14

The integer following E is the exponent of the number. The following are examples of E form: The number $265.0 = 0.265 \times 10^3$ will appear as 0.265 E3; the number $0.00000107 = 0.107 \times 10^{-5}$ will appear as $0.107\,E - 5$. Note that sometimes, especially in calculator displays, the E is omitted and the exponent is instead spaced over from the mantissa. Thus, 0.672 4 means 0.672 E4.

How is floating-point notation implemented in a computer? We have already noted that numbers are manipulated by a computer in registers. We can store only a finite number of numbers in a register because a register is fixed in size. To aid our discussion, we shall consider a hypothetical computer that does decimal arithmetic.* Its one and only register, F, contains five decimal digits. An example of the contents of F is F: 06315. No matter how we represent floating-point numbers in F, the numbers themselves will have a limited precision. How can the real number system, which contains numbers with infinite decimal expansions, be implemented on a computer with only finite precision? The answer: It can't! We must be content with only approximating the real number system. This limitation has several implications, as given in Note 1.21.

Note 1.21

1. In a floating-point register, certain numbers, such as irrationals and certain rationals, cannot be represented exactly.

2. Because some numbers cannot be represented exactly, some of the results of floating-point arithmetic will be in error.

We now show how to represent floating-point numbers in F.

Since floating-point numbers can be positive or negative, we need a way of storing the sign of the number. We store the sign in the left-most digit position. A digit of 0 denotes a positive number, and a digit of 1 denotes a negative number. Any other digit in the left-most position is considered invalid. The next position to the right holds the exponent, and the three remaining positions hold the mantissa. Thus, F stores numbers with 3-digit precision. F is depicted in Figure 1.14.

Since we are restricted to exponents between 0 and 9 inclusive when storing floating-point numbers in this form, how can we store negative exponents? We do so by using a device that is employed in most actual floating-point systems and that

---

* We consider decimal rather than binary arithmetic to take advantage of the reader's familiarity with the decimal system. However, most computers represent the mantissa and exponent in binary, octal, or hexadecimal rather than in decimal.

F:

Sign   Characteristic   Mantissa

Figure 1.15

enables us to represent exponents in the range -5 to $+4$. To store an exponent in this range in F, add 5 to the exponent, obtaining a value between 0 and 9. If an exponent is actually -3, we store $-3 + 5 = 2$ in the exponent position of F. If the exponent is $+1$, we store $1 + 5 = 6$ in the exponent position of F. The number we obtain by adding 5 to the exponent is called the **characteristic** of the number. The exponent is said to be stored using **excess-5** notation because the characteristic is 5 more than the exponent it represents. Thus, a more accurate representation of the contents of F is shown in Figure 1.15.

This form of floating-point representation—namely, sign, characteristic, and mantissa—is called **normal form.** We shall refer to the "mantissa times a power of 10" form as the **exponent form** of a floating-point number. In the remainder of this chapter we shall refer to the hypothetical 3-digit mantissa floating-point system as simply the **3-digit system.**

EXAMPLE 1.39

Show how $+265$, $+0.00107$, and -3.49 are stored in normal form in F.

SOLUTION

Number	Exponent Form	Normal Form in F
$+265$	0.265×10^3	08265
$+0.00107$	0.107×10^{-2}	03107
-3.49	-0.349×10^1	16349

When converting to normal form we add 5 to the exponent to get the characteristic. When going in the opposite direction, that is, when converting from normal form to exponent form, we subtract 5 from the characteristic to get the exponent.

EXAMPLE 1.40

Given the following contents of F in normal form, express the contents in both exponent and decimal forms.

a. F: 12345   b. F: 05399

SOLUTION

a. F: 12345. The sign is $-$; the characteristic is 2. Hence the exponent is $2 - 5 = -3$. The mantissa is 345. Hence, the number is $-0.345 \times 10^{-3} = -0.000345$.

b. F: 05399. The sign is $+$; the characteristic is 5. Hence the exponent is $5 - 5 = 0$. The mantissa is 399. Hence, the number is $+0.399 \times 10^{0} = +0.399$.

What range of numbers can we store in F? The largest characteristic is 9, which represents an exponent of 4. The largest mantissa is 999. Thus, the largest floating-point number that can be stored in F is $0.999 \times 10^4 = 9990$. The smallest positive number that can be stored is 0.100×10^{-5} or 0.000001 since the smallest exponent is -5 (a characteristic of 0) and the smallest nonzero mantissa is 0.100. Note that the negatives of all these numbers can also be stored in F.

Suppose we take a typical floating-point number stored in F. For example, 0.123×10^2, or in normal form F: 07123. What is the next larger number that can be stored in F? Obviously, the number is 0.124×10^2. In decimal form these are 12.3 and 12.4 and are depicted on the number line in Figure 1.16. These numbers are one-tenth unit apart. Now, how can we represent a number such as 12.37, which falls between 12.3 and 12.4? The answer is we can't. To represent 12.37 on our hypothetical computer, we must approximate it to either 12.3 or 12.4 because the precision of our computer is only three decimal digits. We can approximate 12.37 by either of two methods. First, we can **truncate** or drop anything more than three significant digits. Therefore, we approximate 12.37 to 12.3. Second, we can **round** the number. By rounding we approximate 12.37 to 12.4.

We saw, in the remark following Note 1.17, that although the rational number line has holes in it, the rational numbers can be used to approximate any real number to any degree of accuracy. The floating-point number line corresponding to the numbers representable in F also has holes in it. However, as Note 1.22 points out, there may be *no* floating-point number between a given pair of floating-point numbers.

Note 1.22

The floating-point number system has gaps in it; that is, it is sometimes not possible to find a floating-point number between a given pair of floating-point numbers.

12.0	12.3	12.4	13.0

Figure 1.16

For example, there is no 3-digit floating-point number between 12.3 and 12.4. However, there are infinitely many rational and real numbers between 12.3 and 12.4.

Also, the magnitude of the gap between two consecutive floating-point numbers increases when the numbers are larger in absolute value. For example, 12.3 and 12.4 are consecutive 3-digit floating-point numbers. The size of the gap is 0.1. The floating-point numbers 78200 and 78300 are also consecutive 3-digit floating-point numbers. The size of the gap between this pair, however, is 100.

The gaps in the floating-point system impose an intrinsic limitation on the accuracy of calculations performed in that system. We must, therefore, carefully analyze the numeric errors the gaps cause in arithmetic calculations. In Case Study 1C, we discuss floating-point arithmetic and briefly discuss some of the errors that arise when we use it.

EXERCISES: CASE STUDY 1B

1. Write the following numbers in exponent form:
   a. 51.273   **b.** 86400000.0   **c.** -0.0000000893
2. Write the following numbers in normal form as they would appear in the floating-point register F.
   a. -0.00392 **b.** 92.5   **c.** -8940
3. Given the following contents of F, express the contents in both exponent and decimal form:
   a. 07239   **b.** 01458   **c.** 19269
4. What are the largest and smallest negative numbers that can be stored in F?
5. What are the two smallest positive numbers that can be stored in F? How far apart are they?
6. In order of magnitude, what are the two next larger numbers after 9.99 that can be stored in F?
7. How can zero be stored in F?

Problems and Projects

8. Floating-point numbers may be defined in Pascal by using the variable type REAL. (Floating-point numbers are frequently called real numbers in programming languages.) The precision and range of exponents that can be used in a Pascal program depend on the specific implementation of Pascal. Find both the precision and range of exponents for the implementation of Pascal you use.
9. Because you are dissatisfied with the limited range of numbers in the hypothetical 3-digit computer described in the text, you decide to build a computer with a larger floating-point register called FL. FL can store seven decimal digits. You decide on the following format for the storage of floating-point numbers:

FL: [ ]   [|]   [| |]

   Sign   Characteristic   Mantissa

You thus have available a 4-digit mantissa.

a. The characteristic can store numbers in the range 00 to 99. What exponents will these represent?

b. Write the following decimal numbers as they will be stored in FL:

   i. -8723000000000 **ii.** 0.008435 **iii.** 0.000000000000001234

c. What is the largest number that can be stored in FL?

d. What are the two smallest numbers that can be stored in FL? How far apart are they?

e. Given the following contents of FL, express the contents in both exponent and decimal form:

   i. 0263795   **ii.** 1757890   **iii.** 0513141

f. What advantages does your computer have over the one described in the text?

g. How would 374.577 be approximated in FL if truncation is used? If rounding is used?

h. How would zero be represented in FL?

CASE STUDY 1C   COMPUTER ARITHMETIC: FLOATING-POINT ARITHMETIC

In this case study we discuss floating-point arithmetic and the associated error analysis. We do all arithmetic in the case study on the hypothetical computer with floating-point register F that was described in Case Study 1B. To make the arithmetic easier, we shall write all floating-point numbers in exponent form. Also, we assume that answers to arithmetic calculations are truncated, as they are on most computers. The important points to remember are that our floating-point numbers have 3-digit precision and that the exponents can range from -5 to $+4$ inclusive.

Addition

> **Note 1.23**
>
> To add two floating-point numbers having the same exponent, add the mantissas. If the sum of the mantissas is greater than 1, adjust the exponent accordingly. Otherwise, the exponent is the same as that of the numbers being added.

EXAMPLE 1.41 _____

Add: a. $0.123 \times 10^2 + 0.456 \times 10^2$ b. $0.567 \times 10^3 + 0.678 \times 10^3$

SOLUTION

a. $0.123 \times 10^2 + 0.456 \times 10^2 = 0.579 \times 10^2$

b. $0.567 \times 10^3 + 0.678 \times 10^3 = 1.245 \times 10^3$
$$\approx 0.124 \times 10^4$$

Note that the answer to (b) is truncated to 0.124×10^4. Thus, we obtain only an approximate answer. Therefore, computing with limited precision numbers may give rise to errors in our computations.

Note 1.24

To add two floating-point numbers having different exponents, adjust the number with the smaller exponent before adding so that both numbers have the same exponents. Then add the numbers according to Note 1.23.

EXAMPLE 1.42

Add: $0.346 \times 10^2 + 0.567 \times 10^3$

SOLUTION

First adjust the exponent of 0.346×10^2 to make it a 3: $0.346 \times 10^2 = 0.0346 \times 10^3$. Since we are allowed only 3-digit precision, we must truncate the right-most digit. Thus, $0.346 \times 10^2 = 0.0346 \times 10^3 \approx 0.034 \times 10^3$. Now we add: $0.034 \times 10^3 + 0.567 \times 10^3 = 0.601 \times 10^3$.

At this point let us consider the ordinary decimal addition version of Example 1.42. The number $0.346 \times 10^2 = 34.6$ and $0.567 \times 10^3 = 567$, and their sum is exactly 601.6. The answer obtained in the example was $0.601 \times 10^3 = 601$. Thus, there is an error of 0.6. Before proceeding, we introduce some concepts of error analysis.

In this discussion we shall use letters such as a, b, c to denote exact numbers and a^*, b^*, c^* to denote approximations to those numbers. In the context of Example 1.42, the exact answer to the summation is $a = 601.6$, but the 3-digit approximation we obtained was $a^* = 601$. The **absolute error** in approximating a by a^* is defined as $\varepsilon = |a - a^*|$. Thus, the absolute error in the answer to the addition of Example 1.42 is $|601.6 - 601| = 0.6$.

Frequently, the absolute error is of little use. An absolute error of 1 in a number such as 1,000,000 is not as important as an absolute error of 1 in the number 5. Most millionaires would not be too upset if they lost a dollar. However, a boy whose weekly allowance is five dollars would have good reason to be upset if he lost a dollar. A better way to measure the significance of an error is to measure it relative to the number it approximates. The **relative error** in approximating a by a^* is $r = |a - a^*|/|a| = \varepsilon/|a|$. Thus, the relative error in approximating 601.6 by 601 is $r = |601.6 - 601|/601 = 0.6/601 \approx 0.001$. This means that the error is 0.001 per unit, or 1 part in 1000. To make the relative error more intuitively satisfying, we frequently translate it into an equivalent percent. The **relative percent error,** or simply **percent error,** in approximating a by a^* is $r \times 100$. Thus, the percent error in approximating 601.6 by 601 is $p = (0.001)(100) = 0.1\%$, or about one-tenth of 1 percent.

EXAMPLE 1.43 _____

Find the absolute, relative, and percent errors if we approximate:

a. 87.38 by 87.3   b. π by $\dfrac{22}{7}$

SOLUTION

a. $\varepsilon = |87.38 - 87.3| = 0.08$

$r = \dfrac{0.08}{87.38} = 0.0009$

$p = (0.0009)(100) = 0.09\%$

b. Correct to ten places, $\pi = 3.1415926536$, and $22/7 = 3.\overline{142857}$. Thus,

$$\varepsilon = \left| \pi - \frac{22}{7} \right| \approx 0.0013$$

$$r = 0.0013/\pi \approx 0.0004$$

$$p \approx (0.0004)(100) = 0.04\%$$

Thus, 22/7 is a reasonably good approximation of π; it is off by only four one-hundredths of 1 percent.

_____

Since all the arithmetic answers we obtain on our hypothetical computer must be truncated to three digits, we might ask what is the largest error possible in truncating a number to three digits? We drop all digits after the third, so the mantissa can be incorrect by at most one in the last digit. Thus, $\varepsilon = |a - a^*| < 0.001 \times 10^k$. Let $a = m \times 10^k$. Then an upper bound on the relative error is

$$r = \varepsilon/|a| < (0.001 \times 10^k)/(|m| \times 10^k) = 0.001/|m| \leq 0.001/0.1 = 0.01$$

since the smallest possible mantissa is 0.100. It follows that an upper bound for p in the 3-digit system is

$$p < (0.01)(100) = 1\%$$

EXAMPLE 1.44 _____

Find the maximum percent error in an 8-digit floating-point system.

SOLUTION

Since we truncate to the eighth digit, the approximation can be off by at most one in the eighth decimal place. Hence, the absolute error is bounded by $\varepsilon < 0.00000001 \times 10^k$ and

$$r < (0.00000001 \times 10^k)/(|m| \times 10^k) = (0.00000001)/|m| \leq 0.0000001$$

Thus, $p < 0.00001\%$ (that is, one one-hundred-thousandth of 1 percent)! This degree of accuracy will suffice for most calculations.

_____

Now we consider adding the four floating-point numbers: 0.868×10^2, 0.749×10^2, 0.946×10^3, and 0.877×10^3. Since addition is a binary operation, we can add only two numbers at a time. The answer to the first addition must be truncated. Then we add the next number to the truncated sum of the first two, and so on. To begin:

$$0.868 \times 10^2 + 0.749 \times 10^2 = 1.617 \times 10^2 = 0.1617 \times 10^3$$
$$\approx 0.161 \times 10^3$$

We now add the third number to the approximation, 0.161×10^3:

$$0.161 \times 10^3 + 0.946 \times 10^3 = 1.107 \times 10^3 = 0.1107 \times 10^4$$
$$\approx 0.110 \times 10^4$$

The sum of the first three numbers is approximate for two reasons. First, to find their sum we used a number, namely 0.161×10^3, which is itself only an approximate answer for the first addition. Second, we needed to truncate the right-most digit (7) of the second sum. The error in the first calculation is, therefore, brought forward into the second calculation. This is called the **propagation of errors.** To finish the summation, we add the fourth number to the sum of the first three:

$$0.110 \times 10^4 + 0.877 \times 10^3 = 0.110 \times 10^4 + 0.0877 \times 10^4$$
$$\approx 0.110 \times 10^4 + 0.087 \times 10^4$$
$$= 0.197 \times 10^4$$

In decimal form, our final answer is 1970. The exact answer, however, is 1984.7. Hence, the absolute error is $\varepsilon = 14.7$, the relative error is 0.007, and the percent error is $p = 0.7\%$.

Subtraction

Floating-point subtraction is analogous to floating-point addition. Example 1.45 illustrates this operation.

EXAMPLE 1.45 _____

Subtract:

a. $0.829 \times 10^{-1} - 0.726 \times 10^{-1}$ b. $0.627 \times 10^2 - 0.624 \times 10^2$
c. $0.666 \times 10^2 - 0.333 \times 10^1$

SOLUTION

a. $0.829 \times 10^{-1} - 0.726 \times 10^{-1} = 0.103 \times 10^{-1}$

b. $0.627 \times 10^2 - 0.624 \times 10^2 = 0.003 \times 10^2 = 0.300 \times 10^0$
   In putting the answer into floating-point form we have lost two significant digits: the two zeros between the decimal point and the 3. The final mantissa contains two nonsignificant zeros.

c. $0.666 \times 10^2 - 0.333 \times 10^1 \approx 0.666 \times 10^2 - 0.033 \times 10^2 = 0.633 \times 10^2$
   Since a shift is required in the exponent, a truncation is performed on the mantissa.

Multiplication

> **Note 1.25**
>
> To multiply floating-point numbers, multiply the mantissas and add the exponents.

EXAMPLE 1.46

Multiply:

a. $0.274 \times 10^2 \times 0.465 \times 10^1$ b. $0.274 \times 10^2 \times 0.294 \times 10^2$
c. $0.759 \times 10^{-3} \times 0.511 \times 10^2$

SOLUTION

a. $0.274 \times 10^2 \times 0.465 \times 10^1 = 0.127410 \times 10^3 \approx 0.127 \times 10^3$
   We must truncate to three places as usual.

b. $0.274 \times 10^2 \times 0.294 \times 10^2 = 0.080556 \times 10^4 = 0.80556 \times 10^3$
$$\approx 0.805 \times 10^3$$

c. $0.759 \times 10^{-3} \times 0.511 \times 10^2 = 0.387849 \times 10^{-1}$
$$\approx 0.387 \times 10^{-1}$$

Division

> **Note 1.26**
>
> To divide floating-point numbers, divide the mantissas and subtract the exponents.

EXAMPLE 1.47

Divide:

a. $(0.491 \times 10^3) \div (0.747 \times 10^2)$ b. $(0.813 \times 10^{-2}) \div (0.516 \times 10^{-4})$

SOLUTION

a. $(0.491 \times 10^3) \div (0.747 \times 10^2) \approx 0.6572958 \times 10^{3-2} \approx 0.657 \times 10^1$

b. $(0.813 \times 10^{-2}) \div (0.516 \times 10^{-4}) \approx 1.5755813 \times 10^{-2-(-4)}$
$$= 1.5755813 \times 10^2$$
$$= 0.15755815 \times 10^3$$
$$\approx 0.157 \times 10^3$$

Two problems can occur during any of the four basic elementary arithmetic operations: overflow and underflow. **Overflow** occurs when the answer to a floating-point calculation is too large to be represented in floating-point form. In our 3-digit system, the largest possible exponent is $+4$. So, if we attempt to multiply 0.200×10^3 and 0.600×10^2, we will obtain 0.120×10^5. The exponent $+5$ cannot be represented in the 3-digit system, so overflow has occurred. In a real computer, when overflow occurs, the calculation usually terminates.

Underflow occurs when the answer to a floating-point calculation is too small to be represented in floating-point form. For example, suppose we divide 0.800×10^{-3} by 0.200×10^4. We will obtain 0.400×10^{-6}. The smallest exponent that can be represented in the 3-digit system is -5. Thus, 0.400×10^{-6} is too small to be represented and we have underflow. In the case of underflow, many real computers consider the answer to the calculation to be 0, while other computers just issue a warning message.

In this case study we have merely touched on some of the problems associated with floating-point arithmetic. A more thorough treatment of the subject requires a great deal of mathematical sophistication. For more information, the reader may consult almost any book on numerical analysis.

EXERCISES: CASE STUDY 1C

In Exercises 1 through 7 do all arithmetic as if it were being done in the hypothetical 3-digit system. Note any overflow or underflow.

1. Add: **a.** $0.741 \times 10^2 + 0.642 \times 10^2$ **b.** $0.341 \times 10^{-1} + 0.652 \times 10^{-2}$
2. Subtract: **a.** $0.741 \times 10^2 - 0.642 \times 10^2$ **b.** $0.621 \times 10^3 - 0.621 \times 10^2$
3. Multiply: **a.** $0.821 \times 10^3 \times 0.537 \times 10^2$ **b.** $0.921 \times 10^{-3} \times -0.416 \times 10^2$
4. Divide: **a.** $(0.341 \times 10^3) \div (0.911 \times 10^{-1})$ **b.** $(0.274 \times 10^{-3}) \div (0.567 \times 10^3)$
5. Another important difference between real number and floating-point numbers is that some of the usual rules of operation for real numbers do not hold for floating-point numbers. Let $x = 0.267 \times 10^0$, $y = 0.941 \times 10^0$, and $z = 0.175 \times 10^0$.
   a. The associative law for multiplication sometimes does not hold in floating-point. To show this, evaluate both sides of $x(yz) = (xy)z$, using the given values of x, y, and z.
   b. The distributive law sometimes does not hold in floating-point. To show this, evaluate both sides of $x(y + z) = xy + xz$, using the given values of x, y, and z.
6. Find values for x, y, and z in the 3-digit system such that the associative law for addition, $x + (y + z) = (x + y) + z$, does not hold.
7. Frequently, the method used to evaluate an expression affects the accuracy of the answer.
   a. In the 3-digit system, evaluate the expression $a^2 - b^2 - c$, where $a = 0.102 \times 10^0$, $b = 0.101 \times 10^0$, and $c = 0.201 \times 10^{-3}$, using the usual precedence rules of arithmetic.
   b. Evaluate exactly the expression in part a; that is, carry out all answers to as many decimal places as develop in the calculation.

   c. Find the absolute, relative, and percent errors of the approximate value of the expression found in part a.

   d. Use the factorization $a^2 - b^2 = (a + b)(a - b)$ to evaluate $a^2 - b^2 - c$ with the same values given in part a, and note the accuracy of your answer.

8. Referring to the floating-point system described in exercise 9 of Exercises: Case Study 1B, what are the maximum absolute, relative, and percent errors possible in representing real numbers in that floating-point system?

9. In this case study we introduced the notion of propagation of errors. How bad can propagation of errors be?

   a. In the 3-digit system, 1/3 is stored as 0.333. What is the relative error in this approximation?

   b. Using the 3-digit system, add 1/3 to itself nine times. The exact answer is 3. What is the relative error of the approximation obtained in the 3-digit system?

   c. How does the relative error just found in part b of this exercise compare to the relative error found in part a?

   d. Repeat parts b and c for adding 1/3 to itself 27 times.

Problems and Projects

10. To do this exercise, first do exercise 11 in Exercises 1.1. An error that appears in some calculations and often goes unnoticed is conversion error. The mantissa and exponent of floating-point numbers are usually stored in binary (not in decimal as we stored them in Case Studies 1B and 1C). When a number is entered into computer memory, say from a terminal, it is usually entered in decimal. The mantissa (the digits) of the number must, therefore, be converted to binary for storage. Suppose we have a computer with an 8-bit floating-point register in which the characteristic and mantissa are stored in binary.

      Sign   Characteristic   Mantissa

   a. The characteristic holds the exponent in excess-4 form. Why?

   b. What number is stored in the register if its contents are 11101010? If its contents are 00101110?

   c. The decimal number 0.5 will be stored as 01001000. How will 3.25 and 0.8 be stored?

   d. Take the representation for 0.8 just obtained in part c and convert it back to decimal. What number do you get? What is meant by conversion error?

11. a. Write a Pascal program that multiplies two real numbers and, in doing so, causes overflow.

   b. Write a Pascal program that multiplies two real numbers and, in doing so, causes underflow.

CASE STUDY 1D   MEMORY ADDRESSING AND DATA ENCODING

Memory Addressing

The main memory of a computer consists of millions of bits, one after the other, in sequence. For a program to do useful work, it must be able to refer to certain specific locations within the computer's memory. To accomplish this, a computer's memory is divided into cells called **bytes.** Each byte consists of a specific number of bits. The number of bits in a byte varies with the computer. Some computers have 6-bit bytes, some have 8-bit bytes, and some have 9-bit bytes. Each byte can be identified by an **address,** a number unique to that byte. The address is used to locate a byte in much the same way as a house may be located by its address.

Some computers, like the IBM-370 series, use hexadecimal numbers to denote byte addresses in main memory. To see how this is done, we shall look at what is called a **dump** of main memory. Looking at a dump allows us to see what is actually stored in a certain portion of the computer's memory. Reading and interpreting memory dumps is a crucial part of program debugging in assembler language and occasionally in high-level languages such as FORTRAN and COBOL.

Figure 1.17(a) shows a mock-up dump from an IBM-370. Each box represents a byte. Each hexadecimal number on the left is the address of the first byte on that line. Thus, the address of the first byte on the third line is 02A040. Note that there are 32 bytes per line and $32 = 20_{16}$. The bytes on each line are grouped into two sets of 16 and each set of 16 is divided into four groups of four bytes each. The address of the first byte on the second line is, therefore, the address of the first byte on the first line plus 32—but in hexadecimal:

$$02A000_{16} + 20_{16} = 02A020_{16}$$

The address of the first byte on each succeeding line is obtained in a similar manner. The addresses of the other bytes in the dump are obtained by adding the appropriate number to the address of the first byte of the line in which the byte is located. Therefore, the address of the first byte in the 02A000 row is 02A000, the address of the second byte is 02A001, the address of the third byte is 02A002, and so on.

EXAMPLE 1.48

What are the addresses of the two bytes in Figure 1.17(a) that have Xs through them?

SOLUTION

The byte with the X through it in the third line is the fifteenth byte in that row. We thus add 14 to the address at the left of the row:

$$02A040_{16} + E_{16} = 02A04E_{16}$$

The byte with the X through it in the sixth line is the eighteenth byte in that row. We thus add $17 = 11_{16}$ to the address at the left of the row:

$$02A0A0_{16} + 11_{16} = 02A0B1_{16}$$

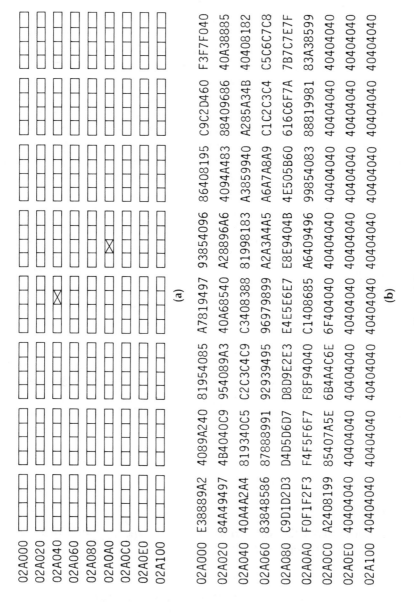

Figure 1.17

Data Encoding

To interpret what is stored in each byte of the dump of Figure 1.17(b), we must be familiar with the data encoding system used on IBM-370 computers, which is the **EBCDIC code** (Extended Binary Coded Decimal Interchange Code).

EBCDIC is an 8-bit code. Special sequences of eight bits are used to represent decimal digits, lower- and uppercase alphabetic characters, and special characters such as the period, blank, and comma. For example, the letter A is represented by 11000001. To make the translation of a given sequence of bits easier for humans, the 8-bit binary is converted into the equivalent hexadecimal number when printed in a dump. Thus, the letter A, which appears in memory as 11000001, is printed as C1 in a memory dump. The contents of one byte, therefore, is represented by two hexits.

EXAMPLE 1.49

How would the following characters appear in a dump: blank (01000000), = (01111110), 8 (11111000), and m (10010100)?

SOLUTION

Character	Binary	Hexadecimal
blank	01000000	40
=	01111110	7E
8	11111000	F8
m	10010100	94

Table 1.5 gives most of the EBCDIC character set.

We now consider the dump illustrated in Figure 1.17(b). When reading the dump, keep in mind that one byte is represented by two hexits.

EXAMPLE 1.50

Use Table 1.5 to describe what characters are stored at the following memory locations:

a. 02A02A   b. 02A078   c. 02A0C1   d. 02A096

SOLUTION

a. The address 02A02A is between 02A020 and 02A040. Since $02A02A_{16} - 02A020_{16} = A_{16} = 10$, the byte we are looking for is the eleventh byte in the 02A020 row. The eleventh byte in this row contains 89. From Table 1.5, 89 represents the letter i.

b. The address 02A078 is between 02A060 and 02A080. Since $02A078_{16} - 02A060_{16} = 18_{16} = 24$, we are looking for the twenty-fifth byte in the 02A060 row; this contains C1, which, from Table 1.5, represents the character A.

c. $02A0C1_{16} - 02A0C0_{16} = 1_{16} = 1$. The second byte in the 02A0C0 row contains 40, which, from Table 1.5, represents the blank character.

Hex	Character	Hex	Character	Hex	Character
40	blank	89	i	D1	J
4A	¢	91	j	D2	K
4B	.	92	k	D3	L
4C	<	93	l	D4	M
4E	+	94	m	D5	N
50	&	95	n	D6	O
5A	!	96	o	D7	P
5B	$	97	p	D8	Q
5E	;	98	q	D9	R
60	−	99	r	E2	S
61	/	A2	s	E3	T
6B	,	A3	t	E4	U
6C	%	A4	u	E5	V
6E	>	A5	v	E6	W
6F	?	A6	w	E7	X
7A	:	A7	x	E8	Y
7B	#	A8	y	E9	Z
7E	=	A9	z	F0	0
7F	"	C1	A	F1	1
81	a	C2	B	F2	2
82	b	C3	C	F3	3
83	c	C4	D	F4	4
84	d	C5	E	F5	5
85	e	C6	F	F6	6
86	f	C7	G	F7	7
87	g	C8	H	F8	8
88	h	C9	I	F9	9

Table 1.5   The EBCDIC Character Set

d. $02A096_{16} - 02A080_{16} = 16_{16} = 22$. The twenty-third byte in the 02A080 row contains 5B, which, from Table 1.5, represents the character $.

EXERCISES: CASE STUDY 1D

1. What is the EBCDIC representation, in both binary and hexadecimal, for each of the following characters? Use Table 1.5.
  a. C   **b.** c   **c.** #   **d.** 4   **e.** ,

2. Write your name in full in EBCDIC (include blanks, initials, and so on). Write each character as it would appear in a dump.

3. What characters are at the following memory locations in the dump of Figure 1.17(b):
    a. 02A094   **b.** 02A10B   **c.** 02A04A

Honeywell-68 computers use a code different from EBCDIC to internally represent data. They use a 9-bit version of the **ASCII code.** Each 9-bit byte is divided into three parts of three bits each. Each set of three bits is represented in a dump by its corresponding octit. For example, the character A in the 9-bit ASCII code is 001000001. If we convert this binary to octal, we obtain 101, which is the way this character would appear in a dump. Table 1.6 gives most of the standard characters in the 9-bit ASCII code in their octal equivalents.

Octal	Character	Octal	Character	Octal	Character
040	blank	101	A	141	a
042	"	102	B	142	b
043	#	103	C	143	c
044	$	104	D	144	d
045	%	105	E	145	e
046	&	106	F	146	f
053	+	107	G	147	g
054	,	110	H	150	h
055	−	111	I	151	i
056	.	112	J	152	j
057	/	113	K	153	k
060	0	114	L	154	l
061	1	115	M	155	m
062	2	116	N	156	n
063	3	117	O	157	o
064	4	120	P	160	p
065	5	121	Q	161	q
066	6	122	R	162	r
067	7	123	S	163	s
070	8	124	T	164	t
071	9	125	U	165	u
072	:	126	V	166	v
073	;	127	W	167	w
074	<	130	X	170	x
075	=	131	Y	171	y
076	>	132	Z	172	z
077	?				
100	@				

Table 1.6   The 9-Bit ASCII Code

Addresses in the Honeywell-68 are also given in octal. Figure 1.18 is a portion of an octal dump from a Honeywell-68.

```
050000 111040150157 160 145040164 150141164040 171157165040
050020 143141156040 162145141144 040164150151 163040144165
050040 155160056012 111146040156 157164054040 164150145156
050060 040171157165 040163150157 165154144040 147157040142
050100 141143153040 141156144040 162145166151 145167040164
050120 150145040143 150141160164 145162040 155 141164145162
050140 151141154056 012107157157 144040154165 143153040167
050160 151164150040 164150145040 162145163164 040157146040
050200 164150145040 142157157153 056012000000 000000000000
050220 000000000000 000000000000 000000000000 000000000000
```

Figure 1.18

As in an IBM-370 dump, the left-most column of numbers gives the octal addresses of the first byte on each line (a byte in a Honeywell-68 is three octits). The memory portion of the dump is divided into four sections of four bytes each. Thus, the address of the byte that contains 160 on the first line is 050004. The address of the byte that contains 155 on the sixth line is 050133.

4. What, in octal, is located at each of the following addresses:
   a. 050174   **b.** 050126   **c.** 050077
5. Translate the first few lines of the dump (all of it, if you feel ambitious).
6. Most microprocessors (a microprocessor is the central processing unit of a microcomputer) have a special register called a **status register.** Each bit of the status register is an **indicator** (or **flag**) that stores the status of a condition of the microprocessor. The 6502 microprocessor, for example, has an 8-bit status register, as shown in Figure 1.19, where S is the sign flag, V the overflow flag, B the break flag, D the decimal mode flag, I the interrupt flag, Z the zero flag, and C the carry flag. The zero flag Z indicates whether the last executed arithmetic operation was zero (1 in bit position 1) or non-zero (0 in bit position 1).

7	6	5	4	3	2	1	0	Bit Number
S	V		B	D	I	Z	C	Flag

Figure 1.19

The contents of the status register are not displayed bit by bit to the assembler programmer. Rather, the status register is shown as two hexits. If the programmer expands the hexits to their binary form, he or she can read each individual bit.

What is in bit positions 2, 4, and 7 of the status register if its contents are A3? 27? 5F?

REFERENCES

Bartee, T. *Introduction to Computer Science.* New York: McGraw-Hill, 1975.

Bohl, M. *Information Processing.* 3d ed. Chicago: Science Research Associates, 1980.

Dudley, U. *Elementary Number Theory.* San Francisco: Freeman, 1978.

Niven, I. *Numbers: Rational and Irrational.* New York: Random House, 1961.

Pennington, R. H. *Introductory Computer Methods and Numerical Analysis.* 2d ed. London: Macmillan, 1970.

We begin this chapter with the study of sets and their properties. Since sets are used in many areas of mathematics, it should not be surprising that sets also arise in a variety of situations in computer science. Some of these situations, such as functions (which we examine in detail in Chapter 5) defined on a set and indexing sets, come from mathematics. Others, such as strings, are more directly related to computer science. In the first four sections, we consider sets, their properties, and some of their uses.

The second topic we consider in this chapter is logic. One of the basic problems studied in logic is determining the truth or falsity of statements. This problem occurs very often in computer programming, where a program will take a particular action (often branch to a different line) if a certain statement (such as "x is greater than 1000") is true but a different action if the statement is false. Logic plays a vital role in examining a computer program for correctness, that is, determining if the program does what it was designed to do.

An understanding of sets makes the study of logic easier. As we shall see, several of the concepts and properties of sets have counterparts in logic. To demonstrate some uses of sets and logic in computer science, we conclude this chapter with a case study on searching and sorting and discuss the logical capabilities of Pascal.

2.1 SETS AND ELEMENTS

A **set** is a well-defined collection of distinct objects. An object within a set is said to be a member of the set or an **element** of the set. The descriptive term "well defined" is explained more fully in Note 2.1.

Note 2.1

By "well defined" we mean there is a rule that enables us to determine whether or not a particular object belongs in the set.

SETS AND LOGIC

For example, we do not speak of the set of "nice colors." This is not a well-defined collection, since it is not clear which colors belong in the set. Two people, acting quite reasonably, could come up with different collections. Thus, when describing sets we do not use vague terms like "nice."

EXAMPLE 2.1

Which of the following are well-defined sets?

a. The set of positive integers less than 25
b. The set of all smart students in the class
c. The set of odd numbers divisible by 10
d. The set of letters in the word *superstars*
e. The set of blonde girls in the school

SOLUTION

a. Well defined
b. Not well defined
c. Well defined, but contains no elements (we will discuss this set soon)
d. Well defined
e. Not well defined (it is not clear where light brown ends and blonde begins)

Although we may be dealing with a well-defined collection of distinct objects, it is not always easy to determine whether a particular object belongs in that set. For example, consider the set of prime numbers. (Recall: a **prime number** is a positive integer greater than 1 that is divisible only by itself and 1.) The list of prime numbers begins 2, 3, 5, 7, 11, 13, 17, 19, 23, 29, 31. It takes time, however, to check whether a very large number, for example 355931447, is in the set of primes. Computers are useful in this regard. In fact, very large prime numbers found by computers are used in cryptography, where secret codes based on them are developed. In addition to being used in the world of espionage, codes based on prime numbers are important in the world of finance, where they are used to protect a firm's banking transactions, merger or liquidation plans, and so on, from the watchful eyes of competitors. Incidentally, the number mentioned, 355931447, is not prime. However, the only prime numbers that divide it evenly are 5431 and 65537.

Generally, we use uppercase letters to denote sets and lowercase letters to denote elements. The special notation relating to sets will be introduced when needed. To describe the fact that x is an element of the set A, we write $x \in A$. The symbol $\in$ is read "is an element of," "belongs to," or "is in." To denote that an element y "is not in A," we write $y \notin A$. For example, if P is the set of prime numbers, then $101 \in P$ while $91 \notin P$.

Set Builder Notation

There are various ways of describing a set. The two most common ways are: (1) listing the elements of the set, and (2) describing the elements in terms of a property that characterizes them. Still another way is by using intersections, unions, or complements of sets. We examine this third method in Section 2.4. Regardless of whether we use a list or a property to describe a set, the description is generally enclosed in braces, { }.

When the list format is used, the elements are separated by commas—for example, $\{3, 6, 8, 12, 17\}$. If there are either infinitely many elements or a large finite number of elements, an **ellipsis** $(\dots)$ is used to indicate that the pattern of the listed elements continues, as in $\{3, 6, 9, 12, 15, \dots, 60\}$. When the ellipsis is used, it should be clear what the pattern is.

When the property format is used, the set description generally begins with a variable and then a colon, which is read "such that" or "where." (*Note:* Some authors use a vertical bar ($|$) instead of a colon.) The colon is followed by the property (or list of properties separated by commas) that characterizes the variable. An example is $\{x : x$ is an even integer$\}$, read "the set of all x such that x is an even integer." If there is a list of properties, the variable must satisfy all of the properties. In this case, the separating commas are read "and." Consider the following sets:

$$P = \{x : x \text{ is a prime number}\}$$
$$Z = \{0, 1, -1, 2, -2, 3, -3, \dots\}$$
$$B = \{2, 4, 6, 8, 10\}$$
$$C = \{y : y \text{ is an integer}, 4 \leq y < 9\}$$

P and C use the property format; Z and B use the list format. In Z an ellipsis is used to describe the infinite set using the list format. We can also write B as $\{8, 4, 6, 10, 2\}$, by Note 2.2.

> **Note 2.2**
>
> The order of the elements in the listing of a set is not important. When the set is infinite, however, the order helps the reader to detect the pattern.

When the ellipsis is used for a large finite set, the pattern must be clear. In this case, we list enough elements so the pattern can be detected, then insert the ellipsis, and finally show where the pattern ends. We illustrate this procedure in Example 2.2.

EXAMPLE 2.2 _____

Write the following sets using the list format, but without writing all the elements:

a. $\{x : x$ is an odd integer, $5 \le x \le 93\}$
b. $\{x : x$ is a perfect square, $x \le 225\}$
c. $\{y : y$ is a positive integer, y is a multiple of 3$\}$

SOLUTION

a. $\{5, 7, 9, 11, \ldots, 91, 93\}$
b. $\{0, 1, 4, 9, 16, 25, \ldots, 225\}$
c. $\{3, 6, 9, 12, 15, \ldots\}$

Note that (c) is an infinite set, while (a) and (b) are finite.

_____

Keep in mind the distinction between a colon and a comma in a set description. The description for part a is read "the set of all x such that x is an odd integer and x is between 5 and 93 inclusive."

EXAMPLE 2.3 _____

Use the sets $P = \{x : x$ is a prime number$\}$, $Z = \{0, 1, -1, 2, -2, 3, -3, \ldots\}$, $C = \{y : y$ is an integer, $4 \le y < 9\}$, and $B = \{2, 4, 6, 8, 10\}$ for the following:

a. List the 15 smallest elements of P.
b. What are the next four elements in the listing for Z?
c. Describe Z using the property format.
d. Describe B using the property format.
e. Describe C with a list format.
f. Which of the sets are finite and which are infinite?

SOLUTION

a. 2, 3, 5, 7, 11, 13, 17, 19, 23, 29, 31, 37, 41, 43, 47

b. 4, -4, 5, -5

c. $Z = \{x:x \text{ is an integer}\}$

d. $B = \{y:y \text{ is an even integer, } 2 \leq y \leq 10\}$

e. $C = \{4, 5, 6, 7, 8\}$

f. P and Z are infinite sets. B and C are finite sets.

Not all infinite sets can be described using the list format. An example is the set $X = \{x:x = 6n - 1, n \text{ is an integer, } x \text{ is not prime}\}$.

At the opposite extreme from an infinite set is the set containing no elements, which we encountered in Example 2.1(c). This set can be described in many ways and comes up often enough that it has been given a special symbol. The set containing no elements is called the **empty set** (or **null set**) and is denoted $\emptyset$. An example is:

$$\{\text{Humans who can run the mile in less than 2 minutes}\} = \emptyset$$

EXERCISES 2.1

1. Give three examples of collections that are not well defined and therefore not sets.

2. Show how the descriptions of the following collections might be altered to describe well-defined collections:
   a. The collection of all elderly people
   b. The collection of all expensive cars
   c. The collection of all tricky questions

3. Let $N = \{1, 2, 3, 4, \ldots\}$. Use the list format to describe each of the following sets:
   a. $\{x:x \in N, x \text{ is a multiple of 5}\}$
   b. $\{y:y \in N, y < 7\}$
   c. $\{x:x \in N, 20 < x \leq 75\}$

4. Express each of the following sets in a different way:
   a. $\{3, 4, 5, 6, 7, 8, 9, 10\}$
   b. $\{x:x \text{ is a positive binary integer, } x < 1011\}$
   c. $\{n:n \text{ is a state in the United States, } n \text{ begins with "C"}\}$
   d. $\{x:x = 2k + 1, k \text{ is an integer}\}$
   e. $\{10, 20, 30, 40, \ldots\}$
   f. $\{x:x \text{ is a real number, } x^2 \text{ is negative}\}$

5. Which of the sets in exercise 4 are finite?

Problems and Projects

6. Write a program that will list the elements in the following sets:
   a. $\{x:x$ is a perfect square, $x < 5000\}$
   b. $\{y:y = 6k + 1, k$ is an integer, $0 \leq k \leq 250\}$
7. Write a program that reads a set A of integers, and reads an integer x. Use the Pascal operator IN to test whether $x \in A$, and have your program print an appropriate message.

2.2 SUBSETS

Occasionally two sets containing precisely the same elements are described differently. We saw many examples of this in Section 2.1. When two sets A and B contain precisely the same elements, we say that A and B are **equal** and write $A = B$.

Another relation also often arises when comparing two sets. We often find that each element of a set A is also an element of set B, but B may happen to have additional elements. If every element of set A is also an element of set B, we say that set A is a **subset** of set B. In this case, we write $A \subset B$. (*Note:* Some authors use $A \subseteq B$). If $A \subset B$ but $A \neq B$, that is, each element of A is an element of B but B has additional elements, then A is said to be a **proper subset** of B. The subset relation is related to equality of sets, as shown in Note 2.3.

Note 2.3

If $A \subset B$ and $B \subset A$, then $A = B$.

Note also that for any set A, we have $\varnothing \subset A$. (Why?)

EXAMPLE 2.4 _____

Which of the following are subsets of other sets described?

a. $A = \{2, 4, 6, 8, 10, \ldots\}$
b. $X = \{x:x$ is a prime number$\}$
c. $E = \{y:y$ is a positive multiple of 8$\}$
d. $S = \{x:x = n^2, n$ is an integer$\}$
e. $T = \{16, 64, 196\}$

SOLUTION

We leave verification of the following to the reader: $E \subset A$, $T \subset A$, and $T \subset S$.

If A is a subset of B, then B is said to be a **superset** of A. In Example 2.4, A is a superset of both E and T, and S is a superset of T.

Four important sets in mathematics are the sets of natural numbers, integers, rational numbers, and real numbers. These are commonly represented by the letters N, Z, Q, and R. We describe these sets in Note 2.4.

Note 2.4

$$N = \{1, 2, 3, 4, \ldots\}$$
$$Z = \{0, 1, -1, 2, -2, 3, -3, \ldots\} = \{\ldots, -3, -2, -1, 0, 1, 2, 3, \ldots\}$$
$$Q = \{x : x = a/b, \, a \in Z, \, b \in Z, \, b \neq 0\}$$
$$R = \{x : x \text{ corresponds to a point on a number line}\}$$

Note the relationship $N \subset Z \subset Q \subset R$.

In some applications, we need to examine many (sometimes all) subsets of a given set. The **power set** of a set A is the collection of all subsets of A and is denoted by $\mathscr{P}(A)$. The power set is, therefore, a set whose elements are themselves sets. To distinguish this fact, we generally refer to the power set as a **collection** (or **family**) of sets.

EXAMPLE 2.5 _____

Find the power set of $A = \{a, b, c\}$.

SOLUTION

$$\mathscr{P}(A) = \{\varnothing, \{a\}, \{b\}, \{c\}, \{a,b\}, \{a,c\}, \{b,c\}, \{a,b,c\}\}$$

Note that A is in $\mathscr{P}(A)$ since $A \subset A$. Also, $\varnothing$ is always in $\mathscr{P}(A)$ no matter what the set A is. A method for generating $\mathscr{P}(A)$ utilizes our knowledge about generating binaries with a given number of bits. If A has n elements, we generate the table of n-bit binaries. Above each column of the table we list a different element of A. Each row of the table then corresponds to one set within $\mathscr{P}(A)$. For any given row of the table, a "1" in a given position indicates that the element at the top of that column is included in the set for that row. However, if the entry is "0," the element is not included. For example, the case for $A = \{a, b, c\}$ is done as follows: There are three elements in A, so we generate all 3-bit binaries and label the three columns with a, b, and c. See Table 2.1.

From this procedure, we see that there is a correspondence between n-bit binaries and subsets of a set of n elements. Note 2.5 follows from Note 1.19.

Note 2.5

If the set A has n elements, $\mathscr{P}(A)$ has 2^n elements.

---

a	b	c	Set from $\mathscr{P}(A)$
0	0	0	$\varnothing$
0	0	1	$\{c\}$
0	1	0	$\{b\}$
0	1	1	$\{b,c\}$
1	0	0	$\{a\}$
1	0	1	$\{a,c\}$
1	1	0	$\{a,b\}$
1	1	1	$\{a,b,c\}$

Table 2.1

EXAMPLE 2.6

Let $B = \{a,b,c,d,e,f\}$.

a. How many elements are in $\mathscr{P}(B)$?

b. To which set in $\mathscr{P}(B)$ does the number 1101_2 correspond?

c. Write the binary that corresponds to $\{b,c,e,f\}$.

d. Which binary corresponds to $\varnothing$?

SOLUTION

a. $\mathscr{P}(B)$ contains $2^6 = 64$ subsets of B.

b. The leading zeros of the 6-bit binary are left out. Thus $1101_2 = 001101_2$, which corresponds to $\{c,d,f\}$.

c. The set corresponds to $011011_2 = 11011_2$.

d. $\varnothing$ corresponds to $000000_2 = 0$.

In Chapter 3, we will develop the techniques for proving Note 2.5.

EXERCISES 2.2

1. Let $A = \{a,b,c\}$, $B = \{c,d\}$, $C = \{a,b\}$, and $D = \{c\}$. Which of the following are true:

    a. $C \subset A$ **b.** $B \subset C$ **c.** $D \subset A$ **d.** $\varnothing \subset B$ **e.** $B \subset A$

2. Which of the following are subsets of the other sets (here $N = \{1,2,3,4,\ldots\}$)? Are any of the sets equal to one another?

    a. $Y = \{y : y \text{ is prime}, y > 2\}$ **b.** $W = \{1,3,5,7,9,\ldots\}$

    c. $X = \{x : x \in N, 1 \le x \le 6\}$ **d.** $F = \{f : f^2 = 25, f > 7\}$

    e. $G = \{g : g = 2n - 1, n \in N\}$

3. Let $C = \{1, 2, 3, 4\}$. How many sets are there in $\mathscr{P}(C)$? List all of them.

4. For the set $B = \{a, b, c, d, e\}$, list all of the subsets having exactly two elements.

5. Why is $\{3, 6, 15\}$ not a subset of $\{2, 15, 3, 10, 60\}$?

6. Is every subset of a finite set a finite set? Is every subset of an infinite set an infinite set? Explain.

7. If $A \subset B$, do you think that $\mathscr{P}(A) \subset \mathscr{P}(B)$? Why or why not?

Problems and Projects

8. Let $A = \{1, 2, 3, 4, 5, 6, 7\}$. Write a program that will generate $\mathscr{P}(A)$.

9. Let $A = \{1, 2, 3, 4, 5, 6, 7\}$. Write a program that will print all the subsets of A that contain precisely:

a. two elements   **b.** three elements

10. Write a program that reads two nonempty sets A and B. Use the set relational operator $<=$ to test whether one of the sets is a subset of the other and print an appropriate message.

11. The null set is represented by [ ] in Pascal. Write a program that reads a nonempty set A, and uses the set relational operators $<=$ and $<>$ to verify the following:

a. A is an improper subset of itself.

b. $\varnothing$ is a proper subset of A.

2.3   VENN DIAGRAMS

In many areas of mathematics, pictures help considerably in solving problems. This is true for sets. A **Venn diagram** (named after the nineteenth-century English mathematician John Venn) is a useful tool for representing relationships among sets.

In many problems involving sets, we restrict our attention to certain types of objects. For example, if $X = \{x : x \in N, x$ is a multiple of 5$\}$, we restrict our attention to the set of natural numbers $N = \{1, 2, 3, 4, \ldots\}$. X is a particular subset of N consisting of those elements in N that are multiples of 5.

The set of objects to which we restrict our attention is called the **universal set** and is denoted by U. In a Venn diagram, U is drawn as a large rectangle, and the sets being considered are drawn as closed regions (usually circles) within U.

EXAMPLE 2.7 _____

Draw a Venn diagram to represent the relationship between the sets $A = \{x : x = y^2, y \in N\}$, that is, the set of perfect squares, and $B = \{z : z \in N, z$ is a multiple of 4$\}$.

SOLUTION

Here we have $U = N = \{1, 2, 3, 4, \ldots\}$. Note that sets A and B overlap since some perfect squares are also multiples of 4 and vice versa. The Venn diagram is displayed in Figure 2.1.

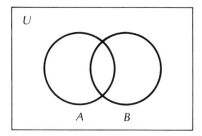

Figure 2.1

When dealing with two sets (whose relationship is unknown), we draw the Venn diagram as in Figure 2.1. This allows for any of the portions of the two circles to be equal to $\varnothing$. If the portion of B that does not overlap with A is $\varnothing$—that is, $B \subset A$—then we have the situation in Figure 2.2(a). If the overlapping portion of A and B in Figure 2.1 equals $\varnothing$, then we could represent the situation as in Figure 2.2(b).

The most general Venn diagram for three sets is displayed in Figure 2.3. Again, we allow for any of the regions to be $\varnothing$.

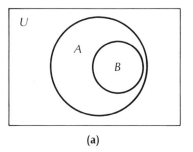
(a)

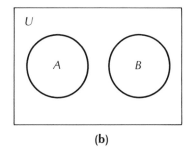
(b)

Figure 2.2

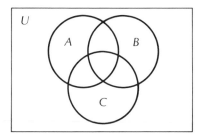

Figure 2.3

If there are n sets, there are 2^n regions in the most general situation. (Note that the area inside of U but outside of all the circles is counted as a region.) Thus, in Figure 2.1 there are four regions. Count the regions in Figure 2.3. You should find eight. When more than three sets are considered simultaneously, the most general Venn diagram cannot be drawn using only circles. In fact, for more than four sets, a Venn diagram can look quite confusing. However, this is not a problem since only rarely are Venn diagrams used for considering more than three sets. We will see in Chapter 4 that Venn diagrams are useful in examining certain problems in probability. In the following sections, we use Venn diagrams to verify various properties for sets and to determine the validity of logical arguments.

EXERCISES 2.3

1. Suppose that $U = \{1, 2, 3, \ldots, 15\}$, $A = \{2, 4, 6, 8, 10\}$, $B = \{5, 10, 15\}$, and $C = \{3, 5, 7, 11\}$. Draw a Venn diagram for these sets. Be careful how they overlap.
2. **a.** Using three infinite sets, give an example where each pair of sets overlap, but all three sets, when taken together, do not overlap. That is, there is no element that is contained in all three sets.
   b. Draw an appropriate Venn diagram for part a.
   c. What is U in your example?
3. For the Venn diagram in Figure 2.4, suppose $U = \{\text{integers}\}$, $N = \{1, 2, 3, \ldots\}$, $P = \{p : p \text{ is a prime}\}$, and $F = \{1, 2, 3, 5, 8, -1, -2, -5\}$. The regions have been labeled with small letters.
   a. Which elements are in region e?
   b. Which elements are in region g?
   c. Which elements are in region h?
   d. Which regions contain no elements, that is, correspond to $\varnothing$?
   e. One of the sets N, P, and F is a subset of another.
   Draw a Venn diagram containing all three sets that will better show the relationship between the three sets.
4. Suppose $U = \{x : x \in N,\ x \le 20\}$, $A = \{3, 6, 9, 12, 15, 18\}$, and $B = \{1, 2, 3, 6, 10, 15\}$.
   a. Draw a Venn diagram for these sets.
   b. Shade in the region that corresponds to $\{1, 2, 10\}$.

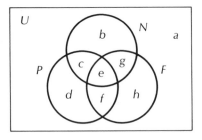

Figure 2.4

c. Describe the set of elements that are within U but not in either of the sets, first by using a list and then in terms of the regions in the Venn diagram.

2.4   INTERSECTIONS, UNIONS, AND COMPLEMENTS

In Section 2.3, we saw how to picture sets with Venn diagrams. We also saw that sets (represented by circles) may overlap. When this happens, we say that the sets intersect. The **intersection** of sets A and B, denoted by $A \cap B$, is the set of all elements that belong to both A and B. Using set notation, $A \cap B = \{x : x \in A, x \in B\}$. The **union** of sets A and B, denoted by $A \cup B$, is the set of all elements that belong to either A or B (or both). "Or both" means that elements contained in both A and B are included in $A \cup B$ along with elements that are contained in only one of the sets A and B. Using set notation, $A \cup B = \{x : x \in A \text{ or } x \in B\}$.

Also, in Section 2.3 we called the set of objects to which we restrict our attention the "universal set." Each set we consider contains elements selected from this universal set U. The **complement** of a set A is the set of all elements that are in U but not in A. We denote the complement of A by A'. Thus $A' = \{x : x \in U, x \notin A\}$.

EXAMPLE 2.8

Suppose $U = \{1, 2, 3, \ldots, 10\}$, $A = \{1, 3, 7, 8, 9\}$, $B = \{2, 3, 6, 7, 10\}$, and $C = \{1, 5, 6, 8\}$. List the elements in each of the following sets:

a. $A \cup B$ b. $A \cap C$ c. B' d. $A \cup C'$ e. $A' \cap B'$

SOLUTION

a. $A \cup B = \{1, 2, 3, 6, 7, 8, 9, 10\}$. Note that although 3 and 7 are in both A and B, we list them each only once in $A \cup B$.

b. $A \cap C = \{1, 8\}$, that is, the set of elements in both set A and set B.

c. $B' = \{1, 4, 5, 8, 9\}$. B' is the set of elements that are in U but not in B.

d. $A \cup C' = \{1, 2, 3, 4, 7, 8, 9, 10\}$. Here, we first find C', which is $\{2, 3, 4, 7, 9, 10\}$, and then find the union of A with this set.

e. $A' \cap B' = \{4, 5\}$. A' is the set of elements in U but not in A. Thus, $A' = \{2, 4, 5, 6, 10\}$. We found B' in part c. $A' \cap B'$ is the set of elements that appear in both set A' and set B'.

EXAMPLE 2.9

Construct Venn diagrams to represent $A \cap B$, $A' \cap B$, and $(A \cup B)'$.

SOLUTION

The Venn diagrams are displayed in Figure 2.5. For $(A \cup B)'$, we first find $A \cup B$, which is the area within either circle, and then take the complement.

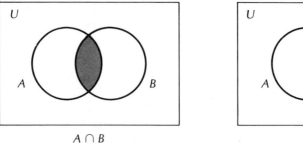

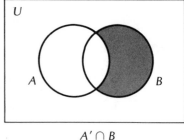

$A \cap B$ $A' \cap B$

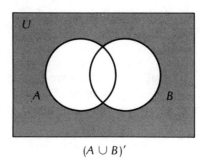

$(A \cup B)'$

Figure 2.5

EXAMPLE 2.10 _____

Suppose $X = \{x : x = y^2, y \in N\}$, $P = \{p : p \text{ is a prime}\}$, and $E = \{s : s = 2t, t \in N\}$.

Determine: a. $P \cap E$ b. $X \cup E$ c. $X \cap P$ d. E'.

SOLUTION

Note that X is the set of all positive perfect squares, and E is the set of all even natural numbers.

a. $P \cap E = \{2\}$. The only even prime number is 2.

b. $X \cup E = \{1, 2, 4, 6, 8, 9, 10, 12, 14, 16, 18, 20, 22, 24, 25, 26, 28, \ldots\}$, the set of positive integers that are either even or perfect squares. Because this set is complicated, it is better to describe $X \cup E$ by properties rather than as a list:

$$X \cup E = \{k : k = y^2 \text{ or } k = 2y, y \in N\}$$

c. $X \cap P = \varnothing$. There are no perfect squares that are also prime.

d. $E' = \{1, 3, 5, 7, \ldots\} = \{w : w = 2x - 1, x \in N\}$. The positive integers that are not even are the positive odd integers. Note that the universal set in this case is N, which is used in the description of E.

_____

Note 2.6 describes several properties of the three set operations $\cap$, $\cup$, and $'$. Recall that two sets are equal if they contain precisely the same elements. One way to check that $X = Y$ is to verify that $X \subset Y$ and that $Y \subset X$ (see Note 2.3).

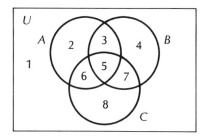

Figure 2.6

Note 2.6   Properties of Set Operations

S1. Associativity of Unions: $(A \cup B) \cup C = A \cup (B \cup C)$

S2. Associativity of Intersections: $(A \cap B) \cap C = A \cap (B \cap C)$

S3. Commutativity of Unions: $A \cup B = B \cup A$

S4. Commutativity of Intersections: $A \cap B = B \cap A$

S5. $(A')' = A$

S6. Distributive Law: $A \cap (B \cup C) = (A \cap B) \cup (A \cap C)$

S7. Distributive Law: $A \cup (B \cap C) = (A \cup B) \cap (A \cup C)$

S8. De Morgan's Law for Sets: $(A \cup B)' = A' \cap B'$

S9. De Morgan's Law for Sets: $(A \cap B)' = A' \cup B'$

To justify S1, an element is in the set described on the left provided that it is in either A or B or it is in C. It is in the set described on the right provided that it is in A or it is in either B or C. But both descriptions define the set consisting of elements contained in any of the three sets A, B, or C. The remaining properties can be verified in a similar manner.

We will now show how Venn diagrams are used to justify certain properties by considering S2. We begin by noting how many sets are involved for the property being examined and draw the general Venn diagram for that number of sets (in this case, three sets). We then number the regions, and compare the regions described by the left side of the equation with those described by the right side. (See Figure 2.6.)

EXAMPLE 2.11 _____

Verify that $(A \cap B) \cap C = A \cap (B \cap C)$.

SOLUTION

For the left side: A corresponds to regions   2, 3, 5, 6

                            B corresponds to regions   3, 4, 5, 7

Therefore, $A \cap B$ corresponds to regions   3, 5

                            C corresponds to regions   5, 6, 7, 8

Therefore, $(A \cap B) \cap C$ corresponds to region   5
For the right side: B corresponds to regions   3, 4, 5, 7
                                       C corresponds to regions   5, 6, 7, 8
Therefore, $B \cap C$ corresponds to regions   5, 7
                                       A corresponds to regions   2, 3, 5, 6
Therefore, $A \cap (B \cap C)$ corresponds to region   5

Since both sides correspond to the same region, S2 is verified.

The other properties can also be justified using Venn diagrams or by using an equality of sets argument. We will give the Venn diagram argument for S7 and the equality of sets argument for S8.

EXAMPLE 2.12

Verify that $A \cup (B \cap C) = (A \cup B) \cap (A \cup C)$ by using a Venn diagram.

SOLUTION

(Refer to Figure 2.6.)

For the left side: B corresponds to regions   3, 4, 5, 7
                                       C corresponds to regions   5, 6, 7, 8
Therefore, $B \cap C$ corresponds to regions   5, 7
                                       A corresponds to regions   2, 3, 5, 6
Therefore, $A \cup (B \cap C)$ corresponds to regions   2, 3, 5, 6, 7
For the right side: $A \cup B$ corresponds to regions   2, 3, 4, 5, 6, 7
                                       $A \cup C$ corresponds to regions   2, 3, 5, 6, 7, 8
Therefore, $(A \cup B) \cap (A \cup C)$ corresponds to regions   2, 3, 5, 6, 7

Since both sides correspond to the same regions, S7 is verified.

EXAMPLE 2.13

Verify that $(A \cup B)' = A' \cap B'$ by using an equality of sets argument.

SOLUTION

We first show that $(A \cup B)' \subset A' \cap B'$. Let x be in $(A \cup B)'$. Then $x \notin A \cup B$, which means that $x \notin A$ and $x \notin B$. Thus, $x \in A'$ and $x \in B'$, which means that $x \in A' \cap B'$. So each element of $(A \cup B)'$ is an element of $A' \cap B'$, that is, $(A \cup B)' \subset A' \cap B'$.

Next, we show that $A' \cap B' \subset (A \cup B)'$. Let y be in $A' \cap B'$. Then $y \in A'$ and $y \in B'$, which means that $y \notin A$ and $y \notin B$. Thus, y is in neither A nor B, that is, $y \notin A \cup B$. Thus, $y \in (A \cup B)'$. So each element of $A' \cap B'$ is an element of $(A \cup B)'$, that is, $A' \cap B' \subset (A \cup B)'$.

Since $(A \cup B)' \subset A' \cap B'$ and $A' \cap B' \subset (A \cup B)'$, we have $(A \cup B)' = A' \cap B'$, by Note 2.3.

EXERCISES 2.4

1. Let $U = \{2, 4, 6, 8, 10, 12, 15\}$, $A = \{2, 6, 12\}$, $B = \{6, 8, 12, 15\}$, and $C = \{4, 8\}$. Determine each of the following:
   a. $A \cup B$ **b.** $B \cap C'$ **c.** $(A \cap B) \cup C$ **d.** $((A \cup B) \cup C)'$

2. Suppose $U = N$, $X = \{x : x = 3y - 1, y \in N\}$, $T = \{t : t = 2n, n < 20, n \in N\}$, and $S = \{s : s = r^2, r \leq 10, r \in N\}$. Determine:
   a. $T \cap S$ **b.** $X \cap T$ **c.** $S \cup T$ **d.** $S \cap X'$

3. Let $A = \{x : x = 2y, y \in N\}$ and let $B = \{z : z = 3w, w \in N\}$. Describe using set notation:
   a. $A \cup B$
   b. $A \cap B$
   c. List the first eight elements for each of the sets from parts a and b.
   d. Give the answers to parts a and b using the congruences studied in Chapter 1.
   e. Describe B' using congruences.

4. Let $A = \{x : x = 7y, y \in N\}$, $B = \{x : x = 4k, k \in N\}$, and $C = \{x : x = 3n + 1, n \in N\}$. List the smallest six elements in each of the following sets:
   a. $A \cap C$ **b.** $B \cup C$ **c.** $A \cap (B \cup C)$ **d.** $A' \cap C$ **e.** $(A \cup B)'$

5. Use Venn diagrams to justify these properties of Note 2.6:
   a. S5   **b.** S6   **c.** S8

6. Use an equality of sets argument to justify these properties of Note 2.6:
   a. S5   **b.** S7   **c.** S9

7. Describe the shaded region in each Venn diagram of Figure 2.7, using an expression involving the sets.

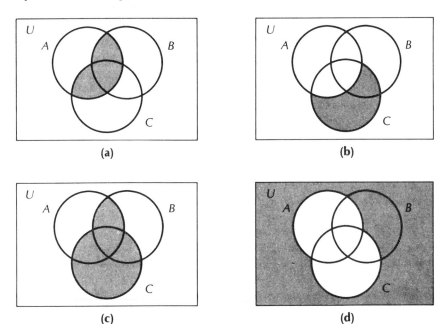

(a)   (b)

(c)   (d)

Figure 2.7

8. Draw a Venn diagram describing each of the following by shading in an appropriate group of regions:
   a. $A \cup (B \cap C)$ **b.** $(A \cap B)'$ **c.** $A' \cup (B \cap C)$
   d. $(A' \cap B') \cup (A \cap C)$ **e.** $A' \cup B$

Problems and Projects

9. Suppose A and B are finite sets having m and n elements. Assume the elements of each set are given in numerical order. Write a program that will find:
   a. $A \cup B$ **b.** $A \cap B$
10. Prove: If $A_1 \subset A_2 \subset A_3 \subset \cdots \subset A_n \subset A_1$, then $A_1 = A_2 = \cdots = A_n$.
11. The elements of a set may be sets themselves. For example, consider $A = \{1, \{1\}, \{1, 2\}\}$.
    a. If $B = \{1\}$ and $C = \{2\}$, which of the following are true: $B \in A$, $B \subset A$, $C \in A$, $C \subset A$?
    b. List the elements of $\mathscr{P}(A)$.
    c. Give an example of a set S such that each element of S is also a subset of S.
12. Prove that the following are equivalent:
    a. $A \subset B$ **b.** $A \cap B = A$ **c.** $A \cup B = B$

The set operations $\cup$ and $\cap$ are represented by $+$ and $*$, respectively, in Pascal. The **difference between sets A and B** is the set of all elements in A that are not in B. The difference is denoted by $A - B$ in Pascal (as well as in mathematics). Let $A = \{1, 2, 5, 8\}$ and $B = \{3, 5, 8, 9\}$. Then $A - B = \{1, 2\}$.

13. Write a program that initializes the universe U to be the set of letters of the alphabet. Your program should then read two sets of letters C and D and determine each of the following:
    a. $C \cup D$
    b. $C \cap D$
    c. $D - C$
    d. C' (*Note:* This is the same as $U - C$.)
    e. $D - (C \cap D)$ (Verify that this is the same answer as in part c.)
14. The **symmetric difference** of two sets A and B, denoted $A \triangle B$, is defined as $A \triangle B = (A \cup B) - (A \cap B)$, or, equivalently, $A \triangle B = (A \cup B) \cap (A \cap B)'$.
    a. Draw a Venn diagram for $A \triangle B$.
    b. Prove: $A \triangle B = B \triangle A$.
    c. Prove: $(A \triangle B) \triangle C = A \triangle (B \triangle C)$.
    d. Prove: $A \triangle A = \varnothing$.
    e. Prove: $A \triangle \varnothing = A$.
    f. Prove: $A = B$ if and only if $A \triangle B = \varnothing$.
15. Write a program to determine the symmetric difference of two sets of integers.

2.5   CONJUNCTION, DISJUNCTION, AND NEGATION

In the next several sections, we will consider various elements of logic. The fundamental building block in logic is the statement. A **statement** is a declarative sentence that is either true or false. In some situations, the statement may correspond to a mathematical statement.

EXAMPLE 2.14 ————————————————————————————

Which of the following are statements:

a. $5 + 7 < 19$

b. How many eggs are there in a dozen?

c. $x + 3 = 7$

d. The capital of California is Los Angeles.

e. This sentence is false.

SOLUTION

a. This is a statement that is true.

b. This is not a statement.

c. This is not a statement. Although we know what value x must have to make the equation into a true statement (that is, 4), we do not know whether x actually has that value. Thus, we cannot tell if the sentence is true or false.

d. This is a statement that is false. (The capital of California is Sacramento.)

e. This is not a statement because it is a declarative sentence that can be neither true and false.

———————————————————————————————————————

In logic we use variables (most commonly p, q, r, and s) to represent statements. Three operations, called **connectives,** are associated with logical statements. These are:

1. **Conjunction**—denoted by $\wedge$

   $p \wedge q$ means "p and q are true" and is read as "p and q."

2. **Disjunction**—denoted by $\vee$

   $p \vee q$ means "p or q, or possibly both are true" and is read as "p or q."

3. **Negation**—denoted by $\sim$

   $\sim p$ means "p is not true" or " it is false that p is true" and is read as "not p."

   Suppose p, q, and r are statements about natural numbers. Statement p is "x is prime," statement q is "x is less than 10," and statement r is "x is divisible by 3." Then $p \wedge q$ means x is prime and less than 10. Thus x is 2, 3, 5, or 7. The statement $q \vee r$ means x is less than 10 or divisible by 3. Thus, x is one of 1, 2, 3, . . . , 8, 9, 12, 15, 18, 21, and so on. The statement $\sim q$ means x is not less than 10. Thus, $\sim q$ is equivalent to the statement "x is greater than or equal to 10."

   We can form **compound statements** by joining statements with connectives. For example, $p \wedge q$, $p \vee q$, and $(p \vee q) \wedge \sim r$ are each compound statements. We then must be concerned with whether the compound statement is true. This depends on whether each (elementary) statement is true or false and which connectives are used. A tool used to examine compound statements is the **truth table.** The truth tables for the simplest statements involving connectives are defined in Note 2.7.

Note 2.7   Truth Tables for Conjunction, Disjunction, and Negation

Conjunction

p	q	$p \wedge q$
T	T	T
T	F	F
F	T	F
F	F	F

Both p and q must be true for $p \wedge q$ to be true.

Disjunction

p	q	$p \vee q$
T	T	T
T	F	T
F	T	T
F	F	F

If either p is true or q is true, or both p and q are true, then $p \vee q$ is true.

Negation

p	$\sim p$
T	F
F	T

The values for $\sim p$ are exactly opposite those of p.

When constructing a truth table, we must consider all possible true-false combinations of the elementary statements. Note 2.8 helps us not to omit any possibilities.

Note 2.8

If there are n elementary statements in a compound statement, the truth table for the compound statement will have 2^n rows.

To begin constructing the truth table, list the n variables across the top and draw 2^n rows. To illustrate, we use the case $n = 3$. In the first column, enter T in the first half of the rows and F in the second half. See Figure 2.8(a). For the second column, in the rows of column 1 just marked T, enter T in the upper half and F in the lower half; in the rows of column 1 marked F, enter T in the upper

(1)	(2)	(3)			(1)	(2)	(3)			(1)	(2)	(3)	
p	q	s	$\cdots$		p	q	s	$\cdots$		p	q	s	$\cdots$
T					T	T				T	T	T	
T					T	T				T	T	F	
T					T	F				T	F	T	
T					T	F				T	F	F	
F					F	T				F	T	T	
F					F	T				F	T	F	
F					F	F				F	F	T	
F					F	F				F	F	F	
(a)					**(b)**					**(c)**			

Figure 2.8

half and F in the lower half. See Figure 2.8(b). Finally in the third column, alternate Ts and Fs. See Figure 2.8(c).

We must now form columns for the parts of the compound statement whose truth table we are constructing. To evaluate the compound statement correctly, we must know which *parts* of the compound statement to evaluate first.

Just as algebra has precedence rules for the order of operations, logic has precedence rules for the connectives. The rules are: Evaluate expressions in parentheses first, then perform negations, then perform conjunctions (from left to right if more than one), and finally do disjunctions (from left to right if more than one). Thus, parentheses have highest precedence, negation has next highest precedence, conjunction next highest, and disjunction has lowest precedence.

We now list as successive column headings the elementary steps involved in evaluating the compound statement. Finally, we fill in the columns by checking the true-false combinations for that row and using the appropriate table from Note 2.7. The following examples illustrate this procedure.

EXAMPLE 2.15

Write the truth table for $p \wedge \sim q$.

SOLUTION

Step 1

p	q
T	T
T	F
F	T
F	F

List the variable names. Below each variable name, list all possible true-false combinations.

Step 2

p	q	$\sim q$
T	T	F
T	F	T
F	T	F
F	F	T

To find $p \wedge \sim q$, we must know p and $\sim q$ individually. We already have p in column 1. Thus, we add a column for $\sim q$ and use the truth table definition for negation along with column 2.

(*Note:* Columns are vertical. The truth table in step 2 has four rows and three columns.)

Step 3

p	q	$\sim q$	$p \wedge \sim q$
T	T	F	F
T	F	T.	T
F	T	F	F
F	F	T	F

We apply the truth table definition of conjunction to columns 1 and 3. For each true-false combination, we determine whether the conjunction is T or F. Thus, for row 1, we see the combination T, F for p and $\sim q$. The combination T, F appears as row 2 of the definition for conjunction in Note 2.7. The conjunction, therefore, has value F, so we record F in column 4.

In some cases, in order to have the statements in the proper order for easy calculation, we recopy a column that is already listed. For example, if we want the truth table for $(a \wedge b) \vee c$, the most convenient listing of the column headings would be $a, b, c, a \wedge b, c, (a \wedge b) \vee c$. When constructing a truth table, we go through the steps described in Example 2.15. However, we list the results within a single table. Therefore, we usually write only the table listed in step 3.

EXAMPLE 2.16

Write the truth table for $\sim(\sim p \wedge q)$.

SOLUTION

p	q	$\sim p$	q	$\sim p \wedge q$	$\sim(\sim p \wedge q)$
T	T	F	T	F	T
T	F	F	F	F	T
F	T	T	T	T	F
F	F	T	F	F	T

EXAMPLE 2.17

Write the truth table for $p \vee (q \wedge s)$.

SOLUTION

p	q	s	$q \wedge s$	$p \vee (q \wedge s)$
T	T	T	T	T
T	T	F	F	T
T	F	T	F	T
T	F	F	F	T
F	T	T	T	T
F	T	F	F	F
F	F	T	F	F
F	F	F	F	F

It is also useful for us to represent verbal expressions using logical symbols and variables.

EXAMPLE 2.18

Suppose that p represents the statement "it is sunny" and q denotes the statement "it is cold." Represent the following using logical symbols:

a. It is sunny and it is not cold.

b. It is cold or it is sunny.

c. It is false that it is cold and it is sunny.

d. It is neither sunny nor cold.

SOLUTION

a. $p \wedge \sim q$

b. $q \vee p$

c. $\sim (q \wedge p)$ Note that the English could also have been interpreted as $\sim q \wedge p$, which is different. This points out that using logical notation is safer than using English.

d. $\sim p \wedge \sim q$ The sentence means "it is not sunny and it is not cold."

Remark   In Chapter 7, logic is used in the more general context of Boolean algebra to analyze *switching networks*. These networks (also called *logic circuits*) are a fundamental technical aspect of computers. By using logic concepts, we can sometimes replace one switching circuit by an equivalent one that is simpler (generally in terms of its structure and/or cost). For details see Chapter 7.

EXERCISES 2.5

1. Which of the following are statements? For those that are statements, tell whether they are true or false.
   a. Help!
   b. What time is dinner?
   c. $(11)(13) = 143$
   d. $11011_2 + 10110_2 = 100001_2$
   e. $CA_{16} > 276_8$
   f. $x^2 + y^2 = 25$
2. Give the negation of each of the following:
   a. My favorite color is blue.
   b. $37 - 19 = 18$
   c. It is six o'clock and all is well.
   d. My ankle is not sore.
   e. $(111_2)(11_2) = 10101_2$ or my name is not Fred.
3. Let p represent the statement "I drive a Chevy" and q represent the statement "I smoke." Describe the following using logical symbols:
   a. I drive a Chevy and I don't smoke.
   b. It is false that I don't smoke or I don't drive a Chevy.
   c. Either I drive a Chevy or I smoke.
   Translate the following logical symbols into English:
   d. $\sim p \vee q$ e. $q \vee \sim q$ f. $\sim (p \wedge q)$
4. Give the truth table for each of the following:
   a. $\sim p \wedge \sim q$ b. $p \vee (\sim q \wedge s)$ c. $s \wedge \sim s$
   d. $\sim p \vee \sim (q \wedge \sim s)$ e. $\sim (p \vee q) \vee s$ f. $(p \vee q) \vee s$
5. Give the truth table for each of the following:
   a. $(p \wedge \sim q) \vee s$ b. $(p \wedge q) \wedge (s \vee t)$ c. $(p \vee q \vee s) \wedge t$

Problems and Projects

6. *Russell's paradox*—In a town with one barber, the barber shaves only those people who do not shave themselves. Who shaves the barber? Put another way, let

$$R = \{x : x \text{ is a set, } x \notin x\}$$

Show $R \in R$. Show $R \notin R$.

2.6   CONDITIONAL AND BICONDITIONAL STATEMENTS

Conditional Statements

We now consider two additional connectives. The first of these arises with statements of the following type: "If Jack fails history, then he will go to summer school." Notice that the sentence does not say whether Jack actually will fail history or whether he will go to summer school. However, if the first condition (Jack fails history) is true, then we are assured that the second condition (he will go to summer school) is also true. Thus, the statement is called conditional. A **conditional statement** (or **implication**) involves two statements p and q and has the form "if p is true, then q is true" or simply "if p, then q." The conditional statement "if p then q" is denoted by $p \rightarrow q$ and is read "p implies q." The truth table for this connective is given in Note 2.9.

Note 2.9   Truth Table for the Conditional

p	q	$p \rightarrow q$
T	T	T
T	F	F
F	T	T
F	F	T

Justifying the truth table for the conditional is more difficult than it was for the other logical connectives we have considered. The truth value assignments for $p \rightarrow q$ are motivated by the desire to have $p \rightarrow q$ true if q is, in some way, deducible from p. If q is true, then it is deducible from p if p is true. This justifies the first row of the table. If q is false, it cannot be deducible from p when p is true (otherwise, q would be true!). This justifies the second row of the table.

The third and fourth rows of the table are somewhat mysterious at first glance. However, they correspond to the situation often described as follows: "If you begin with a false assumption, you can prove anything." That is, you can use a false statement to imply (by logical steps) a true statement, or you can use a false statement to imply (by logical steps) a false statement. For example, suppose we begin with the false statement "$5 = 1$" and we want to prove the true statement "$8 = 8$." We could do this as follows:

1. False statement $5 = 1$
2. Add 1 to both sides $6 = 2$
3. If $a = b$, then $b = a$ $\underline{2 = 6}$
4. Add corresponding sides of last two lines $8 = 8$

Thus, we began with a false statement and got it to imply a true statement: that is, $p \rightarrow q$ in the third row of the truth table is true.

For the fourth line, we want to show that a false statement can be used to prove a false statement. For example, suppose we want to use "$5 = 1$" to prove that "$13 = 29$":

1. False statement $5 = 1$
2. Triple both sides $15 = 3$
3. If $a = b$, then $b = a$ (line 1) $1 = 5$
4. Square both sides of line 3 $\underline{1 = 25}$
5. Add corresponding sides of lines 2, 3, and 4 $17 = 33$
6. Identity $a = a$ $\underline{4 = 4}$
7. Subtract line 6 from line 5 $13 = 29$

Therefore, $p \rightarrow q$ in the fourth row of the truth table is true.

EXAMPLE 2.19

Construct the truth table for $p \to \sim(p \wedge q)$.

SOLUTION

p	q	$p \wedge q$	$\sim(p \wedge q)$	$p \to \sim(p \wedge q)$
T	T	T	F	F
T	F	F	T	T
F	T	F	T	T
F	F	F	T	T

To get each element in the last column, we use the element in column 1 and the element in column 4 and check what that true-false combination produces in the truth table for $p \to q$ (see Note 2.9).

EXAMPLE 2.20

Construct the truth table for $p \vee \sim q \to \sim p \wedge q$.

SOLUTION

p	q	$\sim q$	$p \vee \sim q$	$\sim p$	q	$\sim p \wedge q$	$p \vee \sim q \to \sim p \wedge q$
T	T	F	T	F	T	F	F
T	F	T	T	F	F	F	F
F	T	F	F	T	T	T	T
F	F	T	T	T	F	F	F

Biconditional Statements

The next type of connective arises when a first condition happens precisely when a second condition happens. It is usually expressed in the form: first condition if and only if second condition. Thus, "I sleep if and only if I am tired" means that if I sleep, then I am tired; and if I am tired, then I sleep. Another way to describe this is: If I am tired then I sleep, and if I am not tired then I do not sleep. The connective used in this situation is denoted by $\leftrightarrow$. A statement of the form $p \leftrightarrow q$ is called a **biconditional statement.** The truth value of $p \leftrightarrow q$ is defined to be true if p and q have the same truth value (that is, either both are true or both are false). And $p \leftrightarrow q$ is defined to be false if p and q have different truth values (that is, one is true while the other is false). The truth table for the biconditional is defined in Note 2.10.

Note 2.10   Truth Table for the Biconditional

p	q	$p \leftrightarrow q$
T	T	T
T	F	F
F	T	F
F	F	T

EXAMPLE 2.21

Construct the truth table for $(p \wedge q) \leftrightarrow (p \vee q)$.

SOLUTION

p	q	$p \wedge q$	$p \vee q$	$(p \wedge q) \leftrightarrow (p \vee q)$
T	T	T	T	T
T	F	F	T	F
F	T	F	T	F
F	F	F	F	T

EXAMPLE 2.22

Construct the truth table for $(\sim p \vee q) \leftrightarrow \sim (p \wedge q)$.

SOLUTION

p	q	$\sim p$	q	$\sim p \vee q$	$p \wedge q$	$\sim (p \wedge q)$	$(\sim p \vee q) \leftrightarrow \sim (p \wedge q)$
T	T	F	T	T	T	F	F
T	F	F	F	F	F	T	F
F	T	T	T	T	F	T	T
F	F	T	F	T	F	T	T

Note that at the last step, we focus on columns 5 and 7. If the entries match, the last column entry becomes T. If the entries do not match, the last column entry becomes F.

EXAMPLE 2.23

Construct the truth table for $a \wedge (b \vee c) \leftrightarrow (a \vee b) \wedge c$.

SOLUTION

a	b	c	$b \vee c$	$a \wedge (b \vee c)$	$a \vee b$	c	$(a \vee b) \wedge c$	$a \wedge (b \vee c) \leftrightarrow (a \vee b) \wedge c$
T	T	T	T	T	T	T	T	T
T	T	F	T	T	T	F	F	F
T	F	T	T	T	T	T	T	T
T	F	F	F	F	T	F	F	T
F	T	T	T	F	T	T	T	F
F	T	F	T	F	T	F	F	T
F	F	T	T	F	F	T	F	T
F	F	F	F	F	F	F	F	T

To get the last column entry, we focus on columns 5 and 8. When the entries match, the last column entry is T; otherwise it is F.

EXERCISES 2.6

1. Suppose p represents the statement "the sun is out" and q represents the statement "I am going to the movies." Translate the following into logical symbols:
   a. If the sun is out, then I am not going to the movies.
   b. If it is false that either the sun is out or I am going to the movies, then the sun is not out.
   c. I am going to the movies if and only if the sun is not out.
2. Construct a truth table for each of the following:
   a. $\sim p \rightarrow p \vee \sim q$ b. $p \wedge q \rightarrow \sim p \vee \sim q$
   c. $(p \vee q) \wedge \sim p \rightarrow q$ d. $p \vee \sim p \rightarrow q$
3. Construct a truth table for each of the following:
   a. $a \vee b \leftrightarrow \sim c$ b. $(a \wedge \sim b) \vee c \rightarrow a \vee c$
   c. $a \wedge \sim b \leftrightarrow \sim a \wedge b$ d. $(a \vee b) \wedge (a \vee \sim c) \rightarrow a \vee (b \wedge \sim c)$
   e. $(a \rightarrow \sim b) \leftrightarrow (b \rightarrow a \wedge b)$ f. $\sim(\sim a \wedge b) \rightarrow b \vee \sim a$

2.7 TAUTOLOGIES AND CONTRADICTIONS

Sometimes a compound statement is always true no matter what values are used for the elementary statements. A compound statement that is always true is called a **tautology.** If a compound statement is always false, then the statement is called a **con-**

tradiction. Thus, a statement is a tautology if the last column of its truth table consists entirely of Ts; a statement is a contradiction if the last column of its truth table consists entirely of Fs. The simplest tautology is $p \vee \sim p$. The simplest contradiction is $p \wedge \sim p$.

EXAMPLE 2.24 _____

Show that $(p \vee q) \vee \sim (p \wedge q)$ is a tautology.

SOLUTION

We construct the truth table and verify that the last column consists entirely of Ts.

p	q	$p \vee q$	$p \wedge q$	$\sim(p \wedge q)$	$(p \vee q) \vee \sim(p \wedge q)$
T	T	T	T	F	T
T	F	T	F	T	T
F	T	T	F	T	T
F	F	F	F	T	T

EXAMPLE 2.25 _____

Show that $p \wedge \sim (p \vee q)$ is a contradiction.

SOLUTION

We construct the truth table and verify that the last column consists entirely of Fs.

p	q	$p \vee q$	$\sim(p \vee q)$	$p \wedge \sim(p \vee q)$
T	T	T	F	F
T	F	T	F	F
F	T	T	F	F
F	F	F	T	F

A tautology or contradiction could arise from a conditional or biconditional statement.

EXAMPLE 2.26 _____

Show that $(p \rightarrow q) \leftrightarrow (\sim q \rightarrow \sim p)$ is a tautology.

SOLUTION

p	q	$p \to q$	$\sim q$	$\sim p$	$\sim q \to \sim p$	$(p \to q) \leftrightarrow (\sim q \to \sim p)$
T	T	T	F	F	T	T
T	F	F	T	F	F	T
F	T	T	F	T	T	T
F	F	T	T	T	T	T

At the last step, we see that each element in column 3 matches each element in column 6. Therefore, the last column consists entirely of Ts, and a tautology exists.

Tautologies involving two compound statements joined by the connective $\leftrightarrow$ are of special importance because they show the logical equivalence of two statements. A statement P is said to be **logically equivalent** to a statement Q if the last columns of their truth tables are identical. This is denoted by $\boldsymbol{P \equiv Q}$ (see Note 2.11).

> **Note 2.11**
>
> To say that $P \equiv Q$ is the same as saying that $P \leftrightarrow Q$ is a tautology.

EXAMPLE 2.27

Show that $(a \wedge b) \vee (a \wedge c) \equiv a \wedge (b \vee c)$.

SOLUTION

This is the same as showing that $(a \wedge b) \vee (a \wedge c) \leftrightarrow a \wedge (b \vee c)$ is a tautology.

a	b	c	$a \wedge b$	$a \wedge c$	$(a \wedge b) \vee (a \wedge c)$	$b \vee c$	$a \wedge (b \vee c)$	$(a \wedge b) \vee (a \wedge c) \leftrightarrow a \wedge (b \vee c)$
T	T	T	T	T	T	T	T	T
T	T	F	T	F	T	T	T	T
T	F	T	F	T	T	T	T	T
T	F	F	F	F	F	F	F	T
F	T	T	F	F	F	T	F	T
F	T	F	F	F	F	T	F	T
F	F	T	F	F	F	T	F	T
F	F	F	F	F	F	F	F	T

Finally, we verify that columns 6 and 8 are identical.

De Morgan's Laws

Two very important tautologies are De Morgan's laws, named after a nineteenth-century logician. These laws, given in Note 2.12, are used to simplify logical expressions, to prove other tautologies, and to establish logical arguments. We will study logical arguments in Section 2.8.

> **Note 2.12   De Morgan's Laws for Logic**
>
> $$\sim(p \vee q) \equiv \sim p \wedge \sim q$$
> $$\sim(p \wedge q) \equiv \sim p \vee \sim q$$

We prove the first of De Morgan's laws and leave the second as an exercise.

EXAMPLE 2.28 _____

Prove De Morgan's first law: $\sim(p \vee q) \equiv \sim p \wedge \sim q$.

SOLUTION

We show that $\sim(p \vee q) \leftrightarrow \sim p \wedge \sim q$ is a tautology.

p	q	$p \vee q$	$\sim(p \vee q)$	$\sim p$	$\sim q$	$\sim p \wedge \sim q$	$\sim(p \vee q) \leftrightarrow \sim p \wedge \sim q$
T	T	T	F	F	F	F	T
T	F	T	F	F	T	F	T
F	T	T	F	T	F	F	T
F	F	F	T	T	T	T	T

De Morgan's laws for sets can be proved using set theory arguments or (more easily) by taking advantage of the relationship between $\cap$, $\cup$, $'$ and $\wedge$, $\vee$, $\sim$, respectively. Let p be the statement "$x \in A$" and let q be the statement "$x \in B$." Then $p \wedge q$ means $x \in A \cap B$, $p \vee q$ means $x \in A \cup B$, and $\sim p$ means $x \in A'$. The tautology of Example 2.28 corresponds to the set identity $(A \cup B)' = A' \cap B'$. This is De Morgan's first law for sets. A similar translation of De Morgan's second law into set theory language gives De Morgan's second law for sets. These are stated symbolically in Note 2.13.

> **Note 2.13   De Morgan's Laws for Sets**
>
> $$(A \cup B)' = A' \cap B'$$
> $$(A \cap B)' = A' \cup B'$$

These laws were stated as Properties S8 and S9 in Note 2.6.

EXERCISES 2.7

1. Show that the following are tautologies:
   a. $p \vee \sim p$
   b. $p \vee \sim (p \wedge q)$
   c. $((p \leftrightarrow q) \wedge \sim p) \rightarrow \sim q$
   d. $(p \wedge q) \rightarrow (p \vee \sim q)$
   e. $p \vee \sim q \leftrightarrow \sim (\sim p \wedge q)$

2. Show that the following are contradictions:
   a. $p \wedge \sim p$
   b. $p \wedge (\sim p \wedge q)$
   c. $(p \wedge q) \wedge \sim (p \vee q)$
   d. $(p \rightarrow q) \wedge (\sim q \wedge p)$

3. Example 2.27 established the distributive property for $a \wedge (b \vee c)$. Prove the distributive property for $a \vee (b \wedge c)$ (part a), and prove the associative property for $\vee$ (part b) and for $\wedge$ (part c).
   a. $a \vee (b \wedge c) \equiv (a \vee b) \wedge (a \vee c)$
   b. $(a \vee b) \vee c \equiv a \vee (b \vee c)$
   c. $(a \wedge b) \wedge c \equiv a \wedge (b \wedge c)$

4. **a.** Show that $p \vee q \equiv \sim (\sim p \wedge \sim q)$ by using a truth table.
   b. Show that $p \vee q \equiv \sim (\sim p \wedge \sim q)$ by using one of De Morgan's laws and the property that $\sim (\sim a) \equiv a$.

5. Prove De Morgan's law $\sim (p \wedge q) \equiv \sim p \vee \sim q$.

6. Prove De Morgan's laws for sets by using set theory arguments. Prove part a by showing that $(A \cup B)' \subset A' \cap B'$ and $A' \cap B' \subset (A \cup B)'$. Prove part b by showing that $(A \cap B)' \subset A' \cup B'$ and $A' \cup B' \subset (A \cap B)'$.
   a. $(A \cup B)' = A' \cap B'$
   b. $(A \cap B)' = A' \cup B'$.

7. Prove the following:
   a. $(p \rightarrow q) \equiv (\sim p \vee q)$
   b. $(p \rightarrow q) \equiv \sim (p \wedge \sim q)$
   c. $(p \leftrightarrow q) \equiv (p \rightarrow q) \wedge (q \rightarrow p)$
   d. $(p \wedge q) \vee \sim p \equiv \sim p \vee q$

8. Translate the tautologies from exercises 1(d), 3(a), 4(b), and 7(d) into set theory language, thereby getting several set theory identities.

The **inverse** of $p \rightarrow q$ is $\sim p \rightarrow \sim q$.
The **converse** of $p \rightarrow q$ is $q \rightarrow p$.
The **contrapositive** of $p \rightarrow q$ is $\sim q \rightarrow \sim p$.

9. State the inverse, the converse, and the contrapositive of each of the following statements:
   a. If x is odd, then x^2 is odd.
   b. He's 60, if he's a day.
   c. $x \leq y \rightarrow f(x) \leq f(y)$

10. **a.** Show that the inverse is the contrapositive of the converse.
    b. Show that the contrapositive of the contrapositive is the conditional.

2.8 LOGICAL IMPLICATION AND DECISION TABLES

Logical Implication

One use of logic is analyzing the validity of an argument. An **argument** is an assertion that the conjunction of a set of statements (called **premises**) implies another statement (called the **conclusion**). Thus, an argument has the form $p_1 \wedge p_2 \wedge \cdots \wedge$

$p_n \to q$. Note that q and each of the p_i's may be compound statements. The argument $p_1 \wedge p_2 \wedge \cdots \wedge p_n \to q$ is also written $p_1, p_2, \ldots, p_n \vdash q$. We will write an argument in this second way to stress that we are thinking of it as an argument rather than simply as a conditional. An argument $p_1, p_2, \ldots, p_n \vdash q$ is **valid** if $p_1 \wedge p_2 \wedge \cdots \wedge p_n \to q$ is a tautology. Otherwise, it is **invalid** (also called **fallacious**). If the left side of a conditional is false, the conditional is automatically true no matter what the right side is. In considering the validity of an argument, therefore, we need test only the situation where the left side is true. In this case, the right side must also be true for the conditional to be true. The left side of an argument is the conjunction of $p_1, p_2, \ldots, p_n$. By the definition of the truth table for conjunction, $p_1 \wedge p_2 \wedge \cdots \wedge p_n$ is true only when each of the p_i's is true. Thus we have Note 2.14.

Note 2.14

$p_1, p_2, \ldots, p_n \vdash q$ is a valid argument if q is true whenever all of the p_i's are true.

EXAMPLE 2.29 _____

One of the simpler valid arguments is $(p \to q)$, $p \vdash q$. Show this using a truth table.

SOLUTION

We must show that $((p \to q) \wedge p) \to q$ is a tautology.

p	q	$p \to q$	p	$(p \to q) \wedge p$	q	$((p \to q) \wedge p) \to q$
T	T	T	T	T	T	T
T	F	F	T	F	F	T
F	T	T	F	F	T	T
F	F	T	F	F	F	T

To get column 5, apply the definition of conjunction to columns 3 and 4. The conjunction is true only when both columns 3 and 4 are true. Thus, row 1 is T and the rest are F. Then recopy column 2 and apply the definition of the conditional to columns 5 and 6 to get column 7. Since we have a tautology, $(p \to q)$, $p \vdash q$ is valid.

EXAMPLE 2.30 _____

Analyze the validity of the following argument: An interesting teacher keeps me awake. I stay awake in sociology class. Therefore, my sociology teacher is interesting.

SOLUTION

First translate the premises and the conclusion into logical symbols. Let t = my teacher is interesting; let a = I stay awake; and let s = I am in sociology class. Then translate the argument as follows:

$$(t \rightarrow a), (a \wedge s) \vdash (s \wedge t)$$

To check whether $((t \rightarrow a) \wedge (a \wedge s)) \rightarrow (s \wedge t)$ is a tautology, use the following truth table:

t	a	s	$t \rightarrow a$	$a \wedge s$	$(t \rightarrow a) \wedge (a \wedge s)$	$s \wedge t$	$((t \rightarrow a) \wedge (a \wedge s)) \rightarrow (s \wedge t)$
T	T	T	T	T	T	T	T
T	T	F	T	F	F	F	T
T	F	T	F	F	F	T	T
T	F	F	F	F	F	F	T
F	T	T	T	T	T	F	F
F	T	F	T	F	F	F	T
F	F	T	T	F	F	F	T
F	F	F	T	F	F	F	T

Since the last column contains an F, the implication is not a tautology. Thus, the argument is invalid.

_____

Sometimes the word *or* is used in the exclusive sense. For statements a and b, the **exclusive or** means a or b is true but not both. The exclusive or is translated into logical symbols as $(a \vee b) \wedge \sim (a \wedge b)$.

EXAMPLE 2.31 _____

Analyze the validity of the following argument: Tom is either a carpenter or a plumber (but not both). If he carries a wrench, he's a plumber. Tom is a carpenter. Therefore, he does not carry a wrench.

SOLUTION

Let c = Tom is a carpenter; let p = Tom is a plumber; and let w = Tom carries a wrench. Then translate the argument as:

$$((c \vee p) \wedge \sim(c \wedge p)), w \rightarrow p, c \vdash \sim w$$

Therefore, we must check whether $(((c \vee p) \wedge \sim(c \wedge p)) \wedge (w \rightarrow p) \wedge c) \rightarrow \sim w$ is a tautology. Since there are so many steps in this problem, we will separate the truth table into two parts:

c	w	p	$c \vee p$	$c \wedge p$	$\sim(c \wedge p)$	$(c \vee p) \wedge \sim(c \wedge p)$	$w \rightarrow p$
T	T	T	T	T	F	F	T
T	T	F	T	F	T	T	F
T	F	T	T	T	F	F	T
T	F	F	T	F	T	T	T
F	T	T	T	F	T	T	T
F	T	F	F	F	T	F	F
F	F	T	T	F	T	T	T
F	F	F	F	F	T	F	T

c	$((c \vee p) \wedge \sim(c \wedge p)) \wedge (w \rightarrow p) \wedge c$	$\sim w$	$(((c \wedge p) \wedge \sim(c \wedge p)) \wedge (w \rightarrow p) \wedge c) \rightarrow \sim w$
T	F	F	T
T	F	F	T
T	F	T	T
T	T	T	T
F	F	F	T
F	F	F	T
F	F	T	T
F	F	T	T

Since the final column contains all Ts, the implication is a tautology and the argument is valid.

Decision Tables

In many situations, we make decisions based on whether certain conditions are true or false. For example, we may decide to carry an umbrella based on a weather forecast or a gloomy sky. Two conditions in this example might be "rain is predicted" and "the sky is gloomy." If the first condition is true and the second condition is false, we may decide not to carry an umbrella. The particular combination "true-false" for the two conditions results in an action: "Do not carry an umbrella." The true-false combination together with the action is called a **rule.** Certain occupations such as decision manager, systems analyst, or operations researcher require making decisions in more complex situations. Techniques have been developed to organize the decision process. In this section we discuss one of the tools used for decision analysis—decision tables.

A **decision table** displays in the upper rows of the first column conditions to be tested, and in the lower rows of the first column the actions that may be taken. The upper and lower parts of the table are separated by a double horizontal line. In successive columns of the table, we list the rules. To the right of the conditions, we list all possible true-false combinations for the conditions. For each true-false combination, the action to be taken is indicated by placing a check mark (✓) in the appropriate row. Thus, the decision table lists all the rules in the decision process. See Figure 2.9.

	Conditions and Actions for Bills Overdue ≥ 30 Days	Rules			
Conditions	Highly rated customer	T	T	F	F
	Payment overdue ≥ 60 days	T	F	T	F
Actions	Send second notice	✓			✓
	Send "legal" letter			✓	
	Wait for payment		✓		

Figure 2.9   Decision Table for Bills Overdue ≥ 30 Days

EXAMPLE 2.32

A student wants to set up his course schedule. If he is on probation, he can take a maximum of 12 credits. Otherwise, he can take up to 18 credits. However, if his average is at least B+, he can take up to 21 credits. If his average is less than B+ (and he is not on probation) he may take up to 21 credits if he gets the dean's approval. The program of a student with an appropriate number of credits will be processed. The program of a probationary student who wants more than 12 credits will be rejected. A student with an average of less than B+ who wants between 19 and 21 credits will be sent to the dean. Any program for more than 21 credits will be rejected. Construct a decision table for this problem.

SOLUTION

(An * at the bottom of a column means the case is impossible. Can you justify each *?)

Conditions and Actions																																
On probation	T	T	T	T	T	T	T	T	T	T	T	T	T	T	T	T	F	F	F	F	F	F	F	F	F	F	F	F	F	F	F	F
Credits ≤12	T	T	T	T	T	T	T	T	F	F	F	F	F	F	F	F	T	T	T	T	T	T	T	T	F	F	F	F	F	F	F	F
Credits ≤18	T	T	T	T	F	F	F	F	T	T	T	T	F	F	F	F	T	T	T	T	F	F	F	F	T	T	T	T	F	F	F	F
Credits ≤21	T	T	F	F	T	T	F	F	T	T	F	F	T	T	F	F	T	T	F	F	T	T	F	F	T	T	F	F	T	T	F	F
Average ≥B+	T	F	T	F	T	F	T	F	T	F	T	F	T	F	T	F	T	F	T	F	T	F	T	F	T	F	T	F	T	F	T	F
Process	✓												✓	✓											✓	✓						
Reject									✓						✓	✓																✓
Send to dean																													✓	✓		
			*	*	*	*	*	*			*	*							*	*	*	*	*	*			*	*				

Note that if there are n conditions in a decision table, there are 2^n columns. (Does this sound familiar? Where have we seen this before and in what contexts?) Thus, a decision table with only eight conditions will have 256 columns. This is clearly too cumbersome.

However, it is general practice to simplify decision tables. First, we eliminate all impossible cases. These sometimes occur because of conflicting conditions, as in Example 2.32. When the condition "credits ≤ 12" is true and the condition "credits ≤ 18" is false, there is a conflict, because the number of credits would have to be at most 12 and greater than 18 simultaneously. This is impossible. The second step in simplifying is to combine certain cases where the action taken is determined only by the truth values of some subset of the conditions. For example, suppose there are five conditions C_1, C_2, C_3, C_4, and C_5 and three possible actions A_1, A_2, and A_3. Suppose further that for a certain true-false combination of C_1 and C_3, action A_2 is taken regardless of the truth values for C_2, C_4, and C_5. Then we simplify the table by eliminating all but one of the columns where C_1 and C_3 have the given true-false combination and replace the true-false entries for C_2, C_4, and C_5 by dashes (—). The dashes are referred to as **"don't cares."**

EXAMPLE 2.33 _____

Construct the simplified table for the problem in Example 2.32.

SOLUTION

Conditions and Actions

On probation	T	T	F	—	F	—
Credits ≤ 12	T	F	—	—	—	—
Credits ≤ 18	—	—	T	—	F	—
Credits ≤ 21	—	—	—	T	T	F
Average $\geq B+$	—	—	—	T	F	—
Process	✓		✓	✓		
Reject		✓				✓
Send to dean					✓	

EXERCISES 2.8

1. Determine the validity of the following arguments:
   a. $(p \to \sim q), (\sim r \to p), q \vdash r$ **b.** $p \vee q, \sim p \vdash q$
   c. $p \to q, q \to r, \sim r \vdash \sim p$ **d.** $p \to \sim q, \sim r \to q \vdash \sim p \to r$

2. Analyze the following argument: I will go to school if I do not feel sick. I sleep 12 hours a day and feel sick if and only if I have had too much exercise. I did not sleep 12 hours today. Therefore, I did not have too much exercise and I will go to school.

3. Analyze the following argument: If he does not have an explanation, then he will be found guilty. He either has an explanation or he has been framed. Therefore, if he has been framed, then he will be found guilty.

4. Analyze the following argument: A dog does not bark a lot if and only if he is friendly. A chihuahua barks a lot. My dog is friendly. Therefore, my dog is not a chihuahua.

5. Construct a decision table for the following problem: If it is raining, Judy will go to the movies. If it is sunny, Judy will go to the beach (assume that there are no sun showers). However, if it is above 95°, Judy will go to the movies because she cannot deal with the heat.

6. Construct a simplified decision table for the following: A man would like to make a purchase using his credit card. If he has special credit approval, process the purchase. If his credit limit has been exceeded, reject his request unless he has a gold card in which case he should be sent to the credit manager. If his card has expired, reject his request no matter what color the card is. If his credit

limit has not been exceeded but he already has three transactions today, then send him to the credit manager. Otherwise, process the purchase.

CASE STUDY 2A   SEARCHING AND SORTING

In many situations in computer science, we want to check through a set of data that is stored in computer memory to locate a particular item. Our purpose in looking for the item might be to: (1) alter the item, (2) remove the item from the set, (3) insert the item into the set if it is not already there, or (4) obtain additional information about the item.

Checking through a set for a particular item is called **searching.** Searching a set A for an element x corresponds to determining whether $x \in A$. If a search were repeated for every element in a set B of objects, we would be testing whether $B \subset A$. There are various methods of searching a set, such as sequential (or linear) search, binary search, ternary search, and Fibonacci search. In this case study we examine the sequential and the binary search techniques.

Each search method has its own measure of efficiency. A method's efficiency depends on how the elements of the set are arranged. In one situation a particular method might be very efficient. However, if the elements of the set were arranged differently, that same method might not be as efficient. For example, suppose we are given a well-shuffled deck of cards (the set to be searched) and asked to find the jack of diamonds (the element to be searched for). One way to find the card would be to turn over each card until we locate the jack of diamonds. However, if we are given a brand new, unshuffled deck (a new deck of cards is arranged so that all cards of the same suit are grouped together; and within a suit, the cards are arranged in order A, 2, 3, . . . , J, Q, K), the same technique would not be as efficient. Suppose, instead, we try the following: With three "cuts" of the deck, we locate the diamonds. From there we simply count to locate the jack. By using the second method with the new deck, we reduce the average number of cards we must turn over from 26 for the first method to about 4.

The new deck can be searched more quickly than the well-shuffled deck because the new deck has been (pre-)sorted. **Sorting** is the procedure that begins with an unordered set and ends with a completely ordered set. Some sorted sets that we use everyday are dictionaries, telephone directories, library card catalogs, and house numbers on any street. Recall that the order of the elements within a set does not matter. That is, $\{2, 3, 6, 9\} = \{3, 2, 9, 6\}$. By sorting, however, we rearrange the elements of a set into some particular order. The most commonly desired order for sets with numeric data is numerical order, and for sets with nonnumeric data, alphabetical order.

For the rest of this case study, we restrict our attention to sets of real numbers stored as one-dimensional arrays (see Chapter 6). We first consider searching techniques and then a sorting procedure.

Sequential Search

Sequential search is the simplest search technique and was, in fact, the technique described for the well-shuffled deck of cards. This technique is good for small, randomly ordered sets. Two common situations in which we use a sequential search are: (1) when we are searching to see *if* a particular element is in the set, and (2) when we are searching for an element in the set having some special property.

We assume the elements of set A are stored in random order and that A has n elements. For example, let $A = \{3, 1, 2, 5, 8, 6, 13\}$. We also assume the elements of set A are stored, in the order given, in a one-dimensional array (or list variable) A. We can use subscripts, therefore, to specify a particular number in the array. i_I represents the *i*th number in the array. So $A_1 = 3$, $A_2 = 1$, $A_3 = 2$, and so on.

EXAMPLE 2.34

Write a program that will read a set A of at most 50 elements, its size N, a number K, and test whether $K \in A$.

SOLUTION

```
program element(input,output);
var n,i: integer;
    k:real;
    a:array[1..50]of real;
begin
   writeln('input a positive integer');
   read(n);
   if n>50 then
      writeln('array is too large')
   else
      begin
         writeln('input',n,' real numbers');
         for i:=1 to n do
            read(a[i]);
         writeln('input the real number being searched for');
         read(k);
         i:=1;
         while(a[i]<>k) and (i<n) do
            i:=i+1;
         if a[i]=k then
            writeln(k,'is in the set')
         else
            writeln(k,'is not in the set')
      end
end.
```

In Example 2.34, we do not have to search through the whole set if we find the element K before i is incremented to N. However, situations do occur when we must search the entire set—for example, in the second type of sequential searching, when we are searching for an element having a particular property. For instance, we may want to find the largest element in the set.

EXAMPLE 2.35

Write a program that will read a set A of at most 50 elements, its size N, and determine the largest element in A.

SOLUTION

```
program largest(input,output);
var n,i: integer;
    large: real;
    a: array[1..50] of real;
begin
    writeln('input a positive integer');
    read(n);
    if n>50 then
       writeln ('array is too large')
    else
       begin
          writeln('input',n,'  real numbers');
          for i:=1 to n do
             read (a[i]);
          large:= a[1];
          for i:=2 to n do
             begin
                if a[i] > large then
                   large:=a[i]
             end;
          writeln ('large=',large)
       end
end.
```

Binary Search

If our set has already been sorted, then performing a sequential search wastes time. For example, if we were looking for the word *jodhpurs* in a dictionary, we would not search through the dictionary, one word at a time, beginning with "a, aardvark." A more effective way would be to open the dictionary halfway and determine whether the word we want is before or after the page to which we have opened. In this way we can eliminate half the dictionary in one step. We then open halfway the part of the dictionary that is still of interest and determine if our word is before or after that page. We can now eliminate half of this part; that is, we eliminate an additional one-fourth of the original dictionary. We continue in this manner and eventually converge on the word we want.

Note 2.15

In a binary search, we continually eliminate from our set half the remaining elements we are considering until we find the element we want.

As another example, suppose a friend picks a number between 1 and 500 and writes it down. Our task is to guess the number. The set we are searching is $A = \{1, 2, 3, \ldots, 499, 500\}$, which is in numerical order. After each guess, our friend will tell us if the number written is higher or lower than (or perhaps the same as) our guess. If we use the binary search, we will find the number in at most nine guesses. This is because the size of the set of possible values will be shrunk on successive guesses to 250, 125, 63, 32, 16, 8, 4, 2, 1. At each stage, the procedure for determining our next guess is to average the largest and smallest numbers in the remaining set and then truncate (that is, delete) any decimal part.

EXAMPLE 2.36

Suppose our friend writes 171. Write the succession of events in our search.

SOLUTION

Set	(Largest + Smallest)/2	Guess	Response
$\{1, 2, 3, \ldots, 500\}$	$(500 + 1)/2 = 250.5$	250	lower
$\{1, 2, 3, \ldots, 249\}$	$(1 + 249)/2 = 125$	125	higher
$\{126, 127, \ldots, 249\}$	$(126 + 249)/2 = 187.5$	187	lower
$\{126, 127, \ldots, 186\}$	$(126 + 186)/2 = 156$	156	higher
$\{157, 158, \ldots, 186\}$	$(157 + 186)/2 = 171.5$	171	correct

In this case, it took 5 guesses. Compare this to the 171 guesses it would have taken by a sequential search!

In most situations, the set we are searching does not consist of consecutive integers. However, we can still use a binary search provided the elements of the set have been ordered. A set in which the elements have been ordered is called an **ordered set.**

Keep in mind that the set is stored using a list (that is, subscripted) variable and the *subscripts* are consecutive integers. Thus, when performing a binary search using the computer, we use the same procedure as that described for the last example except that we average the largest and smallest *subscripts*, truncate, and then examine the element corresponding to that truncated subscript. Also note that the element for which we are searching may not be in the set. Our program must handle this situation explicitly.

EXAMPLE 2.37

Write a program that reads a set A (assumed to be sorted) of at most 50 elements, the number of elements N, and an element K, and performs a binary search. If K is found, describe the position within the ordered set where K is located.

SOLUTION

```
program binary(input,output);
var n,small,large,signal,sub,i: integer;
    k: real;
    a: array [1..50] of real;
begin
   writeln('input a positive integer');
   read (n);
   if n>50 then
      writeln ('array is too large')
   else
      begin
         small:=1;
         large:=n;
         signal:=0;
         writeln('input', n,'  numbers in increasing order');
         for i:=1 to n do
            read (a[i]);
         writeln('input the element being searched for');
         read(k);
         while (small <> large) and (signal=0) do
            begin
               sub:=trunc((small+large)/2);
               if a[sub]>k then
                  large:=sub-1
               else
                  if a[sub]<k then
                     small:=sub+1
                  else
                     signal:=1     (* element has been found *)
            end;
         if signal=1 then
            writeln(k,'is item number',sub,' in the list')
         else
            writeln(k,'is not in the list')
      end
end.
```

Sorting

We now consider one simple method of sorting, the method of successive minima. More complex (and more efficient) methods such as bubble sort, quick sort (Chapter 6), and heap sort (Chapter 8) will be considered after we discuss lists and arrays in Chapter 6. In the method of successive minima, we do a sequential search of our unordered set to find the smallest element, which we then move to position 1. We move the element that had been in position 1 to where the smallest element used to be. We then locate the next smallest element by doing a sequential search for the smallest element of the subset beginning with the element in the second position. We move this element to the second position, and move the element that had been in position 2 to where the second smallest element used to be. We continue the process in the same manner until we reach the end of the set.

EXAMPLE 2.38

Write a program that reads a randomly ordered set A of at most 50 elements, its size N, and orders the elements of A using the method of successive minima.

SOLUTION

```
program scsmin(input,output);
var n,i,place,k: integer;
    min: real;
    a: array[1..50] of real;
begin
   writeln('input a positive integer');
   read(n);
   if n>50 then
      writeln ('array is too large')
   else
      begin
         writeln('input',n,' numbers in any order');
         for i:=1 to n do
            read (a[i]);
         for i:=1 to n do
            begin
               min:=a[i];
               place:=i;
               for k:=i+1 to n do
                  if a[k]<min then
                     begin
                        min:=a[k];
                        place:=k
                     end;
               a[place]:=a[i];
               a[i]:=min
            end;
         writeln('the ordered list is as follows:');
         for i:=1 to n do
            writeln (a[i])
      end
end.
```

When we need to search a set, the decision of whether to sort the set before searching depends on various factors, the most important of which is how often we will search the set. Infrequent searches may not warrant the time and effort spent on sorting. Also, in situations where the set is large and/or there are frequent updates (that is, insertions or deletions) done on the set, random access techniques might be more fruitful. Which sorting technique to use, if we decide to sort the set, depends mostly on the form of input (if known) and the purpose of sorting. More will be said about this in Chapter 6. See also Tremblay and Sorenson (1976).

EXERCISES: CASE STUDY 2A

1. Suppose the elements of A are input in the order 5, 7, 12, 8, 4, 6, 9 in Example 2.38. List the elements of A in order after each pass through the outer DO loop, that is, each time i is incremented.

2. Ask a friend to pick a number (integer) between 1 and 200 and to write it down. Using a binary search, you should be able to guess the number within eight guesses, while the friend tells you "lower" or "higher" (or "correct") after each guess.

3. Suppose your friend had picked 118 in exercise 2. Write the succession of events in this search, as we did in Example 2.36.

4. Write a program that will determine the second largest element in a set.

5. Given sets A and B and their sizes n and m, we can test whether $A \subset B$ by testing whether $a_i \in B$, $1 \leq i \leq n$. That is, we successively search through B for each element of A. Write a program to test whether $A \subset B$.

6. In a diving competition there are eight judges, each of whom gives a score selected from the set 0, 0.5, 1, 1.5, 2, 2.5, ..., 9.5, 10. The weighted score for any dive is determined by the following procedure:
   i. Eliminate the highest and lowest scores.
   ii. Add up the remaining six scores.
   iii. Multiply the sum by a degree of difficulty factor, d (d is generally between 1.0 and 3.0).

   Write a program that reads in the set S of eight scores, the degree of difficulty, d, and determines the weighted score for the dive.

Use the following information for exercises 7–11:

We can declare X and Y to be logical variables in Pascal with the statement:

$$\text{VAR } X,Y: \quad \text{BOOLEAN;}$$

The variables X and Y can then take on the logical values TRUE and FALSE. In Pascal the logical connectives $\wedge$, $\vee$, and $\sim$ are translated as AND, OR, and NOT, respectively.

7. Following are several expressions in Pascal. Write the corresponding expressions using logical symbols.
   a. P AND (NOT (Q OR (NOT S)))
   b. ((P OR Q) OR (NOT S)) AND (NOT Q)
   c. (P AND Q) OR ((NOT Q) AND S)

8. Suppose that for exercise 7, P, Q, R, and S have the values TRUE, TRUE, FALSE, and TRUE. Evaluate the expressions (a), (b), and (c) from exercise 7.

9. Translate the following expressions into Pascal:
   a. $\sim p \vee q$ **b.** $\sim p \wedge \sim q$
   c. $(p \wedge q) \wedge (s \vee t)$ **d.** $p \vee (\sim q \wedge r)$

10. Suppose the registration problem of Example 2.32 is to be computerized. Use the simplified decision table of Example 2.33 to write a Pascal program fragment for the logical-if tests involved.

11. Write a program that will generate and print the truth table for
    a. $p \vee q$ **b.** $(p \wedge \sim q) \vee s$ **c.** $\sim (p \vee \sim q) \wedge s$

REFERENCES

Beckman, F. S. *Mathematical Foundations of Programming*. Reading, Mass.: Addison-Wesley, 1980.

Belford, G. G., and C. L. Liu. *Pascal*. New York: McGraw-Hill, 1984.

Bergman, M., J. Moor, and J. Nelson. *The Logic Book*. New York: Random House, 1980.

Bittinger, M. L. *Logic, Proof, and Sets*. 2d ed. Reading, Mass.: Addison-Wesley, 1982.

Graham, N. *Introduction to Pascal*. 2d ed. New York: West, 1983.

Hurley, R. B. *Decision Tables in Software Engineering*. New York: Van Nostrand Reinhold, 1983.

Scheid, F. *Computers and Programming*. New York: McGraw-Hill, 1982.

Tremblay, J. P., and P. G. Sorenson. *An Introduction to Data Structures with Applications*. New York: McGraw-Hill, 1976.

CHAPTER 3

In Chapter 2 we studied logic and examined its relationships to computer science, particularly to decision tables and logical-if statements. We also studied sets and their properties and saw how sets arise in various contexts in computer science and data processing. Important applications involving sets deal with searching and sorting algorithms, which are of fundamental importance for efficient programming. Recall that sorting orders a set, and thus deals with the rearrangement of objects. The subject of this chapter, **combinatorics,** is a branch of mathematics concerned with the arrangement of elements from a set. A typical problem in this field is: Can the elements of a set be arranged in a particular manner (described by certain conditions) and, if so, in how many ways can they be arranged? Another problem may be the following. New York Telephone just installed an additional area code for New York City because it was running out of telephone numbers for the 212 area code. How many phone numbers can there be within one area code?

We begin by studying mathematical induction, a primary technique for proving theorems, especially those involving counting problems. Mathematical induction is also the basis for recursively defined functions and recursive procedures used in programming. In Case Study 3A we shall see how induction can be used as a technique for proving programs correct.

For most of this chapter, we consider a variety of problems dealing with the arrangement of objects from a set. Several useful formulas will be developed for counting these arrangements. These formulas will be used throughout the rest of the book to:

1. Determine probabilities by calculating the proportion of items having a specific property (Chapter 4)
2. Analyze algorithms and recursive functions (Chapters 3 and 5)
3. Study stacks and queues (Chapter 6)
4. Solve graph theory problems (Chapter 8)

COMBINATORICS

3.1 MATHEMATICAL INDUCTION

Sometimes, to discover properties of the natural numbers, we examine special cases, look for a pattern, and make an inference. We illustrate this procedure with several examples.

The inequalities $2^1 > 1$, $2^2 > 2$, $2^3 > 3$, $2^4 > 4$ suggest $2^n > n$ for each natural number n. We show this is so in Example 3.1. The equations $1 = 1^2$, $1 + 3 = 2^2$, $1 + 3 + 5 = 3^2$, $1 + 3 + 5 + 7 = 4^2$ suggest that the sum of the first n odd natural numbers is n^2. We prove this in Example 3.7.

Consider the polynomial $n^4 - 10n^3 + 35n^2 - 50n + 24$. If we substitute 1, 2, 3, or 4 for n, the value of the polynomial is zero. This might suggest that $n^4 - 10n^3 + 35n^2 - 50n + 24$ is zero for each $n \in N$. This is not true, however, as we see by substituting $n = 5$: $5^4 - 10(5)^3 + 35(5)^2 - 50(5) + 24 = 24$.

The number 12252240 is divisible by 1, 2, 3, ..., 18. This might lead us to conclude that 12252240 is divisible by all natural numbers (a foolish conclusion since 12252240 is not divisible by 12252241; it is also not divisible by 19).

Consider the polynomial $n^2 + n + 41$. The value of $n^2 + n + 41$ is a prime number for $n = 1, 2, 3, \ldots, 40$. Can we conclude that $n^2 + n + 41$ is always a prime number? No. In fact, $n^2 + n + 41$ is not prime for $n = 41$.

Using observations alone, we can never conclude a statement is true for all natural numbers because there are infinitely many natural numbers. We need a method to prove such statements true without examining them for all natural numbers. Mathematical induction is such a method. To understand how this method works, imagine an infinitely long row of dominoes standing on end, spaced close to one another. If we knock down the first domino, our intuition tells us that all the dominoes will eventually fall. Why? First, we start the falling process by knocking over the first domino. Call this action

$$S_1: \text{the first domino falls}$$

The action guiding our intuition is that if a domino anywhere down the line should fall, it will knock over the next domino in line (this is a familiar disaster to every domino enthusiast). Call this action

S_k: if a domino (say the kth one) falls, it knocks down
the next (or $k + 1$st one)

We believe statements S_1 and S_k are true, so we must conclude that all the dominoes fall.

To be more specific, S_1 is true because we knock over the first domino on purpose. Since S_k is always true, it must be true for $k = 1$. Thus, if the first domino falls, it knocks over the second. S_k is also true for $k = 2$. Thus, if the second falls, it knocks over the third, and so on. In this way every domino must eventually fall.

To further test our intuition, suppose we want to convince ourselves that the 1000th domino will fall. We purposely knock over the first (this is statement S_1). Then domino 1 knocks over domino 2 (S_k with $k = 1$), domino 2 knocks over domino 3 (S_k with $k = 2$), ..., domino 998 knocks over domino 999 (S_k with $k = 998$), and domino 999 knocks over domino 1000 (S_k with $k = 999$). Clearly, the argument will apply to any domino, no matter how far down the line it may be.

We now use the analogy of a row of dominoes to discuss proofs of mathematical statements by induction. Suppose S_n is a statement that in some way depends on the positive integer n. For example

$$S_n: 2^n > n$$

or

$$S_n: 1 + 3 + 5 + \cdots + (2n - 1) = n^2$$

or

$$S_n: n^4 - 10n^3 + 35n^2 - 50n + 24 = 0$$

Suppose further that we wish to prove S_n true for all $n \in N$. Let the statement "S_n is true" be equated with the statement "domino n falls."

Note 3.1   The Principle of Mathematical Induction

The Principle of Mathematical Induction states that S_n is true for all positive integers n if

1. S_1 is true (that is, the first domino falls), and

2. If S_k is true, then S_{k+1} is true (that is, if the kth domino falls, then the $k + 1$st domino falls).

Step 1 is called the **initial step,** and step 2 is called the **inductive step.** In step 2, the statement "S_k is true" is referred to as the **inductive hypothesis.** It is so called because we hypothesize (or assume) that S_k is true, and we must show that this implies S_{k+1} is true. Thus, whenever S_n is true about $n = k$ objects, S_n is true about $n = k + 1$ objects.

As in the row of falling dominoes, successive application of step 2 implies that all the statements S_n are true. The first four cases in this process are shown in Table 3.1.

Application of Step 2	Value of k	Result
1st	1	S_1 implies S_2 is true
2nd	2	S_2 implies S_3 is true
3rd	3	S_3 implies S_4 is true
4th	4	S_4 implies S_5 is true
⋮	⋮	⋮

Table 3.1

Now let us consider an example.

EXAMPLE 3.1

Prove by induction: $2^n > n$ for all $n \in N$.

SOLUTION

Step 1 Show that S_1 is true. $2^1 > 1$ is trivially true.

Step 2 Inductive hypothesis: Assume S_k: $2^k > k$ is true.

We must show that S_{k+1}: $2^{k+1} > k + 1$ is true. We begin by rewriting 2^{k+1} as $2(2^k)$ so we can use the inductive hypothesis:

$$2^{k+1} = 2(2^k) > 2(k) = k + k \geq k + 1$$

(The inductive hypothesis $2^k > k$ is used for the first inequality.) Thus, assuming S_k: $2^k > k$, we have shown that S_{k+1}: $2^{k+1} > k + 1$. The principle of mathematical induction then implies S_n is true for all n.

In all inductive proofs, we must write the expression for S_{k+1} (2^{k+1} in Example 3.1) in terms of an expression used in S_k so we can use the inductive hypothesis. This usually involves factoring a product or grouping the first k terms of a sum.

Before considering more examples of inductive proofs, we shall discuss a notational device that will prove useful in this and subsequent chapters, namely, summation notation.

Many of the formulas we shall establish involve sums of integers. It is useful, therefore, to have a special notation for sums. Generally, the terms we add are described in a way similar to the way elements of sets are described. For example, the set X of even numbers between 5 and 17 is described by $X = \{x : x = 2y, y \in N,$

$3 \le y \le 8\}$. Using **summation notation**, adding the numbers in X is described by

$$\sum_{y=3}^{8} 2y \quad \text{or} \quad \sum_{y=3}^{8} 2y$$

Note 3.2

The symbol $\sum$ (the Greek letter "sigma") is used to indicate a sum.

The summation $\sum_{y=3}^{8} 2y$ is read "the sum of $2y$ as y goes from 3 to 8" or "the sum as y goes from 3 to 8 of $2y$." This means y initially has a value of 3 and is successively incremented to 4, 5, 6, 7, and 8. Thus,

$$\sum_{y=3}^{8} 2y = 2(3) + 2(4) + 2(5) + 2(6) + 2(7) + 2(8)$$

$$= 6 + 8 + 10 + 12 + 14 + 16 = 66$$

In the sum $\sum_{y=3}^{8} 2y$, y is called the **index**. A sum may involve more than one variable, as in $\sum_{i=1}^{30} (5i + 3x - 2)$. However, only the variable being incremented (in this case, i) is the index. The index is sometimes called a **dummy variable** because its name is unimportant. So

$$\sum_{y=3}^{8} 2y = \sum_{k=3}^{8} 2k = \sum_{t=3}^{8} 2t = 66$$

The numbers 3 and 8 are called the **initial** and **final values** (or the **lower** and **upper limits**) of the sum.

Summation notation does not calculate the sum for us, but rather makes the sum appear more compact. This is particularly evident when we are adding many numbers.

EXAMPLE 3.2 ———————————————————————————————

Express the following sums using summation notation:

a. The sum of the first 1000 positive integers

b. The sum of the squares of natural numbers less than 15

c. The sum of the elements of the set $Y = \{y : y = 3n - 1, n \in N, 20 \le n \le 50\}$

SOLUTION

a. $\displaystyle\sum_{i=1}^{1000} i$ b. $\displaystyle\sum_{x=1}^{14} x^2$ c. $\displaystyle\sum_{n=20}^{50} (3n - 1)$

In the last case, we use parentheses so that $\sum_{n=20}^{50} (3n - 1)$ will not be confused with $\sum_{n=20}^{50} 3n - 1 = (60 + 63 + 66 + \cdots + 150) - 1$.

EXAMPLE 3.3 _____

Calculate the following sums:

a. $\displaystyle\sum_{j=1}^{8} j$ b. $\displaystyle\sum_{i=5}^{15} (2i + 1)$ c. $\displaystyle\sum_{x=1}^{6} x^2 - \sum_{x=1}^{6} x$

SOLUTION

a. $\displaystyle\sum_{j=1}^{8} j = 1 + 2 + 3 + 4 + 5 + 6 + 7 + 8 = 36$

b. $\displaystyle\sum_{i=5}^{15} (2i + 1) = 11 + 13 + 15 + 17 + 19 + 21 + 23 + 25 + 27 + 29 + 31 = 231$

c. $\displaystyle\sum_{x=1}^{6} x^2 - \sum_{x=1}^{6} x = (1^2 + 2^2 + 3^2 + 4^2 + 5^2 + 6^2) - (1 + 2 + 3 + 4 + 5 + 6)$
$$= (1 + 4 + 9 + 16 + 25 + 36) - 21 = 91 - 21 = 70$$

The next three examples illustrate useful properties of summation notation.

EXAMPLE 3.4 _____

Evaluate each of the following:

a. $\displaystyle\sum_{i=5}^{10} 3i$ b. $\displaystyle\sum_{j=1}^{6} 3(j + 4)$ c. $\displaystyle\sum_{x=8}^{13} 3(x - 3)$ d. $3\displaystyle\sum_{i=5}^{10} i$

SOLUTION

a. $\displaystyle\sum_{i=5}^{10} 3i = 3(5) + 3(6) + 3(7) + 3(8) + 3(9) + 3(10)$
$$= 15 + 18 + 21 + 24 + 27 + 30 = 135$$

b. $\displaystyle\sum_{j=1}^{6} 3(j + 4) = 3(1 + 4) + 3(2 + 4) + 3(3 + 4) + 3(4 + 4) + 3(5 + 4) + 3(6 + 4)$
$$= 3(5) + 3(6) + 3(7) + 3(8) + 3(9) + 3(10)$$
$$= 15 + 18 + 21 + 24 + 27 + 30 = 135$$

c. $\displaystyle\sum_{x=8}^{13} 3(x - 3) = 3(8 - 3) + 3(9 - 3) + 3(10 - 3) + 3(11 - 3) + 3(12 - 3) + 3(13 - 3)$
$$= 3(5) + 3(6) + 3(7) + 3(8) + 3(9) + 3(10)$$
$$= 15 + 18 + 21 + 24 + 27 + 30 = 135$$

d. $3\displaystyle\sum_{i=5}^{10} i = 3(5 + 6 + 7 + 8 + 9 + 10) = 3(45) = 135$

The answer is 135 in all cases! We can show that the sums in parts a, b, and c are equivalent by a **change of variables.** In the sum $\sum_{i=5}^{10} 3i$, let $i = j + 4$. Then

$$\sum_{i=5}^{10} 3i = \sum_{j=?}^{?} 3(j + 4)$$

To get the lower and upper limits, we note that when $i = 5$, $5 = j + 4$. Thus, $j = 1$. When $i = 10$, $10 = j + 4$ gives $j = 6$. Therefore,

$$\sum_{i=5}^{10} 3i = \sum_{j=1}^{6} 3(j + 4)$$

Comparing part b to part a, we see that in part b the initial and final values of the index are decreased by 4 while the index is increased by 4 in the expression being summed.

In the sum $\sum_{i=5}^{10} 3i$, let $i = x - 3$. Then

$$\sum_{i=5}^{10} 3i = \sum_{x=?}^{?} 3(x - 3)$$

To get the upper and lower limits, we note that when $i = 5$, $5 = x - 3$ gives $x = 8$. When $i = 10$, $10 = x - 3$ gives $x = 13$. Therefore,

$$\sum_{i=5}^{10} 3i = \sum_{x=8}^{13} 3(x - 3)$$

Comparing part c to part a, we see that in part c the initial and final values of the index are increased by 3 while the index is decreased by 3 in the expression being summed. In part a, 3 is a common factor of each term. Factoring out the 3, we obtain part d. Thus,

$$\sum_{i=5}^{10} 3i = 3 \sum_{i=5}^{10} i$$

EXAMPLE 3.5 _____

Evaluate each of the following:

a. $\displaystyle\sum_{i=1}^{5} (3i + i^2)$ b. $\displaystyle\sum_{i=1}^{5} 3i + \sum_{i=1}^{5} i^2$

SOLUTION

a. $\displaystyle\sum_{i=1}^{5} (3i + i^2) = 3(1) + 1^2 + 3(2) + 2^2 + 3(3) + 3^2 + 3(4) + 4^2 + 3(5) + 5^2$

$\qquad = 3 + 1 + 6 + 4 + 9 + 9 + 12 + 16 + 15 + 25 = 100$

b. $\displaystyle\sum_{i=1}^{5} 3i + \sum_{i=1}^{5} i^2 = 3(1) + 3(2) + 3(3) + 3(4) + 3(5) + 1^2 + 2^2 + 3^2 + 4^2 + 5^2$

$\qquad = (3 + 6 + 9 + 12 + 15) + (1 + 4 + 9 + 16 + 25) = 100$

_____

The answer for part a and for part b is 100, that is,

$$\sum_{i=1}^{5} (3i + i^2) = \sum_{i=1}^{5} 3i + \sum_{i=1}^{5} i^2$$

This equality is a result of the commutative and associative properties for addition.

EXAMPLE 3.6 _____

Evaluate the sum $\sum_{j=1}^{4} 12$.

SOLUTION

$$\sum_{j=1}^{4} 12 = 12 + 12 + 12 + 12 = 48$$

$\sum_{j=1}^{4} 12$ means add 12 to itself four times. More generally, if c is any constant, $\sum_{i=1}^{n} c = nc$.

Summarizing the results of Examples 3.4, 3.5, and 3.6 are Properties 3.1–3.6.

Property 3.1

$$\sum_{i=a}^{b} i = \sum_{i=a+k}^{b+k} (i - k) \qquad\qquad k \in N$$

Property 3.2

$$\sum_{i=a}^{b} i = \sum_{i=a-k}^{b-k} (i + k) \qquad\qquad k \in N$$

Property 3.3

$$\sum_{i=a}^{b} cx_i = c \sum_{i=a}^{b} x_i \qquad\qquad$$ c is a constant, x_i is an expression involving i.

Property 3.4

$$\sum_{i=a}^{b} (x_i + y_i) = \sum_{i=a}^{b} x_i + \sum_{i=a}^{b} y_i \qquad$$ x_i and y_i are expressions involving i.

Property 3.5

$$\sum_{i=1}^{n} c = nc$$

Property 3.6

$$\sum_{i=a}^{a} x_i = x_a \qquad\qquad$$ x_i is an expression involving i.

Note that Property 3.5 implies $\sum_{i=1}^{1} c = c$. This generalizes to $\sum_{i=a}^{a} c = c$, which in turn generalizes to Property 3.6.

An example of the use of Property 3.6 is

$$\sum_{i=10}^{10} (2i^2 + 5i) = 2(10)^2 + 5(10) = 250$$

We now return to our discussion of mathematical induction.

Most of the formulas obtained for summations were discovered by noticing patterns, guessing a formula, and then trying to prove that the formula was always true. We illustrate this process in Example 3.7.

EXAMPLE 3.7 _____

Suppose we want to add the first n odd natural numbers. Display several of the sums and make a guess at the correct formula.

SOLUTION

The results are displayed in Table 3.2.

n	Sum of the First n Odd Natural Numbers	Total
1	1	$1 = 1^2$
2	$1 + 3$	$4 = 2^2$
3	$1 + 3 + 5$	$9 = 3^2$
4	$1 + 3 + 5 + 7$	$16 = 4^2$
5	$1 + 3 + 5 + 7 + 9$	$25 = 5^2$
6	$1 + 3 + 5 + 7 + 9 + 11$	$36 = 6^2$

Table 3.2

From the table, the sum of the first n odd natural numbers appears to be n^2. Using summation notation, we guess that the formula

$$\sum_{i=1}^{n} (2i - 1) = n^2$$

is always true.

_____

We now use mathematical induction to prove the formula guessed at in Example 3.7.

EXAMPLE 3.8 _____

Use mathematical induction to prove the formula guessed at in Example 3.7:

$$\sum_{i=1}^{n} (2i - 1) = n^2$$

SOLUTION

Step 1 Show that $S_1: \sum_{i=1}^{1}(2i - 1) = 1^2$ is true. Using Property 3.6,

$$\sum_{i=1}^{1}(2i - 1) = 2(1) - 1 = 2 - 1 = 1 = 1^2$$

Step 2 Inductive hypothesis: Assume that $S_k: \sum_{i=1}^{k}(2i - 1) = k^2$ is true. Showing that S_{k+1} is true involves "splitting off" the last term. It is important to understand this technique because it is used in many inductive proofs. We must show that S_{k+1}: $\sum_{i=1}^{k+1}(2i - 1) = (k + 1)^2$ is true. We begin with $\sum_{i=1}^{k+1}(2i - 1)$ and split off the last term of the sum:

$$\sum_{i=1}^{k+1}(2i - 1) = \sum_{i=1}^{k}(2i - 1) + \underbrace{2(k + 1) - 1}_{\text{the last term}}$$

$$= \sum_{i=1}^{k}(2i - 1) + 2k + 2 - 1$$

$$= \boxed{\sum_{i=1}^{k}(2i - 1)} + 2k + 1$$

By the inductive hypothesis, the sum in the box is k^2. Thus, we have

$$\sum_{i=1}^{k+1}(2i - 1) = \boxed{\sum_{i=1}^{k}(2i - 1)} + 2k + 1$$

$$= k^2 + 2k + 1 = (k + 1)(k + 1) = (k + 1)^2$$

Thus, S_{k+1} is true. Since S_k is true implies S_{k+1} is true, by the principle of mathematical induction, the formula holds for all values of $n \in N$.

Formula 3.1

$$\sum_{i=1}^{n}(2i - 1) = n^2$$

EXAMPLE 3.9

Prove the following formula by mathematical induction:

$$\sum_{i=1}^{n} i = \frac{n(n + 1)}{2}$$

SOLUTION

Step 1 Show that $S_1: \sum_{i=1}^{1} i = 1(1 + 1)/2$ is true:

$$\sum_{i=1}^{1} i = 1 = \frac{1(1 + 1)}{2}$$

Step 2 Inductive hypothesis: Assume that S_k: $\sum_{i=1}^{k} i = k(k+1)/2$ is true. We must show that S_{k+1}: $\sum_{i=1}^{k+1} i = (k+1)(k+2)/2$ is true, so we split off the last term and use the inductive hypothesis on the boxed sum:

$$\sum_{i=1}^{k+1} i = \boxed{\sum_{i=1}^{k} i} + (k+1) = \frac{k(k+1)}{2} + (k+1)$$

$$= \frac{k(k+1)}{2} + \frac{2(k+1)}{2}$$

$$= \frac{k(k+1) + 2(k+1)}{2} = \frac{(k+1)(k+2)}{2}$$

At the final step we notice that $(k+1)$ is a factor common to both terms in the numerator, so we factor out $(k+1)$. Since S_k is true implies S_{k+1} is true, by the principle of mathematical induction, the formula holds for all $n \in N$.

Formula 3.2

$$\sum_{i=1}^{n} i = \frac{n(n+1)}{2}$$

Formula 3.1 and Formula 3.2 should be memorized since they make certain calculations simpler. At first sight, it appears that to use these formulas, the sums must begin with $i = 1$. This is not the case, however, as we indicate in the next example. Note also that the index name is unimportant.

EXAMPLE 3.10

Calculate each of the following sums:

a. $\sum_{i=1}^{1000} i$ b. $\sum_{j=1}^{50} (2j-1)$ c. $\sum_{a=20}^{100} a$ d. $\sum_{i=56}^{95} 2i$

SOLUTION

a. This is Formula 3.2, with $n = 1000$. Thus,

$$\sum_{i=1}^{1000} i = \frac{1000(1000+1)}{2} = 500(1001) = 500,500$$

b. This is Formula 3.1, with $n = 50$. Thus,

$$\sum_{j=1}^{50} (2j-1) = 50^2 = 2500$$

c. We can rewrite this sum and apply Formula 3.2 twice

$$\sum_{a=20}^{100} a = \sum_{a=1}^{100} a - \sum_{a=1}^{19} a = \frac{100(100 + 1)}{2} - \frac{19(20)}{2}$$

$$= 50(101) - 190 = 5050 - 190 = 4860$$

d. Using Property 3.3, we can factor out 2:

$$\sum_{i=56}^{95} 2i = 2\sum_{i=56}^{95} i$$

We calculate the sum as we did in part c:

$$2\sum_{i=56}^{95} i = 2\left[\sum_{i=1}^{95} i - \sum_{i=1}^{55} i\right] = 2\left[\frac{95(96)}{2} - \frac{55(56)}{2}\right]$$

$$= 2[95(48) - 55(28)] = 2(4560 - 1540)$$

$$= 2(3020) = 6040$$

Another useful formula deals with the sum of the first n perfect cubes. In Table 3.3, each total is a perfect square. To get a formula in terms of n, we must determine how each of the perfect squares is related to the corresponding value of n. The number that is squared in the Total column corresponds to the sum of the first n positive integers. (Verify that this is true. That is, check $\sum_{i=1}^{1} i = 1$, $\sum_{i=1}^{2} i = 3$, $\sum_{i=1}^{3} i = 6$, and so on.) Thus, by Formula 3.2 the number that is squared is $n(n + 1)/2$. Therefore, the sum of the first n perfect cubes should be

$$\left[\frac{n(n + 1)}{2}\right]^2 = \frac{n^2(n + 1)^2}{4}$$

n	Sum of the First n Perfect Cubes		Total
1	1^3	$= 1$	$1 = 1^2$
2	$1^3 + 2^3$	$= 1 + 8$	$9 = 3^2$
3	$1^3 + 2^3 + 3^3$	$= 1 + 8 + 27$	$36 = 6^2$
4	$1^3 + 2^3 + 3^3 + 4^3$	$= 1 + 8 + 27 + 64$	$100 = 10^2$
5	$1^3 + 2^3 + 3^3 + 4^3 + 5^3$	$= 1 + 8 + 27 + 64 + 125$	$225 = 15^2$

Table 3.3

EXAMPLE 3.11

Prove the following formula by mathematical induction:

$$\sum_{i=1}^{n} i^3 = \frac{n^2(n + 1)^2}{4}$$

SOLUTION

Step 1   Show that S_1: $\sum_{i=1}^{1} i^3 = 1^2(1+1)^2/4$ is true:

$$\sum_{i=1}^{1} i^3 = 1^3 = 1 = \frac{1^2(1+1)^2}{4} = \frac{2^2}{4} = \frac{4}{4} = 1$$

Step 2   Inductive hypothesis: Assume that S_k: $\sum_{i=1}^{k} i^3 = k^2(k+1)^2/4$ is true. We must show that S_{k+1}: $\sum_{i=1}^{k+1} i^3 = (k+1)^2(k+2)^2/4$ is true, so we split off the last term and use the inductive hypothesis:

$$\sum_{i=1}^{k+1} i^3 = \boxed{\sum_{i=1}^{k} i^3} + (k+1)^3 = \frac{k^2(k+1)^2}{4} + (k+1)^3$$

$$= \frac{k^2(k+1)^2}{4} + \frac{4(k+1)^3}{4}$$

$$= \frac{k^2(k+1)^2}{4} + \frac{4(k+1)(k+1)^2}{4}$$

$$= \frac{(k+1)^2[k^2 + 4(k+1)]}{4} = \frac{(k+1)^2(k^2 + 4k + 4)}{4}$$

$$= \frac{(k+1)^2(k+2)(k+2)}{4} = \frac{(k+1)^2(k+2)^2}{4}$$

Since S_k is true implies S_{k+1} is true, by the principle of mathematical induction, the formula is true for all $n \in N$.

Formula 3.3

$$\sum_{i=1}^{n} i^3 = \frac{n^2(n+1)^2}{4}$$

Induction is also used to prove results other than formulas, as seen in Example 3.12.

EXAMPLE 3.12

Prove by induction: If set A contains n elements, then the power set $\mathscr{P}(A)$ contains 2^n sets.

SOLUTION

Step 1   Show S_1: If set A contains one element, then $\mathscr{P}(A)$ contains 2^1 sets is true. If $A = \{a\}$, then $\mathscr{P}(A) = \{\emptyset, \{a\}\}$. Therefore, if A contains one element, then $\mathscr{P}(A)$ contains $2^1 = 2$ sets.

Step 2 Inductive hypothesis: Assume that S_k: If set B contains k elements, then $\mathcal{P}(B)$ contains 2^k sets is true. We must show that S_{k+1} is true: If A contains $k + 1$ elements, then $\mathcal{P}(A)$ contains 2^{k+1} sets. Let $A = \{a_1, a_2, \ldots, a_k, a_{k+1}\}$. In previous examples, we split off the last term of a sum; now we split off the last element of a set. Let $B = \{a_1, a_2, \ldots, a_k\}$. Then $A = B \cup \{a_{k+1}\}$. Since B contains k elements, the inductive hypothesis implies that $\mathcal{P}(B)$ contains 2^k sets. Each set X in $\mathcal{P}(B)$ corresponds to 2 sets in $\mathcal{P}(A)$: X and $X \cup \{a_{k+1}\}$. Therefore, there are twice as many sets in $\mathcal{P}(A)$ as there are in $\mathcal{P}(B)$, giving $2(2^k) = 2^{k+1}$ sets in $\mathcal{P}(A)$. Thus, S_k is true implies S_{k+1} is true. Therefore, by the principle of mathematical induction S_n is always true: If A has n elements, then $\mathcal{P}(A)$ contains 2^n sets.

EXERCISES 3.1

1. Use Properties 3.1–3.6 to express the following sums more simply:

   a. $\displaystyle\sum_{i=10}^{17} (i - 8)$ b. $\displaystyle\sum_{j=3}^{10} 25j$ c. $\displaystyle\sum_{k=1}^{50} 9$

   d. $\displaystyle\sum_{i=2}^{10} (2i + 5) + \sum_{i=2}^{10} (i - 4)$ e. $\displaystyle\sum_{i=36}^{50} 4(i - 30)$ f. $\displaystyle\sum_{t=8}^{8} (5t - 3)$

2. Write each of the terms in the following sums and add:

   a. $\displaystyle\sum_{i=1}^{8} (5i - 2)$ b. $\displaystyle\sum_{i=4}^{10} i^2$ c. $\displaystyle\sum_{i=1}^{5} (i^2 + 3i)$

3. Write each of the following using summation notation:
   a. $6 + 12 + 18 + 24 + \cdots + 120$
   b. $10 + 13 + 16 + 19 + \cdots + 31$
   c. $4 + 16 + 36 + 64 + 100 + \cdots + 900$ (*Hint:* Factor out 4.)

4. Prove by induction: $1^2 + 2^2 + 3^2 + \cdots + n^2 = n(n + 1)(2n + 1)/6$.

5. Prove by induction: $2^2 + 4^2 + 6^2 + \cdots + (2n)^2 = 2n(n + 1)(2n + 1)/3$. (*Hint:* Factor out 4 from each term and then use the result of exercise 4.)

6. Prove by induction: The number of n-bit binaries is 2^n.

7. Prove by induction: $1/(1 \cdot 2) + 1/(2 \cdot 3) + \cdots + 1/n(n + 1) = n/(n + 1)$.

8. Recall that $x^0 = 1$ when $x \neq 0$. Prove by induction:

$$\sum_{i=0}^{n} 2^i = 2^{n+1} - 1$$

9. Prove by induction: $2n + 1 \leq 3^n$ for all $n \in N$.

10. Prove by induction: $n^3 + 2n$ is divisible by 3. (*Hint:* Use congruences mod 3.)

11. Prove by induction: $3 + 7 + 11 + \cdots + (4n - 1) = n(2n + 1)$.

12. Find the flaw in the following proposed proof that everyone in the world is the same age: To use induction, we must first define a statement S_n. Define S_n: If a set contains n people, they are all the same age.

    Step 1 Show S_1 is true: If a set contains one person, then everyone in that set is the same age. Consider any set P containing a single person. That person is the same age as himself, so everyone in P is the same age.

Step 2   Inductive hypothesis: Assume S_k is true: If a set contains k people, then all those people are the same age. We must show that S_{k+1} is true: If a set contains $k + 1$ people, then they are all the same age. Let A be a set of $k + 1$ people. Let A_1 and A_2 each be subsets containing k people from A so that $A_1 \cup A_2 = A$ and $A_1 \cap A_2 \neq \varnothing$ (that is, A_1 and A_2 overlap). Since A_1 contains k people, the inductive hypothesis implies they are all the same age x. Since A_2 contains k people, they are all the same age. $A_1 \cap A_2 \neq \varnothing$, so there is at least one person a in A_1 who is also in A_2. Since a's age is x and everyone in A_2 is of the same age, everyone in A_2 must be age x. Therefore, everyone in A_1 and A_2 is age x. Since $A_1 \cup A_2 = A$, everyone in A (containing $k + 1$ people) is the same age.

Problems and Projects

13. Prove by induction the following generalizations of De Morgan's laws to n sets:
  a. $(A_0 \cup A_1 \cup \cdots \cup A_{n-1})' = A_0' \cap A_1' \cap A_2' \cap \cdots \cap A_{n-1}'$
  b. $(A_0 \cap A_1 \cap \cdots \cap A_{n-1})' = A_0' \cup A_1' \cup A_2' \cup \cdots \cup A_{n-1}'$
  (*Hint:* Use the associative property to group the first k sets together at the inductive step.)

14. The FOR DO loop facilitates the summation process in Pascal. To determine the sum $\sum_{y=3}^{8} 2y$, use the sequence of statements

$$\text{ANSWER} := 0;$$
$$\text{FOR Y} := 3 \text{ TO } 8 \text{ DO}$$
$$\text{ANSWER} := \text{ANSWER} + 2*\text{Y};$$

within your program.
  a. Write a program that reads in two integers a and b, tests for the larger of the two, and prints the sum

$$\sum_{\text{smaller of } a \text{ and } b}^{\text{larger of } a \text{ and } b} i$$

  b. Run the program using the following input:
     i. $a = 48, b = 17$ **ii.** $a = 6, b = 35$

15. Formula 3.2 says that $1 + 2 + 3 + \cdots + n = n(n + 1)/2$, and Formula 3.3 says that $1^3 + 2^3 + 3^3 + \cdots + n^3 = n^2(n + 1)^2/4 \ [= (n(n + 1)/2)^2]$. Thus, $1^3 + 2^3 + 3^3 + \cdots + n^3 = (1 + 2 + 3 + \cdots + n)^2$. Write a program that calculates the left side and the right side separately and prints each answer for all $n \leq 20$. (You should get 20 pairs of numbers (a, b) such that $a = b$ in each case.)

3.2   PERMUTATIONS

It is easy to determine how many natural numbers there are between 17 and 50 or how many 3-digit numbers there are. It is more difficult to count how many 6-element subsets can be formed from a 23-element set or how many license plate numbers there are consisting of two letters followed by four digits. In this

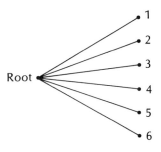

Figure 3.1

section we develop techniques for counting arrangements of objects. We begin, in Note 3.3, with a basic principle.

Note 3.3   Multiplication Principle

Suppose a procedure consists of n stages. At the first stage there are a_1 choices, at the second stage there are a_2 choices, and so on. If the choice at each stage is independent of the choice at any other stage, then the number of possible sequences of choices is the product $a_1 \cdot a_2 \cdot a_3 \cdot \cdots \cdot a_n$.

We can justify the multiplication principle by using a tree diagram. For example, in how many ways can a schoolboy select a shirt and tie combination if he has six shirts and three ties? Each selection of a shirt is represented by a branch from the root, as in Figure 3.1 (the shirts are numbered from 1 to 6).

Each selection of a shirt may be followed by selecting any of three ties (labeled A, B, and C). Therefore, we grow three branches from each of the shirt branches. See Figure 3.2. Consequently, there are $6 \cdot 3 = 18$ branches in all, each representing one shirt-tie combination.

The multiplication principle tells how to determine the number of possible sequences of choices for a procedure involving n stages, where the choice at each stage is independent of the choices at the other stages. The formula says to multiply $a_1 \cdot a_2 \cdot a_3 \cdot \cdots \cdot a_n$. The multiplication principle is also called the **fundamental counting principle**.

EXAMPLE 3.13 ──

The dinner special at a restaurant consists of an appetizer, an entree, one vegetable, and a beverage. One may choose from two appetizers, three entrees, five vegetables, and three beverages. How many different dinners are possible?

SOLUTION

We apply the multiplication principle and get $2 \cdot 3 \cdot 5 \cdot 3 = 90$ different dinners.

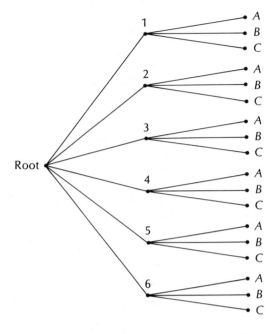

Figure 3.2

EXAMPLE 3.14 _____

How many 7-digit phone numbers are there if the first digit cannot be 0 or 1?

SOLUTION

There are eight choices for the first digit and ten choices for the other digits. Therefore, the number of 7-digit phone numbers is $8 \cdot 10 \cdot 10 \cdot 10 \cdot 10 \cdot 10 \cdot 10 = 8{,}000{,}000$.

We have seen that sorting is important in computer science. By sorting a set, we rearrange the elements to produce an ordered set. If a set contains n distinct elements, in how many ways can they be arranged? A **permutation** is an arrangement or ordering of objects. Our question then becomes: How many permutations are there of n distinct elements? We can answer this question by using the multiplication principle. There are n possible choices for the first element in the set, $n - 1$ choices for the second element, $n - 2$ for the third, and so on. The elements, therefore, can be arranged in $n(n - 1)(n - 2) \cdots 3 \cdot 2 \cdot 1$ ways. This product occurs often in counting problems and is denoted by the symbol $n!$ (read "**n factorial**").

Note 3.4

There are $n!$ permutations of n distinct objects.

EXAMPLE 3.15 ⎯⎯⎯⎯⎯⎯⎯⎯⎯⎯⎯⎯⎯⎯⎯⎯⎯⎯⎯

a. Five people arrive simultaneously at a theater. In how many ways can they line up at the ticket booth?

b. In how many ways can the letters of the word *computer* be arranged?

SOLUTION

a. Since we have five distinct people, there are $5! = 5 \cdot 4 \cdot 3 \cdot 2 \cdot 1 = 120$ different lines.

b. There are eight distinct letters to arrange in the word *computer*. There are, therefore, $8! = 8 \cdot 7 \cdot 6 \cdot 5 \cdot 4 \cdot 3 \cdot 2 \cdot 1 = 40{,}320$ permutations.

⎯⎯

As noted in the introduction to this chapter, combinatorics is concerned with methods for counting the number of elements in sets. Suppose we must count the number of elements in a set A. One of the simplest techniques for doing so is to divide A into nonoverlapping subsets $A_1, A_2, \ldots, A_k$, and then add the number of elements in $A_1, A_2, \ldots,$ and A_k.

To formally state our next counting principle, we require two definitions. A collection of sets $A_1, A_2, \ldots, A_k$ is **pairwise disjoint** if no two of them have an element in common. That is, $A_i \cap A_j = \varnothing$ if $i \neq j$. For example, if $A =$ The set of cards in a standard deck and $A_1 = \{$All spades$\}$, $A_2 = \{$All hearts$\}$, $A_3 = \{$All clubs$\}$, and $A_4 = \{$All diamonds$\}$, then A_1, A_2, A_3, and A_4 are pairwise disjoint. This situation is depicted in Figure 3.3.

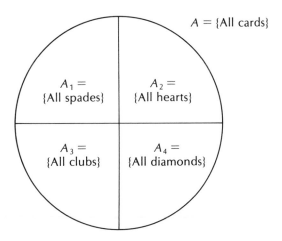

Figure 3.3

A **partition** of a set A is a collection of pairwise disjoint sets $A_1, A_2, \ldots, A_k$ whose union is A; that is, $A_1 \cup A_2 \cup \cdots \cup A_k = A$. In the card example, A_1, A_2, A_3, A_4 is a partition of the set A of all cards in the deck. Note 3.5 states our next principle.

Note 3.5   The Addition Principle

Let $A_1, A_2, \ldots, A_k$ be a partition of A, and let A_1 contain a_1 elements, A_2 contain a_2 elements, and so on. Then A contains $a_1 + a_2 + \cdots + a_k$ elements.

EXAMPLE 3.16 _____

Suppose I plan to buy one dessert item. There are five kinds of cake, three kinds of pie, three flavors of Jell-o, and four kinds of fruit available. In how many ways can I make my selection?

SOLUTION

Let $A = \{\text{All desserts}\}$. The number of ways of choosing an element from A is the number of elements in A—one choice for each element. In this problem, the way to partition A is forced upon us by the way in which we must make our choice of a dessert. Let $A_1 = \{\text{Cake items}\}$, $A_2 = \{\text{Pie items}\}$, $A_3 = \{\text{Jell-o items}\}$, and $A_4 = \{\text{Fruit items}\}$. Clearly, A_1, A_2, A_3, and A_4 are pairwise disjoint, and their union is A. By the addition principle, A contains $5 + 3 + 3 + 4 = 15$ elements. There are, therefore, 15 ways in which I can make my selection.

In some cases, when the choices overlap, we can reformulate the problem so that the addition principle applies. This is done in Example 3.17.

EXAMPLE 3.17 _____

In how many ways can I select a picture card (jack, queen, or king) or a diamond from a deck of cards?

SOLUTION

Let $A = \{\text{Picture card or a diamond}\}$. Again, the number of ways of choosing an element from A is the number of elements of A. In this problem, however, the partition of A is not so obvious. Let $A_1 = \{\text{Picture card}\}$, and let $A_2 = \{AD, 2D, 3D, 4D, 5D, 6D, 7D, 8D, 9D, 10D\}$, that is, the set of diamonds that are not picture cards. Then A_1 and A_2 are a partition of A. By the addition principle, A contains $12 + 10 = 22$ elements. Thus I can select a picture card or a diamond in 22 ways.

EXAMPLE 3.18

How many sequences of three distinct letters can be formed from the letters a, b, c, d, e if the sequence must contain a or b (or both)?

SOLUTION

We separate the problem into three pairwise disjoint cases, so we can apply the addition principle. Define:

$$A_1 = \{\text{3-letter sequences containing } a \text{ but not } b\}$$
$$A_2 = \{\text{3-letter sequences containing } b \text{ but not } a\}$$
$$A_3 = \{\text{3-letter sequences containing both } a \text{ and } b\}$$

Clearly, A_1, A_2, and A_3 are pairwise disjoint and, together, they include all possible 3-letter sequences containing a or b. Note that here we are using "or" in the inclusive sense. (In Chapter 2, we saw that "or" is sometimes used in the exclusive sense, in which case "a or b" would mean "a or b but not both.") The addition principle implies that the number of elements in A is the sum of the number of elements in A_1, A_2, and A_3. For A_1, we have three positions to fill and four letters to choose from, but we *must* choose an a. Then we choose two other letters from c, d, e. There are three ways to choose those two letters: cd, ce, or de. Once we determine which three letters we are dealing with, we know that there are $3! = 6$ permutations of those three letters. Since there are three ways to choose the letters and six ways to arrange them, the multiplication principle implies that there are $3 \cdot 6 = 18$ elements in set A_1. Similarly, there are 18 elements in set A_2. For A_3, there are three possible choices for the letter used other than a and b. We can then arrange the three letters in $3! = 6$ ways. By the multiplication principle, A_3 contains $3 \cdot 6 = 18$ elements. The addition principle then implies A contains $18 + 18 + 18 = 54$ elements. Therefore, we can form 54 sequences of three distinct letters from a, b, c, d, e containing a or b (or both).

In both parts of Example 3.15, the answer is $n!$, where n is the number of distinct objects involved. However, more analysis is required in counting problems when the objects are not distinct or not all the objects are used.

EXAMPLE 3.19

How many ways are there to arrange the letters of the following words:

a. *sisters*   b. *successes*

SOLUTION

a. *Sisters* has seven letters, three of which are identical. To count the number of arrangements of the letters in *sisters*, we first label the three s's to make them distinct: $s_1is_2ters_3$. If we keep the non-s's fixed in their positions, we see there are $6 = 3!$ ways of arranging the three s's among their positions: $s_1is_2ters_3$, $s_1is_3ters_2$, $s_2is_1ters_3$, $s_2is_3ters_1$, $s_3is_1ters_2$, and $s_3is_2ters_1$. Since each permutation of *sisters* produces six permutations of $s_1is_2ters_3$, there are six times as many arrangements of $s_1is_2ters_3$ as there are of *sisters*. Therefore, we divide the number of arrangements of $s_1is_2ters_3$ by 6 ($= 3!$). There are $7!/3! =$ $7 \cdot 6 \cdot 5 \cdot 4 \cdot 3 \cdot 2 \cdot 1/(3 \cdot 2 \cdot 1) = 7 \cdot 6 \cdot 5 \cdot 4 = 840$ arrangements of the letters of *sisters*.

b. Using reasoning similar to that used in part a, there are 4! ways of arranging the s's, 2! ways of arranging the c's, and 2! ways of arranging the e's in $s_1uc_1c_2e_1s_2s_3e_2s_4$ for the one arrangement *successes*. There are $4! \cdot 2! \cdot 2!$ times as many arrangements of the nine letters of $s_1uc_1c_2e_1s_2s_3e_2s_4$ as there are of the letters of *successes*. There are, therefore,

$$9!/(4! \cdot 2! \cdot 2!) = 9 \cdot 8 \cdot 7 \cdot 6 \cdot 5 \cdot 4 \cdot 3 \cdot 2 \cdot 1/(4 \cdot 3 \cdot 2 \cdot 1 \cdot 2 \cdot 1 \cdot 2 \cdot 1)$$
$$= 9 \cdot 8 \cdot 7 \cdot 6 \cdot 5/(2 \cdot 2) = 9 \cdot 2 \cdot 7 \cdot 6 \cdot 5 = 3780$$

arrangements of the letters of *successes*.

Example 3.19 illustrates the principle given as Formula 3.4.

Formula 3.4

If there are n objects, r_1 of type 1, r_2 of type 2, ..., r_k of type k, then the number of arrangements of the n objects is $n!/(r_1! \cdot r_2! \cdot \cdots \cdot r_k!)$.

In Example 3.19 all the objects were used. However, there are many situations where not all the objects are used.

EXAMPLE 3.20

a. There are ten contestants in a beauty contest. In how many ways can the winner, first runner-up, and second runner-up be chosen?

b. There are 16 youngsters trying out for the nine distinct starting positions on a Little League team. In how many ways can the nine positions be filled?

c. A club with 20 members will elect a president and vice-president. In how many ways can these positions be filled?

SOLUTION

a. There are ten ways a winner can be chosen, nine ways a first runner-up can be chosen, and eight ways a second runner-up can be chosen from the remaining contestants. Using the multiplication principle, there are, therefore, $10 \cdot 9 \cdot 8 = 720$ ways of filling the three positions.

b. Using reasoning similar to that used in part a, the positions can be filled in $16 \cdot 15 \cdot 14 \cdot 13 \cdot 12 \cdot 11 \cdot 10 \cdot 9 \cdot 8 = 4{,}151{,}347{,}200$ ways.

c. Using the multiplication principle, the positions can be filled in $20 \cdot 19 = 380$ ways.

The numbers 720; 4,151,347,200; and 380 from Example 3.20 can be written in terms of factorials as 10!/7!, 16!/7!, and 20!/18!, respectively. In each case, the factorial in the numerator is the number of ways we can arrange the objects, and the factorial in the denominator is the number of ways we can arrange the objects not selected for the given positions.

Formula 3.5

The number of ways of arranging r out of n distinct objects is $_nP_r = n!/(n-r)!$.

We can arrange all n objects in $n!$ ways, that is, $_nP_n = n!$. If we use Formula 3.5, $_nP_n = n!/(n-n)! = n!/0! = n!$. Therefore, 0! is defined as 1, that is, $0! = 1$.

When $n \geq 1$, $n!$ is the product of each positive integer less than or equal to n. In this case, it is useful to write only several of the smaller integers and indicate the rest by using a factorial. For example, $10! = 10 \cdot 9 \cdot 8!$ and $17! = 17 \cdot 16 \cdot 15 \cdot 14!$. We use this method to simplify the fraction we get when applying Formula 3.5.

EXAMPLE 3.21 _____

Evaluate each of the following:

a. $_8P_3$ b. $_{15}P_4$ c. $_{30}P_2$

SOLUTION

a. $_8P_3 = \dfrac{8!}{(8-3)!} = \dfrac{8!}{5!} = \dfrac{8 \cdot 7 \cdot 6 \cdot 5!}{5!} = 8 \cdot 7 \cdot 6 = 336$

b. $_{15}P_4 = \dfrac{15!}{(15-4)!} = \dfrac{15!}{11!} = \dfrac{15 \cdot 14 \cdot 13 \cdot 12 \cdot 11!}{11!}$

   $= 15 \cdot 14 \cdot 13 \cdot 12 = 32{,}760$

c. $_{30}P_2 = \dfrac{30!}{(30-2)!} = \dfrac{30!}{28!} = \dfrac{30 \cdot 29 \cdot 28!}{28!}$

   $= 30 \cdot 29 = 870$

In each case $(n - r)!$ divides the numerator $n!$ and leaves a product of r successively smaller integers beginning with n. An alternate way of calculating $_nP_r$, therefore, is by using Formula 3.6.

Formula 3.6

$$_nP_r = n(n - 1)(n - 2) \cdots (n - r + 1)$$

EXERCISES 3.2

1. A girl has five skirts and eight blouses. How many skirt–blouse outfits does she own?
2. There are five roads from town A to town B, three roads from town B to town C, and two roads from town A to town C that avoid town B. All the roads are two-way streets.
   a. How many ways are there to go from A to C by way of B?
   b. How many ways are there to go from A to C?
   c. In how many ways can we get from A to C and back?
   d. In how many ways can we get from A to C and back if B is passed through at least once?
3. How many permutations are there of the letters a, b, c? List them.
4. How many permutations are there of the letters in the following words:
   a. *build*   b. *sees*   c. *entrepreneur*
   d. *Mississippi*   e. *coincide*   f. *tattletale*
5. A theater will show a new movie called *Middlemen*. If an employee is given the letters (all capitals) for the marquee, in how many ways can he *misspell* the movie title?
6. Evaluate:
   a. 6!   b. $_5P_2$ c. $_9P_4$
7. Write the following using factorials:
   a. $12 \cdot 11 \cdot 10 \cdot 9$ b. $30 \cdot 29$
8. There are eight horses in a race. How many win-place-show finishes are possible?
9. How many 5-digit numbers are there (the first digit cannot be zero)?
10. How many 8-digit numbers are there containing three 6s and five 2s?
11. How many 7-digit numbers are there containing two 3s and three 8s if neither of the other two digits are 3 or 8 (the first digit cannot be zero)?
12. Simplify each of the following:
    a. $12!/8!$ b. $5! \cdot 10!/12!$ c. $_8P_3/3!$ d. $n!/(n - 2)!$
13. In how many ways can five science books, three history books, and four mathematics books be arranged on a shelf if books of the same subject must remain together (all books are distinct)? (*Hint:* First arrange the books for each subject, then arrange the order of the subjects.)

14. How many license plate numbers are there consisting of two letters followed by four digits?

15. A system designer is asked to design an inventory system for a hardware manufacturer. The manufacturer produces 57,826 different kinds of hardware. The system designer must decide on a way to identify the different kinds of hardware. Each id# is to begin with two letters of the alphabet. Any letter can be used except I, O, and Z. The rest of the id# is to consist of digits. How many digits must the designer include in the id# format to ensure that each kind of hardware has a unique id#?

Problems and Projects

16. Write a program that prints the first fifteen factorials.

17. Write a program that inputs N and R and prints all values of $_nP_r$, where $1 \le r \le R$ and $r \le n \le N$.

18. Write a program that reads in an integer K and determines the first value of N for which $N!$ exceeds K^3.

19. What would be the output of the following Pascal fragment:

```
BEGIN
    N := 0;
    FOR I := 1 TO 50 DO
        FOR J := 1 TO 100 DO
            FOR K := 1 TO 50 DO
                N := N + 1;
    WRITELN(N)
END.
```

3.3 COMBINATIONS

In each example of Section 3.2, we observed the order of the objects chosen. In many situations, however, order is not important. As mentioned in Chapter 2, for example, the order of the elements of a set is not important: $\{3, 4, 8, 10\} = \{10, 4, 3, 8\}$. Thus, we do not use permutations to count the number of 6-element subsets of a 23-element set.

A **combination** is an unordered selection or subset of objects from a given set of objects. The number of combinations of r objects selected from a set of n objects is denoted by $\binom{n}{r}$, read "n choose r" or "n above r." Some authors use $C(n, r)$ or $_nC_r$ instead of $\binom{n}{r}$, but we use only $\binom{n}{r}$, the most commonly used of the three notations.

To find a formula for $\binom{n}{r}$ we can proceed as follows: Find the permutations of r out of n objects in two steps: (1) Select the r objects, and (2) arrange the r objects. By the definition of combinations, we can select the r objects in $\binom{n}{r}$ ways. We can arrange the r objects in $r!$ ways (see Note 3.4). By the multiplication principle, therefore, $_nP_r = \binom{n}{r} \cdot r!$. Solving for $\binom{n}{r}$ gives Formula 3.7.

Formula 3.7

$$\binom{n}{r} = \frac{_nP_r}{r!} = \frac{n!}{r!(n-r)!}$$

EXAMPLE 3.22

Evaluate each of the following:

a. $\binom{5}{3}$ b. $\binom{12}{4}$ c. $\binom{6}{6}$ d. $\binom{100}{2}$

SOLUTION

a. $\binom{5}{3} = \dfrac{5!}{3!(5-3)!} = \dfrac{5!}{3! \cdot 2!} = \dfrac{5 \cdot 4 \cdot 3!}{3! \cdot 2 \cdot 1}$

$= \dfrac{5 \cdot 4}{2} = 10$

There are, therefore, ten 3-element subsets of a 5-element set.

b. $\binom{12}{4} = \dfrac{12!}{4!(12-4)!} = \dfrac{12!}{4! \cdot 8!} = \dfrac{12 \cdot 11 \cdot 10 \cdot 9 \cdot 8!}{4! \cdot 8!}$

$= \dfrac{12 \cdot 11 \cdot 10 \cdot 9}{4 \cdot 3 \cdot 2 \cdot 1} = 11 \cdot 5 \cdot 9 = 495$

c. $\binom{6}{6} = \dfrac{6!}{6!(6-6)!} = \dfrac{6!}{6! \cdot 0!} = \dfrac{6!}{6! \cdot 1} = 1$

There is only one way to choose six elements from a 6-element set. Choose all of them!

d. $\binom{100}{2} = \dfrac{100!}{2! \cdot 98!} = \dfrac{100 \cdot 99 \cdot 98!}{2 \cdot 98!}$

$= \dfrac{100 \cdot 99}{2} = 4950$

EXAMPLE 3.23 ―――――――――――――――――――――――――――――――

How many 5-element subsets can be formed from a 15-element set?

SOLUTION

$$\binom{15}{5} = \frac{15!}{5!(15-5)!} = \frac{15!}{5! \cdot 10!}$$

$$= \frac{15 \cdot 14 \cdot 13 \cdot 12 \cdot 11 \cdot 10!}{5! \cdot 10!}$$

$$= \frac{15 \cdot 14 \cdot 13 \cdot 12 \cdot 11}{5 \cdot 4 \cdot 3 \cdot 2 \cdot 1} = 7 \cdot 13 \cdot 3 \cdot 11 = 3003$$

―――

EXAMPLE 3.24 ―――――――――――――――――――――――――――――――

A club with 40 members elects a finance committee having 3 members. How many committees are possible?

SOLUTION

Since the order of selection does not matter, this is a combination problem. Selecting the committee corresponds to selecting a 3-element subset from a 40-element set. This can be done in $\binom{40}{3}$ ways.

$$\binom{40}{3} = \frac{40!}{3!37!} = \frac{40 \cdot 39 \cdot 38 \cdot 37!}{3! \cdot 37!} = \frac{40 \cdot 39 \cdot 38}{3 \cdot 2 \cdot 1}$$

$$= 40 \cdot 13 \cdot 19 = 9880$$

―――

EXAMPLE 3.25 ―――――――――――――――――――――――――――――――

A jar contains eight black balls and six white balls. In how many ways can four balls be chosen so that:

a. exactly two black balls are chosen

b. at least three black balls are chosen

SOLUTION

a. The order of selection is not important; *BWWB* is the same as *WBWB*. We can, therefore, consider this a two-stage procedure: (1) Choose two black balls, and (2) choose two white balls. We can choose the black balls in $\binom{8}{2}$ ways and the white balls in $\binom{6}{2}$ ways. By the multiplication principle, the number of 4-ball selections containing exactly two black balls is, therefore

$$\binom{8}{2} \cdot \binom{6}{2} = \left(\frac{8!}{2! \cdot 6!}\right) \cdot \left(\frac{6!}{2! \cdot 4!}\right)$$

$$= \left(\frac{8 \cdot 7 \cdot 6!}{2 \cdot 6!}\right)\left(\frac{6 \cdot 5 \cdot 4!}{2 \cdot 4!}\right)$$

$$= \left(\frac{8 \cdot 7}{2}\right)\left(\frac{6 \cdot 5}{2}\right) = 28 \cdot 15 = 420$$

b. At least three black balls "means that either three black balls and one white ball are chosen or four black balls are chosen. Using reasoning similar to that used in part a, three black balls and one white ball can be chosen in $\binom{8}{3} \cdot \binom{6}{1}$ ways, and four black balls can be chosen in $\binom{8}{4}$ ways. By the addition principle, the number of 4-ball selections containing at least three black balls is:

$$\binom{8}{3} \cdot \binom{6}{1} + \binom{8}{4} = \left(\frac{8!}{3! \cdot 5!}\right)\left(\frac{6!}{1! \cdot 5!}\right) + \frac{8!}{4! \cdot 4!}$$

$$= \left(\frac{8 \cdot 7 \cdot 6 \cdot 5!}{3! \cdot 5!}\right)\left(\frac{6 \cdot 5!}{1 \cdot 5!}\right) + \frac{8 \cdot 7 \cdot 6 \cdot 5 \cdot 4!}{4! \cdot 4!}$$

$$= \left(\frac{8 \cdot 7 \cdot 6}{3 \cdot 2 \cdot 1}\right)(6) + \frac{8 \cdot 7 \cdot 6 \cdot 5}{4 \cdot 3 \cdot 2 \cdot 1}$$

$$= 8 \cdot 7 \cdot 6 + 2 \cdot 7 \cdot 5 = 336 + 70 = 406$$

EXAMPLE 3.26

A jar contains 10 red, 12 white, and 13 blue balls. In how many ways can seven balls be selected so there are at least two balls of each color?

SOLUTION

There are three cases: 3R,2W,2B; 2R,3R,2B; and 2R,2W,3B. Thus the number of ways in which the selection can be made is

$$\binom{10}{3}\binom{12}{2}\binom{13}{2} + \binom{10}{2}\binom{12}{3}\binom{13}{2} + \binom{10}{2}\binom{12}{2}\binom{13}{3}$$

$$= 120 \cdot 66 \cdot 78 + 45 \cdot 220 \cdot 78 + 45 \cdot 66 \cdot 286 = 617{,}760 + 772{,}200 + 849{,}420$$

$$= 2{,}239{,}380.$$

Figure 3.4

Some counting problems at first appear to involve order but are solved using combinations. These problems involve arranging objects, many of which are the same. Suppose, for example, a child wishes to arrange a row of her three black and two red checkers. In how many ways can this be done? We can depict the positions for the five checkers as in Figure 3.4.

To count the number of arrangements of checkers, first we choose three of the five positions in which to place the black checkers. There are $\binom{5}{3} = 10$ ways of doing this. In the two remaining positions, we place the two red checkers. There is $\binom{2}{2} = 1$ way of doing this. By the multiplication principle, therefore, there are $\binom{5}{3} \cdot \binom{2}{2} = 10 \cdot 1 = 10$ ways of placing the checkers in a row.

We use similar reasoning in Example 3.27.

EXAMPLE 3.27 _____

a. Determine the number of 8-digit numbers composed of three 6s and five 2s.
b. Determine how many social security numbers contain four 4s, two 6s, and three 8s.

SOLUTION

a. We can choose the positions for the three 6s in $\binom{8}{3}$ ways. Once these positions are selected, 2s are placed in each of the remaining positions. Therefore, there are $\binom{8}{3} = 8!/(3! \cdot 5!) = 8 \cdot 7 \cdot 6/(3 \cdot 2 \cdot 1) = 8 \cdot 7 = 56$ such 8-digit numbers. Another way to do this problem is to use Formula 3.4. We are arranging eight objects, three of which are of type 1, and five of which are of type 2. Thus, the number of arrangements is $8!/(3! \cdot 5!) = 56$.

b. A social security number is composed of nine digits. We can choose the positions of the 4s in $\binom{9}{4}$ ways. From the remaining five positions, we can choose the positions of the 6s in $\binom{5}{2}$ ways. The 8s go in the remaining three positions. Thus, the number of social security numbers containing four 4s, two 6s, and three 8s is

$$\binom{9}{4} \cdot \binom{5}{2} \cdot \binom{3}{3} = \frac{9!}{4! \cdot 5!} \cdot \frac{5!}{2! \cdot 3!} \cdot \frac{3!}{3! \cdot 0!}$$

$$= \frac{9 \cdot 8 \cdot 7 \cdot 6}{4 \cdot 3 \cdot 2 \cdot 1} \cdot \frac{5 \cdot 4}{2 \cdot 1} \cdot 1$$

$$= 9 \cdot 2 \cdot 7 \cdot 5 \cdot 2 = 1260$$

Alternately, we can use Formula 3.4. We are arranging nine objects, four of which are of type 1, two of type 2, and three of type 3. Thus, the number of arrangements is $9!/(4! \cdot 2! \cdot 3!) = 1260$.

EXERCISES 3.3

1. Evaluate each of the following:

   a. $\binom{10}{3}$ **b.** $\binom{90}{88}$ **c.** $\binom{13}{3} \cdot \binom{12}{2} / \binom{25}{5}$

2. How many 5-card poker hands can be formed from a deck of 52 cards?
3. How many committees of four can be formed from 16 people?
4. I plan to buy 4 shirts from a group of 22 and 2 jackets from a rack of 14. How many different selections are possible?
5. Alphabet soup is popular with children.
   a. How many combinations of five distinct letters can be picked up on a spoon?
   b. Suppose there is one pair of identical letters and three other distinct ones. How many combinations are possible?
6. How many "words" are there consisting of nine distinct letters and using all the vowels *a,e,i,o,u*? (*Hint:* Select the nonvowels, then arrange the letters.)
7. Steve and 10 of his friends are choosing sides for a basketball game. There are five players on a team.
   a. How many teams are possible that exclude Steve?
   b. How many teams are possible that include Steve?
   c. After selecting the members for one team, the members for a second team are selected. Two games are different if the composition of the teams playing is different. How many different games are there in which Steve does not play?
8. A full house in poker consists of three of one kind and a pair of another. For example, JJJ55 is a full house. How many different ways are there to get a full house?
9. A flush in poker consists of five cards from the same suit. How many ways are there to get a flush? (*Hint:* Select the suit, then select the cards from that suit.)
10. A bridge hand contains 13 cards. How many different bridge hands are there?

11. A jar contains six black and nine white balls. How many ways are there of selecting four balls if
    a. two are black?
    b. all four balls are the same color?
12. There are 15 girls and 12 boys in a class. A committee of five is selected for a project. How many possible committees are there containing:
    a. three boys?
    b. at least one boy?
13. A test has 12 questions and you must answer 10. How many choices do you have?
14. How many 6-element subsets are there of a 23-element set?
15. How many sets are there of three distinct integers between 1 and 50 inclusive whose sum is even? (*Hint:* You must count two cases—either all three are even, or one is even and the other two are odd.)

Problems and Projects

16. A woman wants to invite different subsets of r friends to her house on K consecutive nights.
    a. Write a program that reads the numbers r and K, and that determines and prints the smallest number n of friends she must have. [That is, your program should find the smallest n for which $\binom{n}{r} \geq K$].

    b. Run your program with $r = 3$ and $K = 1000$.

17. *Computing Combinations.* Computing the binomial coefficient $\binom{n}{r}$ is not simple when both n and r are large—for example, $\binom{30}{15}$. By Formula 3.7,

$$\binom{30}{15} = \frac{30!}{15!15!} = \frac{30 \cdot 29 \cdot 28 \cdot \cdots \cdot 16}{15!}$$

The products in the numerator and denominator are too large for most computers to compute using integer arithmetic. An accurate floating-point estimate of $\binom{30}{15}$ may be obtained, however, by writing the combination as follows:

$$\binom{30}{15} = \frac{30 \cdot 29 \cdot 28 \cdot \cdots \cdot 16}{15 \cdot 14 \cdot 13 \cdot \cdots \cdot 1} = \frac{30}{15} \cdot \frac{29}{14} \cdot \frac{28}{13} \cdot \cdots \cdot \frac{16}{1}$$
$$= (2.000)(2.071)(2.153) \cdot \cdots \cdot (16.000)$$

    a. Write a program that inputs two positive integers N and R and computes an estimate of $\binom{N}{R}$ using the technique just described.

b. Test your program by obtaining estimates for $\begin{pmatrix} 30 \\ 15 \end{pmatrix} = 155,117,520;$ $\begin{pmatrix} 50 \\ 30 \end{pmatrix};$ and $\begin{pmatrix} 100 \\ 60 \end{pmatrix}.$

18. Write a program that will read an integer n and print all values of $\begin{pmatrix} n \\ r \end{pmatrix}$ for

$0 \le r \le n$ $(n \le 10)$.

3.4 BINOMIAL COEFFICIENTS

The numbers $\begin{pmatrix} n \\ r \end{pmatrix}$ are called **binomial coefficients** because they appear in the expansion of a binomial to an integral power n. Consider $(x + y)^3 = x^3 + 3x^2y + 3xy^2 + y^3$. The expression $x^3 + 3x^2y + 3xy^2 + y^3$ is obtained by expanding $(x + y)^3 = (x + y)(x + y)(x + y)$. To observe more closely where the coefficients 1,3,3,1 come from, we use subscripts to indicate from which set of parentheses the x or y comes:

$$(x_1 + y_1)(x_2 + y_2)(x_3 + y_3) = (x_1x_2 + x_1y_2 + x_2y_1 + y_1y_2)(x_3 + y_3)$$
$$= x_1x_2x_3 + x_1x_2y_3 + x_1x_3y_2 + x_1y_2y_3$$
$$+ x_2x_3y_1 + x_2y_1y_3 + x_3y_1y_2 + y_1y_2y_3$$

Each term contains three factors each with a different subscript, because each term in the expanded expression is formed by multiplying one term from each set of parentheses. The coefficient of x^2y in the expansion of $(x + y)^3$ is 3 because three of the terms in the expansion of $(x_1 + y_1)(x_2 + y_2)(x_3 + y_3)$ contain one y factor: $x_1x_2y_3$, $x_1x_3y_2$, and $x_2x_3y_1$. Three is the number of ways we can choose one y from the three sets of parentheses, that is, $3 = \begin{pmatrix} 3 \\ 1 \end{pmatrix}$. By a similar analysis, we can find the coefficients for the expansion of any binomial.

Note 3.6

The coefficient of the term $x^{n-r}y^r$ in the expansion of $(x + y)^n$ is $\begin{pmatrix} n \\ r \end{pmatrix}$, the number of ways of choosing r y's as factors from the n parentheses.

In the expansion of $(x + y)^6$, we obtain the terms (with coefficients omitted) x^6, x^5y, x^4y^2, x^3y^3, x^2y^4, xy^5, y^6. The exponents always add to 6 because one factor was chosen from each of the six sets of parentheses. The successive exponents of x decrease by 1 and those of y increase by 1. Since the coefficient of $x^{n-r}y^r$ is $\begin{pmatrix} n \\ r \end{pmatrix}$, the coefficients for the terms just given are: $\begin{pmatrix} 6 \\ 0 \end{pmatrix}, \begin{pmatrix} 6 \\ 1 \end{pmatrix}, \begin{pmatrix} 6 \\ 2 \end{pmatrix}, \begin{pmatrix} 6 \\ 3 \end{pmatrix}, \begin{pmatrix} 6 \\ 4 \end{pmatrix}, \begin{pmatrix} 6 \\ 5 \end{pmatrix},$

$\binom{6}{6}$. Hence,

$$(x + y)^6 = \binom{6}{0}x^6 + \binom{6}{1}x^5y + \binom{6}{2}x^4y^2 + \binom{6}{3}x^3y^3$$

$$+ \binom{6}{4}x^2y^4 + \binom{6}{5}xy^5 + \binom{6}{6}y^6$$

$$= x^6 + 6x^5y + 15x^4y^2 + 20x^3y^3 + 15x^2y^4 + 6xy^5 + y^6$$

Theorem 3.1   The Binomial Theorem

Let n be a positive integer. Then

$$(x + y)^n = \sum_{r=0}^{n} \binom{n}{r}x^{n-r}y^r$$

EXAMPLE 3.28

Use Theorem 3.1 to expand $(x + y)^5$.

SOLUTION

$$(x + y)^5 = \binom{5}{0}x^5 + \binom{5}{1}x^4y + \binom{5}{2}x^3y^2 + \binom{5}{3}x^2y^3 + \binom{5}{4}xy^4 + \binom{5}{5}y^5$$

$$= x^5 + 5x^4y + 10x^3y^2 + 10x^2y^3 + 5xy^4 + y^5$$

EXAMPLE 3.29

What is the coefficient of $x^{20}y^4$ in the expansion of $(x + y)^{24}$?

SOLUTION

The coefficient counts the number of ways of choosing four y's from the 24 sets of parentheses:

$$\binom{24}{4} = \frac{24!}{4! \cdot 20!} = \frac{24 \cdot 23 \cdot 22 \cdot 21}{4 \cdot 3 \cdot 2 \cdot 1}$$

$$= 23 \cdot 22 \cdot 21 = 10{,}626$$

Throughout the rest of this section, we explore properties of the binomial coefficients. The first property we obtain comes directly from Theorem 3.1. If we let $x = 1$ and $y = 1$ in Theorem 3.1, we get $(1 + 1)^n = \sum_{r=0}^{n} \binom{n}{r}1^{n-r}1^r$. This gives Formula 3.8.

Formula 3.8

$$\sum_{r=0}^{n} \binom{n}{r} = 2^n$$

Formula 3.8 verifies the following statement, first made as Note 2.5 in Chapter 2 and proved by induction in Example 3.12: If set A contains n elements, the power set $\mathcal{P}(A)$ contains 2^n sets. The left side of Formula 3.8 sums the number of subsets of A that contain $r = 0, 1, 2, \ldots, n$ elements.

Another useful formula states that the number of ways of choosing r out of n elements is the same as the number of ways of choosing the $n - r$ elements we do not want. Each selection of $n - r$ elements we do not want corresponds to one selection of r elements we do want. We thus obtain Formula 3.9.

Formula 3.9

$$\binom{n}{r} = \binom{n}{n-r}$$

We could also use Formula 3.7 to justify Formula 3.9:

$$\binom{n}{r} = \frac{n!}{r!(n-r)!} = \frac{n!}{(n-r)!r!} = \frac{n!}{(n-r)!(n-(n-r))!} = \binom{n}{n-r}$$

Formula 3.9 is the reason for the symmetry in the coefficients in the binomial expansion:

$$(x + y)^5 = x^5 + 5x^4y + 10x^3y^2 + 10x^2y^3 + 5xy^4 + y^5$$
$$(x + y)^6 = x^6 + 6x^5y + 15x^4y^2 + 20x^3y^3 + 15x^2y^4 + 6xy^5 + y^6$$

As we move in from either end of the expansion, the coefficients are equal. The rth coefficient equals the $(n - r)$th coefficient.

Since it is the basis of many inductive proofs, the most important formula involving binomial coefficients is Formula 3.10.

Formula 3.10

$$\binom{n+1}{r+1} = \binom{n}{r} + \binom{n}{r+1}$$

A combinatorial argument that proves Formula 3.10 is the following: Let A be a set containing $n + 1$ elements, and suppose $x \in A$. The left side of Formula 3.10 counts the number of $(r + 1)$-element subsets of A. We can separate these subsets into two groups: (1) those containing x, and (2) those not containing x. If x is in the subset, then we must select r other elements from the remaining n elements of A to form an $(r + 1)$-element subset. There are $\binom{n}{r}$ ways of choosing these r elements. If x is not in the subset, we must select $r + 1$ elements from the remaining n elements of A. There are $\binom{n}{r + 1}$ ways of choosing these elements. By the addition principle (see Note 3.5), the number of $(r + 1)$-element subsets from the $(n + 1)$-element set A equals the number of $(r + 1)$-element subsets containing x plus the number of $(r + 1)$-element subsets not containing x. Thus

$$\binom{n + 1}{r + 1} = \binom{n}{r} + \binom{n}{r + 1}$$

EXAMPLE 3.30

Use Formula 3.7 to prove Formula 3.10.

SOLUTION

We begin with the right side of Formula 3.10.

$$\binom{n}{r} + \binom{n}{r + 1} = \frac{n!}{r!(n - r)!} + \frac{n!}{(r + 1)!(n - (r + 1))!}$$

$$= \frac{n!}{r!(n - r)!} + \frac{n!}{(r + 1)!(n - r - 1)!}$$

$$= \frac{r + 1}{r + 1} \cdot \frac{n!}{r!(n - r)!} + \frac{n!}{(r + 1)!(n - r - 1)!} \cdot \frac{n - r}{n - r} \quad \begin{matrix} \text{Find a common} \\ \text{denominator.} \end{matrix}$$

$$= \frac{n!(r + 1)}{(r + 1)!(n - r)!} + \frac{n!(n - r)}{(r + 1)!(n - r)!}$$

$$= \frac{n!(r + 1) + n!(n - r)}{(r + 1)!(n - r)!} = \frac{n!(r + 1 + n - r)}{(r + 1)!(n - r)!} = \frac{n!(n + 1)}{(r + 1)!(n - r)!}$$

$$= \frac{(n + 1)!}{(r + 1)!(n - r)!} = \frac{(n + 1)!}{(r + 1)!((n + 1) - (r + 1))!} = \binom{n + 1}{r + 1}$$

EXERCISES 3.4

1. Use the Binomial Theorem to expand $(a + 2)^4$.
2. Use the Binomial Theorem to expand $(2a + b)^5$.
3. What is the coefficient of $a^4 b^6$ in the expansion of $(a + b)^{10}$?

4. What is the coefficient of x^3y^{12} in $(x + y^2)^9$?
5. What is the coefficient of a^4b^3 in $(a^2 - 2b)^5$?
6. Expand $(x - y/x)^5$.
7. Use the Binomial Theorem to compute $(101)^5$. (*Hint:* Write 101 as $100 + 1$.)
8. Use the Binomial Theorem to show $(1.1)^{60} > 24$. [*Hint:* Determine the first several terms in the expansion of $(1 + 0.1)^{60}$].
9. Prove by induction: $x + 1$ divides $x^{2n-1} + 1$ for all $n \in N$.

10. Prove by induction: $\sum_{i=1}^{n} i(i!) = (n + 1)! - 1$ for all $n \in N$.

11. Show $\binom{2n}{2} = 2\binom{n}{2} + n^2$.

12. Express $5 \cdot 5^2 \cdot 5^3 \cdot \cdots \cdot 5^n$ as 5 to a power.

13. Show $\binom{n}{1} + 6\binom{n}{2} + 6\binom{n}{3} = n^3$.

Problems and Projects

Exercises 14–17 deal with **multinomial coefficients,** a generalization of the idea of binomial coefficients. For an expression $(x_1 + x_2 + \cdots + x_k)^n$, the coefficient of $x_1^{r_1}x_2^{r_2} \cdots x_k^{r_k}$ is $n!/(r_1!r_2! \cdots r_k!)$, where $\sum_{i=1}^{k} r_i = n$ (see Formula 3.3). For example, in the expansion of $(x + y + z + w)^{12}$, the coefficient of $x^3y^4z^2w^3$ is:

$$\frac{12!}{3! \cdot 4! \cdot 2! \cdot 3!} = \frac{12 \cdot 11 \cdot 10 \cdot 9 \cdot 8 \cdot 7 \cdot 6 \cdot 5}{3 \cdot 2 \cdot 1 \cdot 2 \cdot 1 \cdot 3 \cdot 2 \cdot 1}$$

$$= 11 \cdot 10 \cdot 9 \cdot 8 \cdot 7 \cdot 5 = 277{,}200$$

14. What is the coefficient of $x^4y^2z^3$ in $(x + y + z)^9$?
15. Use multinomial coefficients to expand $(x + y + z)^4$. (*Hint:* There will be 15 terms, and all the coefficients are less than 20. Some of the terms (without coefficients) are x^4, x^3y, x^3z, x^2y^2, x^2yz).
16. What is the coefficient of $a^2b^5c^3d^6$ in $(a + b + c + d)^{16}$?
17. What is the coefficient of $x^6y^3z^{12}$ in the expansion of $(x^3 + 2y + 3z^6)^7$?
18. **Pascal's triangle** is an array that lists the binomial coefficients in a triangular pattern. The binomial coefficients $\binom{n}{r}$, $r = 0, 1, 2, \ldots, n$ appear in the nth row (beginning with $n = 0$). The first six rows of the triangle appear in Figure 3.5. Note that 1s are along the outside of the triangle. To find any internal entry we add the two closest digits from the previous row: one is above to its right, the other is above to its left. For example, the circled 3 is obtained from $2 + 1$, while the circled 10 is obtained from $4 + 6$.
    a. Use the technique described to determine the next row of the triangle. (*Note:* This technique works because of Formula 3.10.)
    b. Write a program that calculates and prints the first 15 rows of Pascal's triangle.

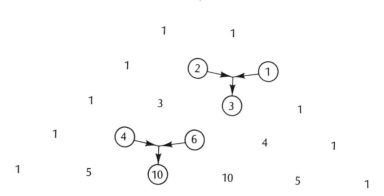

Figure 3.5

c. As an extra challenge, make the computer print the output in a triangular pattern.

19. Use Formula 3.7 to prove $\dbinom{n}{r+1} = \dfrac{n-r}{r+1}\dbinom{n}{r}$.

20. Use Formula 3.7 to prove $r\dbinom{n}{r} = n\dbinom{n-1}{r-1}$.

21. Use the identity of exercise 20 and a change of variables to prove

$$\sum_{r=1}^{n} r\binom{n}{r} = n \cdot 2^{n-1}$$

22. Prove $\dbinom{2n}{r} < \dbinom{2n}{r+1}$ for $0 \le r \le n - 1$.

23. Prove for $2 \le r \le n - 2$: $\dbinom{n}{r} = \dbinom{n-2}{r-2} + 2\dbinom{n-2}{r-1} + \dbinom{n-2}{r}$. (*Hint:* Use Formula 3.10.)

3.5   DISTRIBUTIONS

In this section we find formulas that count the number of ways to distribute objects (either distinct or identical) into boxes (either distinct or identical). The simplest case occurs when both the objects and the boxes are distinct. Suppose we have five distinct objects A, B, C, D, E to distribute into three distinct boxes B_1, B_2, B_3. In how many ways can this be done? For each object there are three possibilities—it can be placed in box B_1, B_2, or B_3. By the multiplication principle, therefore, we get $3^5 = 243$ ways to distribute five distinct objects into three distinct boxes. In general, we have Formula 3.11.

Formula 3.11

The number of ways of distributing n distinct objects into k distinct boxes is k^n.

EXAMPLE 3.31 _____

In how many ways can a penny, nickel, dime, quarter, half-dollar, and silver dollar be distributed among Fred, Mui, John, and Maria?

SOLUTION

There are six distinct objects (coins) to be distributed to four distinct boxes (people). There are, therefore, $4^6 = 4096$ possible distributions. Note that some people might receive no coins.

In Example 3.31, there was no restriction on the number of objects to be placed in each box. We now consider the case where the number of objects placed in each box is specified. Suppose you go Christmas shopping and get carried away. You buy seven toys for your three nieces, Angela, Betty, and Cathy. You decide to give three toys to Angela and two toys each to Betty and Cathy. You place each child's toys in a box labeled with the child's name. In how many ways can this be done? We number the toys 1, 2, 3, 4, 5, 6, 7. Several possible distributions of toys into the three boxes are displayed in Table 3.4.

Another way to describe the distributions in Table 3.4 is to list 7-letter "words" composed of three As (Angela), two Bs (Betty), and two Cs (Cathy). The position of the letters tells who gets which toys. BACCABA means Angela gets toys 2, 5, and 7 because there are As in positions 2, 5, and 7. Likewise, Betty gets toys 1 and 6, and Cathy gets toys 3 and 4.

	Box 1 (Angela)	Box 2 (Betty)	Box 3 (Cathy)
Distribution 1	1, 2, 3	4, 5	6, 7
Distribution 2	2, 4, 7	1, 3	5, 6
Distribution 3	3, 5, 6	2, 7	1, 4
Distribution 4	2, 6, 7	1, 4	3, 5

Table 3.4

EXAMPLE 3.32 _____

List the words composed of three As, two Bs, and two Cs that correspond to the distributions in Table 3.4.

SOLUTION

Distribution 1 becomes AAABBCC; distribution 2 becomes BABACCA; distribution 3 becomes CBACAAB; and distribution 4 becomes BACBCAA.

The number of ways we can distribute three toys to Angela, two to Betty, and two to Cathy is the same as the number of 7-letter words composed of three As, two Bs, and two Cs. By Formula 3.4, this number is $7!/(3! \cdot 2! \cdot 2!) = 7 \cdot 6 \cdot 5 \cdot 4/(2 \cdot 1 \cdot 2 \cdot 1) = 7 \cdot 6 \cdot 5 = 210$. In general, Formula 3.12 applies.

Formula 3.12

The number of ways to distribute n distinct objects into k distinct boxes so there are r_1 objects in box 1, r_2 objects in box 2, ..., r_k objects in box k $(r_1 + r_2 + \cdots + r_k = n)$ is $n!/(r_1! \cdot r_2! \cdot \cdots \cdot r_k!)$.

EXAMPLE 3.33 _____

A housewife goes shopping with her three children and buys 12 items. On the way home the bag breaks. She decides to carry four items, gives two to her youngest child, and three to each of her other children. In how many ways can the 12 items be distributed?

SOLUTION

Here, each person corresponds to a box. Carrying an item is equivalent to placing the item in a box. Therefore, we must place four items in one box, three in each of two others, and two in the last box. By Formula 3.12, we have

$$\frac{12!}{4! \cdot 3! \cdot 3! \cdot 2!} = \frac{12 \cdot 11 \cdot 10 \cdot 9 \cdot 8 \cdot 7 \cdot 6 \cdot 5}{3 \cdot 2 \cdot 1 \cdot 3 \cdot 2 \cdot 1 \cdot 2 \cdot 1} = 11 \cdot 10 \cdot 9 \cdot 8 \cdot 7 \cdot 5$$

$$= 277,200 \text{ ways in which the items can be distributed}$$

Another type of distribution problem is one in which the boxes are not distinct. Suppose, for example, we take the seven toys we bought for Angela, Betty, and Cathy and place them in three *unlabeled* boxes so one box has three toys and the other two have two toys each. We assume that removing the labels makes the boxes indistinguishable. In this case, the distribution 2,4,7; 1,3; 5,6 of toys into unlabeled boxes is the same as the distribution 2,4,7; 5,6; 1,3. What is important is which toys end up together, not which box they are in.

We saw in the preceding discussion that if the boxes are distinct, there are $7!/(3! \cdot 2! \cdot 2!) = 210$ ways of placing the toys into the boxes. If, however, the boxes are unlabeled, the two boxes containing two toys each are considered identical. Since there are 2! ways of arranging these two boxes, we must divide $7!/(3! \cdot 2! \cdot 2!)$ by 2!. There are, therefore,

$$\frac{7!}{3!2!2!(2!)} = \frac{7 \cdot 6 \cdot 5 \cdot 4}{2 \cdot 2 \cdot 2} = 7 \cdot 3 \cdot 5 = 105$$

distributions of seven toys into three unlabeled (that is, indistinguishable) boxes with three toys in one box and two toys in each of the other two boxes.

We now generalize the last example. Suppose we must distribute n objects into k identical boxes. Let r_i be the number of boxes that will contain i objects. In our example, $r_0 = 0$, $r_1 = 0$, $r_2 = 2$, $r_3 = 1$, $r_4 = r_5 = r_6 = r_7 = 0$. There are several things you should note about the r_i's. First, since there are n objects, there will be $n + 1$ r_i's: $r_0, r_1, r_2, \ldots, r_n$. Second, since there are k identical boxes to be filled, $\sum_{i=0}^{n} r_i = k$. This implies that at most k of the r_i's are nonzero. In our example, we had eight r_i's, and clearly $\sum_{i=0}^{7} r_i = 3$, which is the number of boxes to be filled. Finally, since r_i boxes will contain i objects and there are a total of n objects, $\sum_{i=0}^{n} i \cdot r_i = n$. In our example,

$$\sum_{i=0}^{n} i \cdot r_i = (0 \cdot 0) + (1 \cdot 0) + (2 \cdot 2) + (3 \cdot 1) + (4 \cdot 0) + (5 \cdot 0) + (6 \cdot 0) + (7 \cdot 0)$$

$$= 7$$

If we take a box containing i objects, these i objects can be arranged in $i!$ ways. The r_i boxes containing i objects can be arranged in $r_i!$ ways. For each i, therefore, there are $(i!)^{r_i} \cdot r_i!$ ways of arranging the objects and the boxes containing them (the boxes are indistinguishable). This leads to Formula 3.13.

Formula 3.13

The number of ways to distribute n distinct objects into k identical boxes with r_0 empty boxes, r_1 boxes containing one object, r_2 boxes containing two objects, $\ldots$, r_n boxes containing n objects is

$$\frac{n!}{(0!)^{r_0} r_0! \cdot (1!)^{r_1} r_1! \cdot (2!)^{r_2} r_2! \cdots (n!)^{r_n} r_n!}$$

EXAMPLE 3.34 _____

A school gives 15 awards at graduation. Although we do not know which students get awards, we've discovered that two students will get four awards each, three students will get two awards each, and one student will get one award. How many award distributions are possible?

SOLUTION

We must distribute $n = 15$ objects (the awards) among $k = 6$ indistinguishable boxes (the students receiving awards—since we do not know who gets how many awards, the students are indistinguishable to us). Since two boxes (students) get four awards each, three boxes (students) get two awards each, and one box (student) gets one award, $r_4 = 2$, $r_2 = 3$, and $r_1 = 1$. All other r_i's are zero. Formula 3.12, therefore, gives as the number of possible distributions:

$$\frac{15!}{(1!)^1 1! \cdot (2!)^3 3! \cdot (4!)^2 2!} = \frac{15 \cdot 14 \cdot 13 \cdot 12 \cdot 11 \cdot 10 \cdot 9 \cdot 8 \cdot 7 \cdot 6 \cdot 5 \cdot 4!}{2 \cdot 2 \cdot 2 \cdot 3 \cdot 2 \cdot 1 \cdot 4! \cdot 4 \cdot 3 \cdot 2 \cdot 1 \cdot 2}$$

$$= 15 \cdot 7 \cdot 13 \cdot 11 \cdot 5 \cdot 9 \cdot 7 \cdot 5 = 23{,}648{,}625$$

Recall that a partition of a set A is a collection of pairwise disjoint subsets of A whose union is A. In Formula 3.13, if we interpret the n distinct objects as the elements of A and the boxes as subsets of A, the formula gives the number of ways in which a set can be partitioned into k subsets, each subset having a given number of elements.

Formula 3.13′

The number of partitions of a set of n objects into r_1 subsets of size 1, r_2 subsets of size 2, ..., r_n subsets of size n is

$$\frac{n!}{(1!)^{r_1} r_1! \cdot (2!)^{r_2} r_2! \cdots (n!)^{r_n} r_n!}$$

EXAMPLE 3.35

In how many ways can a set of ten elements be partitioned into five 2-element subsets?

SOLUTION

In this example, $n = 10$, $r_2 = 5$, and all other r_i's are zero. Formula 3.13′, therefore, gives as the number of possible partitions into five 2-element subsets:

$$\frac{10!}{(2!)^5 5!} = \frac{10 \cdot 9 \cdot 8 \cdot 7 \cdot 6 \cdot 5!}{2 \cdot 2 \cdot 2 \cdot 2 \cdot 2 \cdot 5!} = 5 \cdot 9 \cdot 7 \cdot 3 = 945$$

Finally, we consider the problem in which we distribute n identical objects into k distinct boxes. Suppose we want to distribute ten identical lollipops among four children, Al, Bea, Carl, and Dan. Take four boxes and label each with a child's name. One distribution of lollipops into the four boxes could be represented by ∗∗/∗∗∗/∗/∗∗∗∗, which means we give Al two, Bea three, Carl one, and Dan four lollipops. This notation allows for the possibility of someone receiving no lollipops.

If we want Al, Bea, Carl, and Dan to receive three, five, zero, and two lollipops, respectively, we would represent this by ∗∗∗/∗∗∗∗∗//∗∗.

Any distribution of the ten lollipops can be represented by an arrangement of 13 symbols, consisting of ten ∗'s and three /'s. Thus, the number of arrangements of the symbols corresponds to the number of ways of choosing 3 positions for the /'s from the 13 available positions. Formula 3.4, therefore gives $13!/10!3! = 13 \cdot 12 \cdot 11/(3 \cdot 2 \cdot 1) = 13 \cdot 2 \cdot 11 = 286$ ways to distribute ten identical lollipops among Al, Bea, Carl, and Dan. In the general case, a distribution of n identical objects into k distinct boxes can be represented by an arrangement of $n + k - 1$ symbols, consisting of n ∗'s and $(k - 1)$ /'s. This observation and Formula 3.4 give Formula 3.14.

Formula 3.14

The number of ways to distribute n identical objects into k distinct boxes is

$$\frac{(n + k - 1)!}{n!(k - 1)!} = \binom{n + k - 1}{n}$$

EXAMPLE 3.36

In how many ways can we distribute eight identical balls into three distinct boxes?

SOLUTION

Applying Formula 3.14 with $n = 8$ and $k = 3$, we get

$$\binom{8 + 3 - 1}{8} = \binom{10}{8} = \frac{10!}{8!2!} = \frac{10 \cdot 9}{2} = 45 \text{ ways}$$

EXAMPLE 3.37

In how many ways can we form a sum of 15 from four nonnegative integers?

SOLUTION

We can represent the four nonnegative integers of the sum by four boxes. We have a supply of 15 (identical) 1s to distribute among the four boxes. Putting three 1s in the first box, two in the second, eight in the third, and two in the fourth would be represented as follows: $15 = 3 + 2 + 8 + 2$. The number of ways the sum can be formed, therefore, is the number of ways the 15 1s can be distributed among the four boxes:

$$\binom{15 + 4 - 1}{15} = \binom{18}{15} = \frac{18!}{15!3!} = \frac{18 \cdot 17 \cdot 16}{3 \cdot 2 \cdot 1} = 3 \cdot 17 \cdot 16 = 816$$

EXAMPLE 3.38

In how many ways can we distribute three identical watches and five distinct radios among Al, Cal, Hal, and Sal?

SOLUTION

By Formula 3.14, we can distribute three identical watches to four distinct people in

$$\binom{4+3-1}{3} = \binom{6}{3} = \frac{6!}{3!3!} = \frac{6 \cdot 5 \cdot 4}{3 \cdot 2 \cdot 1} = 5 \cdot 4 = 20 \text{ ways}$$

By Formula 3.11, we can distribute five distinct radios to four distinct people in $4^5 = 1024$ ways. By the multiplication principle, therefore, we can distribute three identical watches and five distinct radios among Al, Cal, Hal, and Sal in $20 \cdot 1024 = 20{,}480$ ways.

Inclusion–Exclusion

Some combinatorial problems concern counting the number of items that satisfy exactly k properties, where k is a given integer. Typical examples involve inter-sections of sets. Consider set A having 10 elements and set B having 18 elements from a universe of 40 elements. If $A \cup B$ has 22 elements, then how many elements are in $A \cap B$? This problem can be analyzed using a Venn diagram. In Figure 3.6 there are four regions, labeled $X_1, X_2, X_3,$ and X_4. Let x_i be the number of elements in region X_i. Since A corresponds to regions X_2 and X_3, and A has 10 elements we know $x_2 + x_3 = 10$. Similarly, for $B, x_3 + x_4 = 18$. Set $A \cup B$ consists of regions $X_2, X_3,$ and X_4. Thus, $x_2 + x_3 + x_4 = 22$. Since $A \cap B$ corresponds to X_3, we must find x_3. One way of doing this is by finding $x_3 = (x_2 + x_3) + (x_3 + x_4) - (x_2 + x_3 + x_4) = 10 + 18 - 22 = 6$. Thus $A \cap B$ has six elements.

You may ask what this has to do with the properties mentioned earlier. For an element y in the universe U, let P_A be the property of y being in A, and let P_B be the property of y being in B. Finding X_3 corresponds to determining the

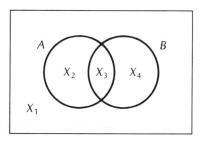

Figure 3.6

number of elements that satisfy the two properties P_A and P_B. Suppose, however, we want to find the number of elements that satisfy neither of the properties P_A or P_B. In which region are these elements? The answer is region X_1.

In many problems we are given information about some of the regions and asked about the others. One type of problem is given in Example 3.39. Let $|A|$ denote the number of elements in A.

EXAMPLE 3.39

A survey of 100 students was taken concerning their programming experience. The following results were obtained:

30 know Pascal
19 know COBOL
16 know PL/I
11 know both Pascal and PL/I
12 know both Pascal and COBOL
6 know both COBOL and PL/I
4 know all three languages

a. How many students know none of the three languages?
b. How many students know Pascal and PL/I but not COBOL?
c. How many students know exactly two languages?

SOLUTION

Again we use a Venn diagram. We have three sets: $S_1 = $ Knows Pascal, $S_2 = $ Knows COBOL, $S_3 = $ Knows PL/I. We begin with the innermost region (the one whose elements are in all three sets) and work our way out. Four people know all three languages, so $|S_1 \cap S_2 \cap S_3| = 4$. Inside each region of the Venn diagram we place the number of elements contained in that region [see Figure 3.7(a)]. From the information given in the question $|S_1 \cap S_2| = 12, |S_1 \cap S_3| = 11$, and $|S_2 \cap S_3| = 6$. By subtracting the number of elements in $S_1 \cap S_2 \cap S_3 (= 4)$ from each of these, we determine the number of elements in the next regions to be 8, 7, and 2 [see Figure 3.7(b)]. We were also given $|S_1| = 30, |S_2| = 19$, and $|S_3| = 16$. Using the Venn diagram, we subtract the numbers already placed within S_i from $|S_i|$ to get 11, 5, and 3 for the next group of regions [see Figure 3.7(c)]. Finally, since 100 students were surveyed, we subtract each of the numbers within the circles of the Venn diagram to find the last region, $100 - (11 - 8 - 4 - 5 - 7 - 2 - 3) = 100 - 40 = 60$ [see Figure 3.7(d)].

a. Sixty students know none of the three languages.
b. Seven students know Pascal and PL/I but not COBOL (this corresponds to $|S_1 \cap S_3 \cap S_2'|$).
c. From the Venn diagram, we see that 17 students $(8 + 2 + 7)$ know exactly two of the languages.

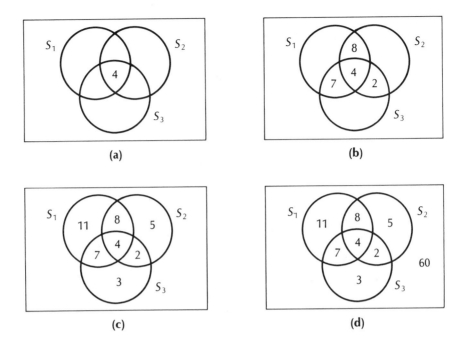

Figure 3.7

In part a of Example 3.39, we are counting the number of elements in none of the sets S_1, S_2, S_3. Thus, $|S_1' \cap S_2' \cap S_3'| = 60$. One way of finding this quantity is by using the **inclusion–exclusion principle**. In a universe with N elements, the inclusion–exclusion formula, for three sets S_1, S_2, S_3 is

$$|S_1' \cap S_2' \cap S_3'| = N - |S_1| - |S_2| - |S_3| + |S_1 \cap S_2| + |S_1 \cap S_3|$$
$$+ |S_2 \cap S_3| - |S_1 \cap S_2 \cap S_3|$$

In Example 3.39, this gives

$$|S_1' \cap S_2' \cap S_3'| = 100 - 30 - 19 - 16 + 11 + 12 + 6 - 4$$
$$= 100 - 65 + 29 - 4 = 60$$

The inclusion–exclusion formula allows us to find the answer for part a of Example 3.39 without referring to a Venn diagram. In the formula, we successively exclude (that is, subtract) and include (add) all combinations of intersections of k of the sets from the universe. The formula generalizes to any number of sets so it can be used even when it is not possible to draw a Venn diagram. The inclusion–exclusion formula for two sets is:

$$|S_1' \cap S_2'| = N - |S_1| - |S_2| + |S_1 \cap S_2|$$

Formula 3.15   Inclusion–Exclusion

Let $S_1, S_2, \ldots, S_n$ be n sets in the universe U of N elements. Let N_k be the sum of the orders of all combinations of intersections of k of the sets S_i. Then

$$|S_1' \cap S_2' \cap \cdots \cap S_n'| = N + \sum_{i=1}^{n} (-1)^i N_i$$

$$= N - N_1 + N_2 - N_3 + \cdots + (-1)^n N_n$$

EXAMPLE 3.40 _____

At the beginning of the semester a professor wants to see if she has figured out which names belong to which students in her class. With their indulgence she returns the papers from the first quiz to each student to whom she thinks that paper belongs. There are 15 students in the class and each student gets one paper back. Find a formula for the number of ways she could distribute the papers so that nobody got his or her own paper back (that is, how many ways could be perfectly wrong?).

SOLUTION

Let P_i mean that person i gets back his or her own paper. We want $|\bigcap_{i=1}^{15} P_i'|$. If k people receive their own papers back, there are $(15 - k)!$ ways the other papers can be distributed among the other students. (Some of those may, in fact, also get their papers back, but they will be excluded in a later term of the formula.) There are $\binom{15}{k}$ ways of selecting the k students who get their own paper back. Thus, there are $\binom{15}{k}(15 - k)!$ ways that (at least) k students get their papers back. By Formula 3.15, with $N = 15!$,

$$\left|\bigcap_{i=1}^{15} P_1'\right| = 15! + \sum_{i=1}^{15} (-1)^i \binom{15}{i}(15 - i)! = \sum_{i=0}^{15} (-1)^i \binom{15}{i}(15 - i)!$$

EXERCISES 3.5

1. In how many ways can a wrench, hammer, plunger, and saw be distributed among Moe, Larry, and Curly?
2. Balls numbered from 1 to 12 are dropped from the top of a maze, flow through the maze and land in slot A, B, C, or D. In how many ways can the balls land if
   a. five, three, one, and three balls fall into slots A, B, C, D?
   b. an equal number of balls fall into each slot?
   c. two slots are empty and an equal number of balls fall into the other two

slots? (*Hint:* First choose the two slots that balls do fall into, then fill them each with half of the balls.)

3. Suppose in exercise 2, the balls are not labeled. In how many ways can the balls land
   a. in part a?
   b. in part b?
   c. in part c?

4. In how many ways can five differently colored candles be arranged on
   a. three differently colored cakes?
   b. three identical cakes?

5. Fifteen identical cups are piled up to be washed. In how many ways can they be separated into three different piles?

6. Three TV shows have been nominated for the same collection of six distinct Emmy Award categories. In how many ways can the winners be selected?

7. Simplify each of the following:

   a. $\dfrac{10!}{6!2!2!}$ b. $\dfrac{14!}{(4!)^2 2!(2!)^3 3!}$ c. $\dbinom{8+5-1}{8}$

8. In how many ways can eight identical peppermint sticks and two distinct toys be distributed among Fay, Jay, and Ray?

9. In how many ways can nine distinct objects be distributed into four distinct boxes so there are four objects in box 1, two objects in box 2, one object in box 3, and two objects in box 4?

10. How many nonnegative integer solutions are there to the equation $x_1 + x_2 + x_3 = 10$?

11. How many solutions (using positive integers) are there to $x_1 + x_2 + x_3 = 8$? (*Hint:* Each $x_i \geq 1$; therefore, make each 1 before you begin.)

12. In how many ways can a 9-element set be partitioned into three 2-element subsets and one 3-element subset?

13. In a holdup, four robbers get 17 $100 bills, a necklace, and a ring. In how many ways can their take be divided up?

14. The bank robbers of exercise 13 are caught and receive a total of ten years in prison. If each robber receives a term that is a nonnegative integer (it is possible some receive zero years, that is, were freed), how many prison term combinations are possible?

15. The bank robbers in exercise 14 are Jesse, Billy, Butch, and Sundance. In how many ways could Jesse, Billy, Butch, and Sundance be sentenced?

16. Write out the inclusion–exclusion formula for four sets A_1, A_2, A_3, A_4 in a universe U of n elements.

17. In a survey of 120 television viewers the following results were obtained:

    40 watch "Dallas"

    28 watch "Love Boat"

    31 watch "Saturday Night Live"

    23 watch "Dallas" and "Love Boat"

19 watch "Love Boat" and "Saturday Night Live"

25 watch "Dallas" and "Saturday Night Live"

17 watch all three programs

a. How many viewers watch none of the three programs?

b. How many viewers watch exactly one of the programs?

c. How many viewers watch "Saturday Night Live" and "Dallas," but not "Love Boat"?

18. How many positive integers are there less than 10,000 that are not divisible by 2, 3, or 5? (*Hint:* Use Formula 3.15.)

CASE STUDY 3A   CORRECTNESS OF ALGORITHMS

People who write computer programs frequently find that their programs do not run. They are burdened with a long list of error messages—clues to their defeat. They then examine and modify their programs until all error messages have been eliminated. Although the program now runs and prints answers, it *still* may be incorrect.

Three major types of errors are found in computer programs: syntax, execution, and logic errors. **Syntax errors,** or errors in the grammar of a computer language, are usually found by the computer when the program is compiled or interpreted (that is, translated from the high-level language you are using into machine language). Execution of the program cannot proceed until corrections are made. For example, in Pascal suppose you want to branch to statement 2 if X equals 5, and you write

IF (X := 5) THEN GOTO 2;

You will receive a syntax error message because you should have written:

IF (X = 5) THEN GOTO 2;

Equals is translated by = in a Pascal logical-if test. Syntax errors are corrected by carefully inspecting the incorrect statements and making sure they are recoded following the syntax of the programming language.

Execution errors occur during execution of a program. They indicate that the computer can no longer continue execution of your program because of some inconsistent internal condition in the computer. Execution errors usually cause a message to be printed that gives a clue to the problem. Execution errors usually occur because the programmer:

1. Divides by zero.

2. Takes the square root of a negative number in a language that does not allow for complex numbers. (PL/I does allow complex valued variables, but they must be declared as such.)

3. Uses a subscript that is outside the declared size of the list.

4. Uses *bad data*—data that is partially missing when a READ statement is being executed or data that is in a form inconsistent with what the program led the computer to expect.

Execution errors are more difficult to correct than syntax errors because we are frequently not told where in the program the error occurred. The programmer must either carefully trace the program by hand, use an interactive debugging program, or try to find the error using a memory dump.

Logic errors are the most difficult to catch because program execution is not interrupted and the computer does not print error messages. **Logic errors** are usually caused by a flaw in the underlying algorithm or by a correct algorithm being incorrectly translated into a programming language.

In the remainder of this case study, we discuss techniques for proving an algorithm correct. An algorithm is **correct** if it performs the tasks it is supposed to and produces the correct output for all possible sets of input data.

Usually the only way beginning programmers "prove" their algorithms correct is by testing them with several sets of input data. In fact, the majority of all programs ever written has been tested only by various sets of input data.

The basic methods for testing an algorithm with data are listed in Note 3.7.

Note 3.7

1. Test simple cases first, preferably ones for which the correct output is known.

2. Do an algorithm trace. In the program implementation, this corresponds to inserting extra print statements so values of variables or the appearance of records can be viewed at intermediate stages of a complex procedure. These print statements would be removed after testing has been completed.

3. Test subroutines separately. This will keep the analysis simple and localize any problems.

4. Be sure that your test data uses all combinations of decision branches in your algorithm.

EXAMPLE 3.41

The following Pascal program calculates the salaries (before deductions) for a certain company. Employees are paid for working overtime (more than 40 hours) at time and a half. If they work on the night shift for that week, they get an extra 10% (differential) added to their gross pay.

a. Do a trace using the following data: 22, 30.5, 5.50, 0, 17, 46, 7.20, 0, 16, 38, 6.30, 1, 21, 25, 5.80, 0, 45, 44.5, 8.00, 0, 10, 36.25, 5.00, 1, 99, 0, 0, 0

b. What code indicates that an employee worked on the night shift?

c. Have all decision branch combinations been used?

d. If an additional data set of 26, 51.5, 6.50, 1 appeared before the last data set, what would the salary be for this employee?

```
program test(input,output);
var emp,night: integer;
    hours,rate,sal: real;
begin
    writeln('input employee #, # of hours, hourly rate, and 1 if a');
    writeln('night employee or 0 if a day employee');
    read(emp,hours,rate,night);
    while emp <> 99 do
        begin
            if hours <=40 then
                sal:=hours*rate
            else
                sal:=40*rate+(hours-40)*1.5;
            if night=1 then
                sal:=1.1*sal;
            writeln('employee # =',emp,'   salary = $',sal);
            writeln('input employee #, # of hours, hourly rate, and');
            writeln('1 if a night employee or 0 if a day employee');
            writeln('input 99   0   0   0    to end');
            read(emp,hours,rate,night)
        end
end.
```

SOLUTION

a.

EMP	22	17	16	21	45	10	99
HOURS	30.5	46	38	25	44.5	36.25	0
RATE	5.50	7.20	6.30	5.80	8.00	5.00	0
NIGHT	0	0	1	0	0	1	0
EMP <> 99?	✓	✓	✓	✓	✓	✓	
HOURS <= 40?	T	F	T	T	F	T	
NIGHT = 1?	F	F	T	F	F	T	
WRITE EMP,	22	17	16	21	45	10	
SAL	167.75	352.80	263.34	145.00	374.00	199.38	

b. If the variable NIGHT has a value of 1, then the employee worked on the night shift.

c. No. The combination F-T never occurred, which indicates that the branch combination corresponding to an employee working overtime while being on the night shift was not tested by the data used.

d. This is a night shift employee on overtime, so her salary is computed by applying the formula following the first ELSE and then increasing that amount by the (differential) formula SAL = 1.1 * SAL. Thus, the salary is 409.34.

Proving Programs Correct

In recent years various methods, many based on mathematical induction, have been developed for proving algorithms correct. Induction is used frequently because (1) recursion and looping are fundamental components of most algorithms, and (2) a computer program can be considered an ordered sequence of n instructions, where $n \in N$.

A difficulty with testing a program using data is that after the problems have been eliminated, other problems may arise when we use a different set of data. This would require further debugging. If we *prove* our program correct, repeated debugging is eliminated (assuming no flaw in our proof—see Exercises 3.1, exercise 12).

Attempting to prove a program correct has additional benefits: (1) We gain a better understanding of our algorithm and, consequently, recognize its flaws more readily, and (2) we gain a better understanding of recursion and looping.

Before we begin to prove a given program correct, we must describe what the program is supposed to do. This description is called the **correctness assertion.** To prove the program correct, we must show that when the program is executed it will eventually terminate, and when it does terminate, the correctness assertion will be true. The correctness proof begins with an **initial assertion,** usually a statement about the input variables (for example: a, b, and c are integers). The initial assertion is usually followed by several **intermediate assertions,** which usually occur at the beginning of loops. The intermediate assertions are of the form "Each time program execution reaches this point then . . . (statement)." The proof of the intermediate assertions is accomplished by induction.

The final step in the correctness proof is to show that the initial assertion and the intermediate assertions together, taken in order of execution, imply the correctness assertion. That is, if the initial assertion is denoted by A_0, the intermediate assertions by $A_1, A_2, \ldots, A_I$, and the correctness assertion by A_C, we must show $A_0 \wedge A_1 \wedge A_2 \wedge \cdots \wedge A_I \rightarrow A_C$.

EXAMPLE 3.42

The following program reads in a positive number X and finds the first natural number N for which 2^N exceeds X.

```
program exceed(input,output);
var x:real;
    n,k:integer;
    function ypowern(y,n: integer) : integer;
    begin
      if n=1 then
         ypowern:= y
      else
         ypowern:= y * ypowern(y,n-1)
    end;

begin
  writeln('input a positive number');
  read(x);
  if x<=0 then
     writeln('x is not positive, error')
  else
     begin
       n:=1;
       k:=ypowern(2,n);
       while k <= x do
          begin
            n:=n+1;
            k:=ypowern(2,n)
          end
     end;
  writeln('the first positive integer n for which 2 to the n exceeds');
  writeln(x,' is', n)
end.
```

a. State the initial, intermediate, and correctness assertions.

b. Prove the program correct.

SOLUTION

Correctness assertion, A_C: The program will terminate and print the first natural number N for which 2^N exceeds X.

Initial assertion, A_0: X is a positive number.

Intermediate assertion, A_1:   Each time execution reaches the statement $N := N + 1$, $2^N \leq X$ is true.

The first time that statement $N := N + 1$ is not executed is the first time the condition $2^N \leq X$ is no longer true. At that point $2^N > X$. The next statement executed is the print statement, which, therefore, prints the first natural number N for which 2^N exceeds X. The only thing remaining to prove is that the program terminates. This is where we use induction. Let Y be the first integer such that $Y \geq X$. Example 3.1 showed $2^Y > Y$. Thus, $2^Y > Y \geq X$. The continual incrementing of N within the WHILE DO loop guarantees we will eventually reach a value of N exceeding X (or such that $2^N > N > X$) and thereby exit the WHILE DO loop. Thus, the program terminates.

The next example is more difficult than Example 3.42 because it contains a counter-controlled FOR DO loop rather than a WHILE DO loop. The intermediate assertion for a WHILE DO loop generally involves the first statement within the loop. In a counter-controlled FOR DO loop, the intermediate assertion usually involves the DO statement and incrementing the counter.

EXAMPLE 3.43

The following program prints the sum of the first T positive integers.

```
program sumint(input,output);
var n,t,sum: integer;
begin
   writeln('input a positive integer');
   read(t);
   sum:=0;
   for n:=1 to t do
      sum:=sum+n;
   writeln('the sum of the first',t,' positive integers is',sum)
end.
```

a. State the initial, intermediate, and correctness assertions.

b. Prove the program correct.

SOLUTION

a. *Correctness assertion, A_C:*   The program will terminate by printing the sum of the first T positive integers.

   Initial assertion, A_0: T is a natural number.

   Intermediate assertion, A_1:   Each time the value of N is incremented at the DO statement, the sum of the first $N - 1$ positive integers will have been accumulated in SUM.

b. The proof of A_1 is by induction. A_1 corresponds to S_N, that is, $A_1 = S_N$: Each time N is incremented, SUM contains the sum of the first $N - 1$ positive integers.

Step 1   Show S_1 is true. When we enter the DO loop for the first time, $N = 1$. Therefore, SUM $= 0$, which is the sum of the first $N - 1 = 1 - 1 = 0$ positive integers.

Step 2   Inductive hypothesis: Assume S_k is true—S_k: When the value of N is incremented to k ($\leq T$), the sum of the first $k - 1$ positive integers will have been accumulated in SUM. We must show S_{k+1} is true—S_{k+1}: When the value of N is incremented to $k + 1$, the sum of the first k positive integers has been accumulated in SUM.

The value of N is incremented to $k + 1$ immediately after the kth pass through the DO loop. By the inductive hypothesis, when N was incremented to k, the value accumulated in SUM was $1 + 2 + \cdots + (k - 1)$. When the statement

$$\text{SUM} = \text{SUM} + N$$

is executed while $N = k$, the resulting value in SUM is $1 + 2 + \cdots + (k - 1) + k$. Thus, when we return to DO to increment N to $k + 1$, the value accumulated in SUM is the sum of the first k positive integers. Since S_k is true implies S_{k+1} is true, by the principle of mathematical induction, A_1 is proven.

To complete the correctness proof, we must show that $A_0 \wedge A_1 \rightarrow A_C$ and that the program terminates. As we saw in Section 2.8, we need only verify that when both A_0 and A_1 are true, A_C is true. WRITELN is reached when we leave the DO loop. This occurs when N is incremented to $T + 1$ and is found to be greater than T. Since A_0 and A_1 are true, when N is incremented to $T + 1$, the sum of the first T positive integers has been accumulated in SUM. Thus WRITELN(SUM) will produce the correct answer. This shows $A_0 \wedge A_1 \rightarrow A_C$.

Finally, since the DO loop is executed exactly T times, it is obvious the program terminates. Note that if the initial assertion A_0 is false (for example, $T = 1/2$), the program may not terminate. This emphasizes the practical importance of verifying that the data input to a program is of the type assumed by the program. Otherwise, the program may abnormally terminate, produce incorrect results, or get caught in an infinite loop.

Example 3.43 shows how much more difficult and tedious it is to prove a program correct than to design and write the program correctly in the first place. The program was only ten lines long, whereas its correctness proof takes a full page.

The two examples we have just considered were simple because each had only one intermediate assertion. More complicated programs could contain several intermediate assertions, as we see in Example 3.44.

EXAMPLE 3.44

Consider the program of Example 2.38, which orders the elements of a set by using the method of successive minima. Describe the initial, intermediate, and correctness assertions.

SOLUTION

Correctness assertion, A_C: The program will terminate, and when it does, the elements of A will be arranged in ascending order.

Initial assertion, A_0: N is a natural number and A is a set of N numbers.

Intermediate assertions, A_1:   Each time i is incremented at the first DO statement, set A has been ordered up to and including position $i - 1$.

A_2:   Each time k is incremented at the second DO statement, the minimum $A[j]$ for $i \leq j \leq k - 1$ is stored in MIN.

Complex problems usually involve nested loops. These problems require a more general approach than the one used in Examples 3.42 and 3.43. The difficulty caused by nested loops is that the proof of an intermediate assertion A_k for an inner loop may depend on the validity of an intermediate assertion A_j in an outer loop. Because the loops are nested, the proof of A_j may in turn depend on the proof of A_k. Correctness proofs of algorithms containing nested loops are done with the method of inductive assertions. The interested reader is referred to Anderson (1979).

EXERCISES: CASE STUDY 3A

1. Consider the program in Example 2.35.
   a. State the initial, intermediate, and correctness assertions.
   b. Prove the program correct.
2. Prove your program from Exercises: Case Study 2A, exercise 4 correct.
3. One way to multiply integers is by repeated addition. If we want to find $5 \cdot 7$, we could add $7 + 7 + 7 + 7 + 7$.
   a. Write a program that reads two integers a and b, and determines and prints the product $a \cdot b$ by repeated addition.
   b. What are the initial, intermediate, and correctness assertions for the program?
   c. Prove your program correct.
4. Prove your program from Exercises 1.3, exercise 17 correct.
5. For each of the following programs, determine for which values of the input variables the program will terminate.

a.
```
program term1(input,output);
var x,j: integer;
begin
    writeln('input an integer');
    read(x);
    j:=100;
    while x<>130 do
        begin
            writeln(x);
            while j>=x do
                j:=j+1;
            x:=x-1
        end;
    writeln(x,j)
end.
```

b.
```
program term2(input,output);
   var a,b: integer;
      small,large,ratio: real;
   begin
      writeln('input two integers');
      read(a,b);
      if a<b then
         begin
            small:=a;
            large:=b
         end
      else
         begin
            small:=b;
            large:=a
         end;
      while small<>large do
         begin
            ratio:=small/large;
            writeln(ratio);
            small:=small+2
         end;
      writeln('finished')
   end.
```

6. State the initial, intermediate, and correctness assertions for the following program (*Note:* The intermediate assertion is a statement about PRODPOS, ZEROS, and SUMNEG, not X.)

```
program mixed(input,output);
var zeros: integer;
   prodpos,sumneg,x:real;
begin
   prodpos:=1;
   zeros:=0;
   sumneg:=0;
   writeln('input a real number');
   read(x);
   while x<>9999 do
      begin
         if x<0 then
            sumneg:=sumneg+x
         else
            if x>0 then
               prodpos:=prodpos*x
            else
               zeros:=zeros+1;
   writeln('input a real number (9999 to end program)');
         read(x)
      end;
   writeln (sumneg,zeros,prodpos)
end.
```

CASE STUDY 3B   INTRODUCTION TO ALGORITHM ANALYSIS

Throughout this chapter we have developed counting techniques and formulas. The type of problem being considered determines which formula we should use. Counting is used in computer science to compare algorithm efficiency by determining how many operations are performed when solving the same problem with different algorithms. A cost is assigned to each operation based on the cost of performing that operation on an idealized computer. In most cases, attention is restricted to particular operations; others are ignored. For example, in comparing algorithms that solve simultaneous linear equations, we might count the number of multiplications and divisions and ignore the number of additions and subtractions (which are less costly to perform). In sorting algorithms, however, we usually count the number of comparisons made of the elements being sorted and ignore the number of assignment and arithmetic operations.

Two types of algorithm analysis are: average case analysis and worst case analysis. In **average case analysis,** the cost of an algorithm is defined as the (weighted) average of the costs of performing the algorithm on sets of n data items. The average is "weighted" by the likelihood of the occurrence of each data set and therefore involves probability. Thus, we will not discuss average case analysis. In **worst case analysis,** the cost of an algorithm is defined as the maximum cost of performing the algorithm on a set of n data items. In this case study, we discuss only worst case analysis. In Chapter 5, we discuss another aspect of algorithm analysis: optimality.

We show how worst case analysis is done by considering the following example.

EXAMPLE 3.45 _____

Following are programs for two algorithms that determine whether a list of n numbers stored in array A is in nondecreasing order. Compare them using worst case analysis.

Algorithm 1

```
program sort1(input,output);
var n,i,sw: integer;
    a: array [1..50] of real;
begin
    writeln('input a positive integer');
    read (n);
    if n>50 then
        writeln ('array is too large')
    else
        begin
            writeln('input the',n,' real numbers');
            for i:=1 to n do
                read (a[i]);
            sw:=0;
            i:=1;
            while (i<n) and (sw=0) do
                begin
                    if a[i]>a[i+1] then
                        sw:=1;
                    i:=i+1
                end;
            if (i=n) and (sw=0) then
                writeln ('ordered')
            else
                writeln ('out of order')
        end
end.
```

Algorithm 2

```
program sort2(input,output);
var  n,i,sw,j: integer;
     a: array[1..50] of real;
begin
   writeln('input a positive integer');
   read (n);
   if n>50 then
      writeln ('array is too large')
   else
      begin
         writeln('input',n,' numbers');
         for i:=1 to n do
            read (a[i]);
         sw:=0;
         i:=1;
         j:=2;
         while i<n do
            begin
               while (j<=n) and (sw=0) do
                  begin
                     if a[i]>a[j] then
                        sw:=1;
                     j:=j+1
                  end;
               if (j=n+1) and (sw=0) then
                  begin
                     i:=i+1;
                     j:=i+1
                  end
               else
                  begin
                     writeln ('out of order');
                     i:=n+2
                  end
            end;
         if (i=n) and (sw=0) then
            writeln ('ordered')
      end
end.
```

SOLUTION

In this example we focus on the maximum number of comparisons that might take place with either algorithm when a list of n numbers is input. Assignment and arithmetic operations are ignored.

In the worst case for Algorithm 1, we do not exit from the DO loop until the whole list has been searched. The conditions $i < n$ and $a[i] < a[i + 1]$ have each been tested for $i = 1, 2, \ldots, n$. Following the DO loop, the value of i is compared to n. Thus, in the worst case, there are $2n + 1$ comparisons.

In the worst case for Algorithm 2, we do not exit from the outer WHILE DO loop until $i = n$. On each pass through the outer WHILE DO loop, we do not exit from the inner DO loop until $j = n + 1$. The conditions $i < n$ and $j = n + 1$ have been tested n times. The conditions $j \leq n$ and $a[i] < a[j]$ are tested n times on the first pass through the outer loop, $n - 1$ times on the second pass, $n - 2$ times on the third pass, and so on. In the final IF clause, the condition $i = n$ is tested once. Thus, in the worst case, there are

$$2n + 2(n + (n-1) + (n-2) + \cdots + 1) + 1 = 2n + 2\sum_{i=1}^{n} i + 1$$

comparisons. By Formula 3.2, the sum $\sum_{i=1}^{n} i$ equals $n(n+1)/2$. Therefore, there are

$$2n + 2(n(n+1)/2) + 1 = n^2 + 3n + 1$$

comparisons in the worst case for Algorithm 2.

Since $2n + 1 < n^2 + 3n + 1$ for all n, Algorithm 1 is a better algorithm for determining if a list of n numbers is in nondecreasing order. In fact, for large values of n, Algorithm 1 is much better. If $n = 100$, Algorithm 1 requires 201 comparisons while Algorithm 2 requires 10,301 comparisons.

In Chapter 5 and Chapter 6 there is a further discussion of algorithm comparison.

Example 3.46 compares two methods for evaluating polynomials. Most readers are somewhat familiar with polynomials. However, for the sake of completeness, we shall state here the definitions we require. Polynomials are considered further in Chapter 5.

A **polynomial** is an expression of the form $a_n x^n + a_{n-1} x^{n-1} + \cdots + a_1 x + a_0$, where a_i are real numbers, n is a nonnegative integer, and $a_n \neq 0$ unless $n = 0$. The number n is called the **degree** of the polynomial, and the numbers a_i are called **coefficients.** The expression $5x^4 + 8x^3 - 2.17x^2 + 4x - 33$ is a polynomial of degree 4. Recall that x^4 means $x \cdot x \cdot x \cdot x$, x^3 means $x \cdot x \cdot x$, and so on. A student asked to evaluate the polynomial for $x = 2$ will usually do the following calculations:

$$5(2)^4 + 8(2)^3 - 2.17(2)^2 + 4(2) - 33$$
$$= 5 \cdot 2 \cdot 2 \cdot 2 \cdot 2 + 8 \cdot 2 \cdot 2 \cdot 2 - 2.17 \cdot 2 \cdot 2 + 4 \cdot 2 - 33$$
$$= 80 + 64 - 8.68 + 8 - 33$$
$$= 144 - 33.68 = 110.32$$

(We use $x = 2$ for simplicity; the student will probably do some steps mentally. However, if x were 372, the student would probably not do any calculations mentally.)

The following is an algorithm for evaluating polynomials:

1. Substitute.
2. Multiply to calculate each term (the terms in our example were $5x^4$, $8x^3$, $-2.17x^2$, $4x$, and -33).
3. Add the terms.

This is the method students learn in high school and continue to use in college. Is it the best method? It is perhaps the simplest method to understand, which is probably why it is taught. However, this method uses many multiplications. (We are concerned about the number of multiplications rather than the number of additions and subtractions because multiplications are much more costly in execution time than are additions and subtractions.)

In our example, there were four multiplications to calculate $5x^4$, three multiplications to calculate $8x^3$, and so on, for a total of $4 + 3 + 2 + 1 = 10$ multiplications.

Another way of evaluating our polynomial takes advantage of factoring to rewrite $5x^4 + 8x^3 - 2.17x^2 + 4x - 33$ as

$$(5x^3 + 8x^2 - 2.17x + 4)x - 33 = ((5x^2 + 8x - 2.17)x + 4)x - 33$$
$$= (((5x + 8)x - 2.17)x + 4)x - 33$$

This is evaluated basically by working backwards. Multiply a coefficient by x and add the next coefficient, multiply that result by x and add the next coefficient, and so on. Evaluating this when $x = 2$, we have

$$(((5 \cdot 2 + 8) \cdot 2 - 2.17) \cdot 2 + 4) \cdot 2 - 33 = ((18 \cdot 2 - 2.17) \cdot 2 + 4) \cdot 2 - 33$$
$$= (33.83 \cdot 2 + 4) \cdot 2 - 33$$
$$= 71.66 \cdot 2 - 33$$
$$= 143.32 - 33 = 110.32$$

Here there are four multiplications instead of ten. The method used here is called **Horner's method.** We compare Horner's method to the previous (standard) method in Example 3.46.

EXAMPLE 3.46 _____

Following are programs for two algorithms that input a natural number n, a list of $n + 1$ real numbers $a_0, a_1, a_2, \ldots, a_n$, and a real number x; and then evaluate the polynomial $a_n x^n + a_{n-1} x^{n-1} + \cdots + a_1 + a_0$. Compare them using worst case analysis.

Algorithm 1 **(Standard Method)**

```
program standard(input,output);
var n,i,j: integer;
  x,y,sum: real;
  a:   array[0..50] of real;
begin
  writeln('input the degree of the polynomial');
  read (n);
  if n>50 then
    writeln ('degree of polynomial too large')
  else
    begin
      writeln('input',n+1,' coefficients with highest powers first');
      for i:=0 to n do
        read (a[i]);
      writeln('input the value of the variable x');
      read (x);
      sum:=0;
      for i:=0 to n-1 do
        begin
          y:=x;
          for j:=2 to n-i do
            y:=y*x;
          sum:=sum + a[i]*y
        end;
      sum:= sum + a[n];
      writeln ('the value of the polynomial at x is ',sum)
    end
end.
```

Algorithm 2 (Horner's Method)

```
program horner(input,output);
var  n,i: integer;
     x,sum: real;
     a: array [0..50] of real;
begin
   writeln('input the degree of the polynomial');
   read (n);
   if n>50 then
      writeln ('polynomial is too large')
   else
      begin
         writeln('input',n+1,' coefficients for highest powers first');
         for i:=0 to n do
            read (a[i]);
         writeln('input the value of the variable x');
         read (x);
         sum:= 0;
         for i:= 0 to n-1 do
            sum:=(sum+a[i])*x;
         sum:=sum+a[n];
         writeln ('the value of the polynomial at x is', sum)
      end
end.
```

SOLUTION

Since arithmetic calculations are the dominant part of this problem, we focus on the number of multiplications or divisions (the more costly operations) and ignore additions and subtractions.

In Algorithm 1, $i - 1$ multiplications take place at $y := y * x$ on each pass through the outer DO loop and one multiplication takes place at $SUM := SUM + a_i * y$. Thus, there are i multiplications on each pass through the outer loop. By Formula 3.2, we have a total of $\sum_{i=1}^{n} i = n(n + 1)/2$ multiplications with Algorithm 1.

In Algorithm 2, there is one multiplication on each pass through the DO loop. There is, therefore, a total of n multiplications with Algorithm 2. Since $n < n(n + 1)/2$ for $n > 1$, Algorithm 2 is more efficient than Algorithm 1.

The preceding discussion is a worst case analysis. When evaluating polynomials, some of the coefficients (the a_i's) may be zero. When $a_i = 0$, the multiplications involved in calculating $a_i x^i$ are generally not done. In the last example, we did all multiplications as if each a_i were not zero and thereby considered the worst case.

Although a physically shorter program is often more efficient than a longer program (this was the case in the two examples just considered), this need not be true. Do not confuse the length of a program with its efficiency. It is quite possible for a physically shorter program to be less efficient than a longer one.

EXERCISES: CASE STUDY 3B

1. We can insert counters in programs to determine how many comparisons or multiplications are performed when the program is run with test data. For example, we can use a counter variable k in Example 3.45. For Algorithm 1 we initialize k to be 2

since two comparisons are made at the DO statement. We increment k by 2 within the DO loop because two comparisons will be made after each pass through the loop. Finally, we increment k by 1 after the loop since one comparison will be made in the IF clause. Thus, we insert the statements $k = 2$, $k = k + 2$, and $k = k + 1$ after the statements $i = 1$, $i = i + 1$, and END DO, respectively.

a. Determine where to insert statements and what statements to insert to count the comparisons made in Algorithm 2.

b. Write a program for Algorithm 1 and a program for Algorithm 2 and run each program with the following test data (remember to insert a PRINT statement at the end of the program to print k):

   i. 7, -11, -3, 10, 11, 5, 2, 17

   ii. 5, 2, 4, 6, 8, 13

   iii. 12, -5, 6, 8.16, 8.25, 12, 13, 17, 19, 22, 19.24, 25, 30

2. a. When using Horner's method, a certain polynomial would be written:

$$(((2x + 5) \cdot x - 3) \cdot x - 7) \cdot x + 4$$

   Multiply to write this polynomial in the usual way.

b. Rewrite the polynomial $5x^6 - 2x^5 + x^4 - 7x^3 + 2x^2 - 11x + 6$ as it would appear using Horner's method.

c. Rewrite the polynomial $10x^7 + 7x^6 - x^4 + x - 5$ as it might appear when using Horner's method. (*Hint:* See Exercise 3.)

3. The polynomial $5x^6 + 2x^5 - 10x^3 + x^2 - x + 3$ would be rewritten for Horner's method as

$$(((((5x + 2)x^2 - 10)x + 1)x - 1)x + 3$$

Evaluate the polynomial in the standard way and then by Horner's method when

a. $x = 3$ **b.** $x = -2$

4. a. Insert counters into each of the algorithms in Example 3.46. Then write programs for the two algorithms (remember to print the value of the counter when the program finishes).

b. Run both programs with the following data:

   i. 5, -7, 3, 8, -2, 1, 34, -6

   ii. 4, 1.2, 7, -2.71, 15, 1.06, 3.2

   iii. 10, 2, 4, -3, 0, 0, -11, 0, 0, 0, 3, 75, -4

5. (Bisection method for finding $\sqrt{x}$). Following is a program that reads in a nonnegative number x and a nonnegative number E (for error) and determines the square root of x, accurate to within E. That is, the true square root and the one calculated differ by no more than E. Do a program trace with the following data:

a. $x = 5$, $E = 0.01$ **b.** $x = 2$, $E = 0.001$

```
program bisect(input,output);
var x,e,l,u,m,d: real;
begin
    writeln('input the positive number and then the error tolerance');
    read (x,e);
    if (e<=0) or (x<0) then
        writeln ('error in x or e')
    else
        begin
            if x>=1 then
```

```
        begin
           l:=1;
           u:=x
        end
     else
        begin
           u:=1;
           l:=x
        end;
     m:=(l+u)/2;
     d:= abs(x-m*m);
     while d>e do
        begin
           if m*m>x then
              u:=m
           else
              l:=m;
           m:=(u+l)/2;
           d:=abs (x-m*m)
        end;
     writeln ('the square root of',x,' with tolerance',e,' is',m)
  end
end.
```

6. Insert counter variables *AS*, *MD*, *C*, and *W* in the program of exercise 5 to count the number of additions and subtractions, multiplications and divisions, and comparisons, and the number of times the WHILE DO loop is executed.

a. Write a program for the algorithm of exercise 5 including these four counter variables and have the program print the values of the counters.

b. Run your program with the following data:

    i. $x = 17$, $E = 0.0001$

   ii. $x = 0.34$, $E = 0.001$

  iii. $x = 85$, $E = 0.0001$

7. (Newton's method for finding $\sqrt{x}$). Following is a program that reads in nonnegative numbers x and E (for error) and determines $\sqrt{x}$ accurate to within E.

```
program newton(input,output);
var   x,e,a,b,d:real;
begin
   writeln('input the positive number and then the allowable error');
   read (x,e);
   if (e<=0) or (x<0) then
      writeln ('error in x or e')
   else
      begin
         a:=1;
         b:=x/a;
         d:=abs(x-a*a);
         while d>e do
            begin
               a:=(a+b)/2;
               b:=x/a;
               d:=abs(x-a*a)
            end;
         writeln(x,e,a)
      end
end.
```

Do a program trace with

a. $x = 5$, $E = 0.01$ **b.** $x = 2$, $E = 0.001$

8. Insert counter variables *AS, MD, C,* and *W* in the program of exercise 7 to count the number of additions and subtractions, multiplications and divisions, and comparisons, and the number of times the WHILE DO loop is executed.

   a. Write a program for the algorithm of exercise 7 including these four counter variables and have the program print the values of the counters.

   b. Run your program with the following data:

      i. $x = 17, E = 0.0001$

      ii. $x = 0.34, E = 0.001$

      iii. $x = 85, E = 0.0001$

9. Using the results of exercises 6 and 8, compare the efficiencies of the bisection method and Newton's method for finding $\sqrt{x}$. Compare on the basis of the number of operations performed by each and the number of times each WHILE DO loop is executed.

REFERENCES

Anderson, R. *Proving Programs Correct.* New York: Wiley, 1979.

Belford, G. G., and C. L. Liu. *Pascal.* New York: McGraw-Hill, 1984.

Bogart, K. P. *Introductory Combinatorics.* Marshfield, Mass.: Pitman, 1983.

Lovasz, L. *Combinatorial Problems & Exercises.* New York: Elsevier, 1979.

Niven, I. *The Mathematics of Choice.* Washington D.C.: The Mathematical Association of America, 1975.

Tucker, A. *Applied Combinatorics.* New York: Wiley, 1980.

Life is full of uncertainty. We do not know if it will rain next 4th of July, or if we will graduate on time, or whom we will marry, or even how long we will live. The mathematical theory of probability attempts to bring some order to this confusion by assigning likelihoods to events that occur at random. If you place a bet on the number 17 in a roulette game, what is the likelihood of your winning? If you play the number 17 over and over again, say 380 times, and bet $1 each time, how much would you expect to be ahead (or behind) at the end of play? Assuming the roulette game is fair, you will be able to answer these questions after you finish studying this chapter.

Many important economic and military decisions must be made on the basis of events whose outcomes are not certain. In some cases a wrong decision can cost a great deal of money or even human life. In such situations, computers are frequently used to simulate the events. How can a computer, which is a deterministic machine, simulate nondeterministic events? We shall see in our case study how a computer, using what is called a pseudorandom number generator, can be made to simulate random experiments.

4.1 FINITE PROBABILITY

Sample Spaces

A **random experiment** is a repeatable action that results in one of a well-defined set of outcomes. Tossing a die is an example of a random experiment. A die may be tossed as many times as we like, and each roll of the die results in one of the numbers 1, 2, 3, 4, 5, 6 appearing on its top face. Another example of a random experiment is flipping a coin. A coin can be flipped as often as we want, with each flip resulting in either a head, H, or a tail, T.

We note several points about the definition of a random experiment. First, the term *well defined* is used in the same sense as in Chapter 2—we can tell precisely which elements are in the set. Second, a random experiment must be random; that is, the outcome of the experiment must not be predetermined. There is no way to

PROBABILITY

predict with certainty the outcome of a toss of a die or the flip of a coin. This is in contrast to what are called **deterministic experiments.** For example, if we take a container of pure water at atmospheric pressure and heat it to 100°C, it will begin to boil. The conditions imposed by the experimentor determine, with relative certainty, the outcome of the experiment. Finally, the set of outcomes of a random experiment must be **exhaustive;** that is, it must contain all possible outcomes of the experiment.

The **sample space** S of a random experiment is the set of all possible outcomes of the experiment. Thus, for the die-tossing experiment $S = \{1, 2, 3, 4, 5, 6\}$. For the coin-flipping experiment $S = \{H, T\}$.

EXAMPLE 4.1

Which of the following are random experiments? Explain why or why not.

a. Observing the number that comes up at a roulette wheel

b. Dropping this book off the roof of your classroom building on a windless day and observing how many seconds it takes to hit the ground

c. Observing the number of student jobs processed by your school's computer system in one day

SOLUTION

a. Observing the number that comes up on a roulette wheel is a random experiment. The roulette wheel can be spun as often as we like, and the result of each spin is unpredictable. The sample space is

$$S = \{0, 00, 1, 2, \ldots, 36\}$$

b. The book-dropping experiment is not a random experiment. Objects fall according to the laws of physics. The number of seconds it will take for this book to hit the ground is predictable with reasonable certainty.

c. Observing the number of student jobs processed by your school's computer system in one day is a random experiment. This number depends on many factors, thus making it impossible to predict. The experiment is repeatable because we can observe the computer system on many days. The sample space of this experiment is not as simply described as in the die, coin, and roulette experiments, however. Clearly, the number of jobs that arrive must be a nonnegative integer, 0, 1, 2, ... But where do we stop? What is the maximum number of jobs the system can process in one day? The maximum number of jobs a given computer system can process in one day is some very large number, which we denote by N. The sample space is then

$$S = \{0, 1, 2, \ldots, N\}$$

It is frequently easier to view certain sample spaces as infinite sets. In the student job experiment, for example, it is not uncommon for the sample space to be $S = \{0, 1, 2, \ldots\}$. Dealing with infinite sample spaces requires mathematical techniques more advanced than those we cover in this book. Our discussion, therefore, will cover only random experiments with finite sample spaces.

It is possible for two people to use different sample spaces to describe the same experiment. Consider, for example, the die-tossing experiment. Person 1 uses our original sample space $S_1 = \{1, 2, 3, 4, 5, 6\}$. Person 2, however, is playing a game in which the parity of the number coming up on the die is important. (The **parity** of an integer is whether it is even or odd.) Thus, person 2 chooses $S_2 = \{E, O\}$ as the sample space of the experiment, where E is the outcome "an even number is rolled" and O is the outcome "an odd number is rolled." We usually choose as our sample space the set having the most detail. Thus, we would normally favor S_1 as the sample space for the die-tossing experiment.

A subset of a sample space is called an **event**. **Simple events** are the 1-element subsets of the sample space. We denote events by capital letters. For the die-tossing experiment, the simple events are $\{1\}$, $\{2\}$, $\{3\}$, $\{4\}$, $\{5\}$, $\{6\}$. Some other events in this sample space are:

> An even number is rolled, or $E = \{2, 4, 6\}$
>
> An odd number is rolled, or $O = \{1, 3, 5\}$
>
> A number between 3 and 6 is rolled, or $G = \{4, 5\}$

An event **occurs** if one of the simple events that it contains results when the experiment is performed. For example, if we toss a die and a 4 results, the event E occurs because $4 \in E$. Note that the event G also occurs if a 4 results.

Since events are subsets of the sample space of an experiment, the empty set, $\varnothing$, and the sample space S itself must be considered events. The empty set, having no elements, can never occur. The empty set, therefore, is referred to as the **impossible event.** The sample space S contains, by definition, all the possible outcomes of the experiment and so must occur on every run of the experiment. For this reason, S is called the **certain event.**

EXAMPLE 4.2

Use set notation to describe the following events:

a. Spin a roulette wheel once, and observe the number that comes up. Let

A = A black number comes up

B = One of the numbers 1 to 12 comes up

b. Observe the number of student jobs processed by your school's computer in one day. Let

C = No student jobs arrive in a day

D = Fewer than 50 student jobs arrive in a day

SOLUTION

a. $A = \{2, 4, 6, 8, 10, 11, 13, 15, 17, 20, 22, 24, 26, 28, 29, 31, 32, 35\}$
$B = \{1, 2, 3, 4, 5, 6, 7, 8, 9, 10, 11, 12\}$
Betting on the set B is called a first-12 bet for obvious reasons.

b. $C = \{0\}$. Note that $C \neq \emptyset$. It is not impossible that no student jobs arrive during a given day. This might occur, for example, during summer vacation.
$D = \{x : x \text{ is an integer}, \quad 0 \leq x \leq 49\}$

Since the sample space of an experiment and its events are sets, they may be depicted in a Venn diagram. The sample space S acts as the universal set, and events are depicted as circles. For the die-tossing experiment with $S = \{1, 2, 3, 4, 5, 6\}$, we illustrate in Figure 4.1 the Venn diagram for the sets $E = \{2, 4, 6\}$ and $G = \{4, 5\}$.

Since a sample space S is a set and events are subsets of S, we can use the set operations introduced in Chapter 2 on the events of S. In Chapter 2 we introduced three basic set operations: union, intersection, and complementation. How can these be interpreted in terms of events?

If A and B are events, they are subsets of a sample space S and their elements are simple events. $A \cup B$ contains those elementary events in A or B or both. Thus, the event $A \cup B$ occurs when A or B or both occur. $A \cap B$ contains those elementary events in both A and B. Thus, the event $A \cap B$ occurs when A and B

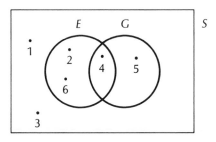

Figure 4.1

both occur. Finally, A' contains those elementary events of S that are not in A. Thus, the event A' occurs when the event A does not occur.

To summarize:

1. The **union** $A \cup B$ of two events A and B occurs whenever A occurs or B occurs, or both occur.

2. The **intersection** $A \cap B$ of two events A and B occurs whenever A and B both occur.

3. The **complement** A' of an event A occurs whenever the event A does not occur.

EXAMPLE 4.3

Describe in English and set notation the following events for the die-tossing experiment, where $E = \{2, 4, 6\}$, $O = \{1, 3, 5\}$ and $G = \{4, 5\}$:

a. $E \cap G$ b. $E \cup O$ c. $O \cup G$ d. $E \cap O$ e. E'

SOLUTION

a. $E \cap G = \{4\}$. $E \cap G$ occurs whenever a 4 is rolled.

b. $E \cup O = \{2, 4, 6, 1, 3, 5\} = S$. $E \cup O$ is the certain event. A roll of the die will certainly result in a number that is even or odd.

c. $O \cup G = \{1, 3, 4, 5\}$. $O \cup G$ occurs whenever a 1, 3, 4, or 5 is rolled.

d. $E \cap O = \varnothing$. $E \cap O$ is the impossible event. We cannot roll a number that is both even and odd.

e. $E' = \{2, 4, 6\}' = \{1, 3, 5\} = O$

Finite Probability Spaces

Each of us has an intuitive idea of what probability means. If we take a balanced coin and flip it, we should be able to express an opinion on the likelihood of a head coming up on the coin. We might say that the odds are even, the possibility of a head is 50%, one out of two times a head will come up, and so on. Each of these notions concerns the frequency with which a head occurs. Let's now discuss how to make these ideas more precise.

Flipping a coin, as we discussed earlier, is an example of a random experiment. One of the properties of a random experiment is that it is repeatable. Suppose we flip a balanced coin a large number of times, say 1000. Most of us would reason as follows: Since the coin is balanced, we would expect about half the flips to result in a head and half to result in a tail. Thus, we expect about 500 heads to occur.

To obtain a more mathematically useful concept of probability, we first require that the probability of an event be a nonnegative number less than or equal to 1. We do so because probability measures the frequency with which an event is likely to occur. The probability of obtaining a head when flipping a balanced coin is obtained as follows: The coin is balanced and there are two possible outcomes: a head, H, and a tail, T. Each of H or T is as likely as the other, making the probability of a head 1/2. This means that we expect a head to appear about half the time the coin is flipped. The important word to remember is *about*. Saying that the probability of a head is 1/2 does not mean that in 100 tosses of a coin we will definitely get 50 heads and 50 tails. It means that the number of heads will be *about* 50. If we obtain 52, 48, or even 45 heads, we should not be surprised. See Note 4.1.

Note 4.1

The probability of an event is an estimate of the frequency of occurrence of that event when the experiment is repeated a large number of times.

The sample space of the coin-flipping experiment is $S = \{H, T\}$, and we denote the probabilities of heads and tails by $P(H) = 1/2$ and $P(T) = 1/2$. Note $P(S) = 1$; that is, the probability of the certain event is 1. Since the impossible event $\varnothing$ cannot happen, we assign to it the probability $P(\varnothing) = 0$.

As a second example, consider the tossing of a balanced die. The sample space is $S = \{1, 2, 3, 4, 5, 6\}$. Since the die is balanced, we expect each of the simple events to be as likely as any other. Each number should occur about 1/6 of the time. The assignment of probabilities is then $P(1) = P(2) = P(3) = P(4) = P(5) = P(6) = 1/6$.

Note 4.2 defines a finite probability space. Let $S = \{e_1, e_2, \ldots, e_n\}$ be the finite sample space of a random experiment.

Note 4.2

A **finite probability space** is S, together with an assignment to each simple event $e_i \in S$ a real number $P(e_i)$, called the probability of e_i, which satisfies:

1. $0 \leq P(e_i) \leq 1$ for all i

and

2. $\displaystyle\sum_{i=1}^{n} P(e_i) = 1$

The **probability of an event** A is defined as the sum of the probabilities of the simple events that constitute A.

Consider the tossing of a balanced die with the probability space described after Note 4.1. Each simple event has probability 1/6, satisfying condition 1 of Note 4.2, and the probabilities sum to 1, satisfying condition 2 of Note 4.2. What are the probabilities of the events E, O, and G? Since $E = \{2, 4, 6\}$, $P(E) = P(2) + P(4) + P(6) = 3/6 = 1/2$. Likewise, $P(O) = 1/2$ and $P(G) = 2/6 = 1/3$.

Remark   A probability space is a mathematical model and, as such, only approximates reality. A toy car is a model of a real car. The toy has some of the characteristics of the real thing (shape, relative proportions, rubber wheels, maybe even headlights and a motor), but it does not have others (size, transmission, air conditioner). The probability space for tossing a balanced die is also a model of reality. It has some of the characteristics of the real thing (six possible outcomes, repeatability), but it does not have others (no die is truly balanced, the die could land on edge). Just as the toy car behaves in some respects like a real car, so the probability space (the mathematical model) behaves in some respects like a real die. The toy is not a real car, however, and so behaves differently from a real car. Likewise, the mathematical model is not a real die and so behaves differently from a die.

Mathematical models are used to approximate reality so predictions of reality can be made. All mathematical models are based on assumptions. The accuracy of the model's predictions depends to a large extent on these assumptions. If the predictions are shown to be accurate, the assumptions made are judged to be reasonable. If the predictions are not accurate, either the assumptions or the entire model should be changed.

In the die-tossing experiment, we assumed the die cannot come to rest on an edge. We also assumed the die is balanced. It has been found that these are reasonable assumptions. One prediction our model makes is that any one face of the die will appear about 1/6 of the time if the die is tossed a large number of times. Try this with a real die and see if the prediction is reasonably accurate (toss the die 120 times).

EXAMPLE 4.4

Consider the following experiment. The spinner depicted in Figure 4.2 is spun. Assume that the spinner comes to rest at a random place and cannot come to rest on a dividing line. Construct a finite probability space for this experiment—

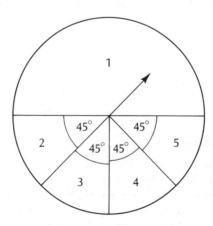

Figure 4.2

that is, a sample space together with an assignment of probabilities. Find the probabilities of $A =$ An odd number comes up, and $B =$ A prime number comes up.

SOLUTION

Region 1 is half the circle and regions 2, 3, 4, and 5 are each one-eighth of the circle. The sample space is $S = \{1, 2, 3, 4, 5\}$, and the assignment of probabilities is $P(1) = 1/2, P(2) = P(3) = P(4) = P(5) = 1/8$. It is easy to see that these probabilities sum to 1.

$$A = \{1, 3, 5\}$$

Thus,

$$P(A) = P(1) + P(3) + P(5) = \frac{1}{2} + \frac{1}{8} + \frac{1}{8} = \frac{6}{8} = \frac{3}{4}$$

$$B = \{2, 3, 5\}$$

Thus,

$$P(B) = P(2) + P(3) + P(5) = \frac{1}{8} + \frac{1}{8} + \frac{1}{8} = \frac{3}{8}$$

The probabilities of each of the simple events in the coin-flipping and the die-tossing experiments were equal to each other. This type of probability space is called an **equiprobable space**; that is, all simple events have the same probability. This was not the case in Example 4.4. Note that if there are n simple events in an equiprobable space, then each of the simple events has probability $1/n$. If, therefore, we take an event A in an equiprobable space, its probability is the number of simple events in A divided by n. For example, in the die-tossing experiment $P(E) = 3/6 = 1/2$, because there are three ways in which an even number can occur, and there are a total of six possible outcomes to the experiment.

Theorem 4.1

If $S = \{e_1, e_2, \ldots, e_n\}$ is an equiprobable space and A is an event in S, then

$$P(A) = \frac{|A|}{|S|} = \frac{|A|}{n}$$

$$= \frac{\text{Number of ways in which } A \text{ can occur}}{\text{Total number of possible outcomes}}$$

Since finding probabilities of events in equiprobable spaces amounts to counting the number of outcomes of those events, we can use Theorem 4.1 and the combinatorial techniques of Chapter 3. We illustrate with several examples.

EXAMPLE 4.5 _____

A standard deck of playing cards is well shuffled, and one card is selected at random. Find the probability of the following events:

a. $A =$ A spade is chosen
b. $B =$ A face card is chosen
c. $C =$ An ace is chosen

SOLUTION

The sample space for this experiment is the set of 52 playing cards. Since the deck is well shuffled and a card is chosen at random, no card is favored over any other, giving us an equiprobable space.

a. The event A can occur in 13 possible ways. Thus,

$$P(A) = \frac{13}{52} = \frac{1}{4}$$

b. The event B can occur in 12 possible ways since each of the four suits contains a jack, queen, and king. Thus,

$$P(B) = \frac{12}{52} = \frac{3}{13}$$

c. The event C can occur in 4 possible ways. Thus,

$$P(C) = \frac{4}{52} = \frac{1}{13}$$

_____

The next two examples deal with selecting more than one item at random from a set S. This can be done in either of two ways: (1) selecting **with replacement**—that is, each item is returned to S before the next is chosen; or (2) selecting **without replacement**—when an item is chosen, it is permanently removed from S.

EXAMPLE 4.6 _____

A standard deck of playing cards is well shuffled and two cards are selected at random without replacement. Find the probabilities of the following events:

a. $A =$ Two spades are chosen
b. $B =$ Two aces are chosen

SOLUTION

Since the deck is shuffled and the cards are chosen at random, we have an equiprobable space. An outcome of this experiment is a pair of cards, making the sample space the set of all possible pairs of cards. Since we are selecting 2 cards out of 52, the total number of such pairs is $\binom{52}{2} = 1326$.

a. The number of ways two spades may be chosen is

$$\binom{13}{2} = 78$$

Thus,

$$P(A) = \binom{13}{2} \bigg/ \binom{52}{2} = \frac{78}{1326} = .0588$$

b. The number of ways two aces may be chosen is

$$\binom{4}{2} = 6$$

Thus,

$$P(B) = \binom{4}{2} \bigg/ \binom{52}{2} = \frac{6}{1326} = .0045$$

EXAMPLE 4.7

A bag contains six green, four red, and five blue balls. If three balls are selected at random without replacement, what are the probabilities of the following events?

a. A = All three balls are of different colors
b. B = Two green and one red ball are selected

SOLUTION

Since the balls are selected at random, we have an equiprobable space. The sample space is the set of all triples of colored balls. Since there are a total of 15 balls in the bag, there are $\binom{15}{3} = 455$ such triples.

a. A is the event that all three balls chosen are of different colors. We must calculate the number of ways of selecting one red, one green, and one blue ball. By the multiplication principle, this number is $6 \times 5 \times 4 = 120$. Thus,

$$P(A) = \frac{120}{455} = .2637$$

b. There are $\binom{6}{2} = 15$ ways of selecting two green balls and 4 ways of selecting one red ball, giving a total of $15 \times 4 = 60$ ways of selecting two green balls and one red ball. Thus,

$$P(B) = \frac{60}{455} = .1319$$

It is frequently useful to be able to calculate the probability of the union and complement of events. Before we state the next theorem, however, we need the following definition. Two events A and B are **mutually exclusive** if they cannot occur together. Put another way, A and B are mutually exclusive if $A \cap B = \emptyset$.

Theorem 4.2

i. If A and B are mutually exclusive, $P(A \cup B) = P(A) + P(B)$.

ii. $P(A) + P(A') = 1$.

If A and B are mutually exclusive, $A \cap B$ is the empty set. Therefore, if $A = \{a_1, a_2, \ldots, a_n\}$ and $B = \{b_1, b_2 \ldots, b_m\}$, then

$$A \cup B = \{a_1, a_2, \ldots, a_n, b_1, b_2, \ldots, b_m\}.$$

Therefore,

$$P(A \cup B) = P(a_1) + P(a_2) + \cdots + P(a_n) + P(b_1) + P(b_2) + \cdots + P(b_m)$$
$$= P(A) + P(B)$$

To justify the second part of the theorem, note that for any event A, the events A and A' are mutually exclusive and $A \cup A' = S$. By the first part of the theorem, $P(A \cup A') = P(A) + P(A')$. Since $A \cup A' = S$, then $P(A \cup A') = P(S) = 1$, which proves the second part of the theorem.

We now illustrate the use of Theorem 4.2.

EXAMPLE 4.8 _____

A standard deck of playing cards is well shuffled, and two cards are chosen at random without replacement. Find the probability of:

a. $X = $ Two aces or two kings are chosen

b. $Y = $ At least one ace is chosen

SOLUTION

This is the same equiprobable space as that of Example 4.6.

a. The event X may be viewed as the union of two mutually exclusive events: $A = $ Two aces are chosen, and $B = $ Two kings are chosen. Thus, $X = A \cup B$. Now, $P(A) = 6/1326$, from Example 4.6(b). Similarly, $P(B) = 6/1326$. Thus,

$$P(X) = P(A) + P(B) = \frac{6}{1326} + \frac{6}{1326} = \frac{12}{1326} = .009$$

b. To calculate the probability of Y, we first calculate the probability of $Y' = $ No ace is chosen. The number of ways in which we can pick two nonaces is

$\binom{48}{2} = 1128$, giving $P(Y') = \binom{48}{2} \Big/ \binom{52}{2} = 1128/1326$. Since $P(Y) + P(Y') = 1$, we obtain

$$P(Y) = 1 - P(Y') = 1 - \frac{1128}{1326} = \frac{198}{1326} = .1493$$

ALTERNATE SOLUTION

The event Y may be decomposed into two mutually exclusive events: $A = $ Two aces are chosen and $C = $ One ace and one nonace are chosen. Since $Y = A \cup C$, $P(Y) = P(A) + P(C)$. We already know that $P(A) = 6/1326$. To count the number of ways C can occur, note that there are 4 ways of selecting an ace and 48 ways of selecting a nonace. Thus, there are $48 \times 4 = 192$ ways in which C can occur, giving $P(C) = 192/1326$. Hence,

$$P(Y) = P(A) + P(C) = \frac{6}{1326} + \frac{192}{1326} = \frac{198}{1326} = .1493$$

If we have more than two events $A_1, A_2, \ldots, A_n$ that are **pairwise mutually exclusive**; that is, if $A_i \cap A_j = \varnothing$ for all $i \neq j$, then the following extension of Theorem 4.2 is true:

Corollary 4.1   The Addition Law of Probability

If $A_1, A_2, \ldots, A_n$ are pairwise mutually exclusive, then

$$P(A_1 \cup A_2 \cup \cdots \cup A_n) = P(A_1) + P(A_2) + \cdots + P(A_n)$$

So far we have been concerned with finite sample spaces in which we can assign probabilities on the basis of the nature of the experiment and several assumptions made about the conditions of the experiment. In the die-tossing experiment or the coin-flipping experiment, we assigned probabilities to the simple events under the assumption that the die and the coin were balanced. When we chose a card from a deck, we assumed the deck was well shuffled, and the card was chosen at random. These assumptions lead to an assignment of $1/52$ as the probability of each elementary event. This way of assigning probabilities is called **a priori** or **conceptual probability.**

However, how can we assign probabilities to the simple events of the sample space of the student job experiment described in Example 4.1(c)? The sample space is not equiprobable. In fact we have no idea how to assign probabilities to the simple events! The only way to proceed in this case is to perform a statistical experiment. Observe how many student jobs come in per day for a large number of days, say 100. Consider the simple event $\{20\}$; that is, 20 jobs arrive on a given day. In observing the number of jobs that arrive during a 100-day span, suppose that on three of those days 20 student jobs arrive. The simple event {20} has occurred 3 out of 100 times. We therefore assign the probability $3/100 = .03$ to the simple event {20}. We assign probabilities to the other simple events in a like manner. This method of assigning probabilities on the basis of the results of many repetitions of the experiment is called **empirical** or **a posteriori probability.** Obtaining reliable empirical probabilities is a difficult problem in statistics, and we will not pursue it further in this text.

The following summarizes the properties of probabilities.

Properties of Probabilities

If S is a finite probability space and A and B are events in S:

a. $0 \le P(A) \le 1$

b. $P(\varnothing) = 0$, $P(S) = 1$

c. If A and B are mutually exclusive, $P(A \cup B) = P(A) + P(B)$.

d. $P(A) + P(A') = 1$

EXERCISES 4.1

1. Spin the spinner depicted in Figure 4.3 and record the number that comes up. Assume the spinner cannot stop on a dividing line.

   a. What is the sample space of this experiment? Assign probabilities to the simple events.

   b. Use set notation to describe the following events:

   A = An even prime comes up

   B = A number greater than 3 comes up

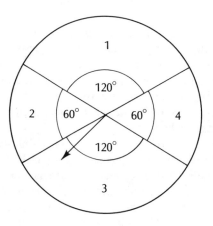

Figure 4.3

    c. Describe the following events in both English and set notation: $A \cap B$, $A \cup B$, B'.

    d. Find the following probabilities: P(A), P(B), P($A \cap B$), P($A \cup B$), P(B').

2. A nickel, a dime, and a quarter are each flipped once. Record the head or tail outcome on each coin. Assume each coin is balanced.

    a. What is the sample space of this experiment? Assign probabilities to the simple events.

    b. Use set notation to describe the following events:

        C = Exactly two heads are flipped

        D = At least two heads are flipped

        E = Three heads are flipped

    c. Describe the following events in both English and set notation: $C \cup E$, $D \cap E$, D'.

    d. Find the following probabilities: P(C), P(D), P(E), P($C \cup E$), P($D \cap E$), P(D').

3. Consider the following experiment: Record the number of cars passing through a particular intersection during the one hour from noon to 1 P.M.

    a. What is the sample space of this experiment? Explain how to find an assignment of probabilities to the simple events of the sample space.

    b. Use set notation to describe the following events:

        F = Fewer than 50 cars pass the intersection

        G = At least 20 cars pass the intersection

    c. Describe the following events in both English and set notation: $F \cup G$, $F \cap G$, F', G'.

4. Which of the following are acceptable assignments of probabilities to the simple events of $S = \{e_1, e_2, e_3, e_4\}$.

    a. $P(e_1) = \dfrac{5}{12}$, $P(e_2) = 0$, $P(e_3) = \dfrac{5}{12}$, $P(e_4) = \dfrac{1}{6}$

    b. $P(e_1) = 0$, $P(e_2) = 1$, $P(e_3) = P(e_4) = 0$

    c. $P(e_1) = \dfrac{3}{4}$, $P(e_2) = P(e_3) = P(e_4) = \dfrac{1}{8}$

    d. $P(e_1) = \dfrac{1}{9}$, $P(e_2) = \dfrac{5}{4}$, $P(e_3) = 0$, $P(e_4) = \dfrac{1}{9}$

5. One hundred cards numbered 00 to 99 are placed in a box, and one card is selected at random. Find the probabilities of the following events:

    a. The number on the card is divisible by 3.

    b. The number on the card is greater than 20.

    c. The number on the card is prime.

    d. The sum of the digits of the number is at least 6.

6. A standard deck of playing cards is well shuffled, and three cards are selected at random without replacement. Find the probabilities of the following events:

    a. A = Three hearts are chosen

    b. B = Exactly two red cards are chosen

    c. C = At least one 10 is chosen (*Hint:* Consider C'.)

  d. D = Three face cards are chosen
  e. $A \cap D$
  f. E = Three aces or three kings are chosen

7. A bag contains 7 white and 13 black balls. Four balls are selected at random without replacement. Find the probabilities of the following events:
  a. A = All four balls are white
  b. B = All four balls are black
  c. C = Two balls are white and two balls are black
  d. D = At least one ball is white
  e. E = The number of black balls exceeds the number of white balls

8. Four cars—A, B, C, and D—are in a drag race. A, B, and C have the same probability of winning and each is three times as likely to win as D. Find each car's probability of winning.

9. A standard deck of playing cards is shuffled and a 5-card poker hand is dealt. What is the probability of being dealt
  a. 4 of a kind?
  b. 4 aces?
  c. A straight flush?

10. The number of projects completed in various programming languages in a computer science class of 73 students is as follows:

Language	Pascal	ALGOL	PL/I
Projects	34	16	23

The instructor chooses one project at random to grade it. What is the probability that the project was completed in Pascal? In ALGOL? In PL/I? In Pascal or PL/I?

Problems and Projects

11. Use Theorem 4.2(i) to prove Corollary 4.1 by mathematical induction.

4.2   CONDITIONAL PROBABILITY AND INDEPENDENCE

Conditional Probability

Suppose you play a card-guessing game with your friend. He deals you one card at random from a well-shuffled standard deck of playing cards. Your job is to guess the card. What is the probability of guessing the correct card? Since you may guess any of the 52 cards, the probability is 1/52. Suppose, however, your friend is a sloppy dealer and you have a quick eye. When he deals you the card you catch a glimpse of it and notice that it is a spade. Now that you have more information, what is the probability of guessing the correct card? You know the card is a spade, so you guess one of the spade cards. The probability of a correct guess is now 1/13.

Thus, if we know some event has occurred (a spade was dealt), our calculation of the probability of another event (guessing the right card) can be affected. The

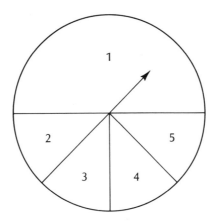

Figure 4.4

probability of event A given that event B has occurred is denoted by $P(A|B)$, read "the probability of A given B," and is called a **conditional probability.**

Referring to our example, let A = Guess the card dealt, B = A spade is dealt. If we have no information about the card that is dealt, $P(A) = 1/52$. If we know that event B has occurred, however, $P(A|B) = 1/13$. Therefore, the occurrence of B has changed our calculation of the probability of A.

EXAMPLE 4.9

a. Toss a balanced die once. Let $O = \{1, 3, 5\}$, $G = \{4, 5\}$. Find $P(O|G)$.

b. Spin the spinner depicted in Figure 4.4. Let

$$A = \text{An odd number comes up}$$
$$B = \text{A prime number comes up}$$

Find $P(B|A)$.

SOLUTION

a. If we know G has occurred, a 4 or a 5 has come up on the die. Since each is as likely as the other, we reason that the probability of the odd number, namely the 5, is 1/2. Thus $P(O|G) = 1/2$.

b. Calculating $P(B|A)$ requires more careful reasoning because the probability space of the spinner experiment is not equiprobable. We are given that event A, an odd number, has occurred. Thus, we restrict our attention to only those elementary events in A, treating A as a new sample space $S^* = A = \{1, 3, 5\}$. We now must assign new probabilities to the simple events of S^* so that the new probabilities add to 1. The original probabilities were $P(1) = 1/2$, $P(3) = 1/8$, $P(5) = 1/8$. We are now restricted to S^* because we know that $A\ (= S^*)$ has occurred. Recalling that probabilities measure the *relative* frequency of occurrence of events, we note that $P(1)$ is four times $P(3)$ or $P(5)$ ($1/2 = 4 \times 1/8$). If x denotes the probability of a 3 occurring in the sample space S^*, and P^*

denotes the probabilities in S^*, then $P^*(3) = x$, $P^*(5) = x$, and $P^*(1) = 4x$. Since the probabilities in S^* must sum to 1, we have $x + x + 4x = 6x = 1$. Thus, $x = 1/6$. Hence, in $S^* P^*(1) = 2/3$, $P^*(3) = 1/6$, $P^*(5) = 1/6$. Now, what is the probability of $B = \{2, 3, 5\}$ in the sample space S^*? We must discard the occurrence 2 because it is not in S^* (if an odd number was spun, the number is surely not 2). Thus,

$$P^*(B) = P(B|A) = P^*(3) + P^*(5) = \frac{1}{6} + \frac{1}{6} = \frac{1}{3}$$

The reasoning used in Example 4.9(b) can be generalized to yield the following useful theorem.

Theorem 4.3

Given events A and B where $P(B) > 0$,

$$P(A|B) = \frac{P(A \cap B)}{P(B)}$$

We require $P(B) > 0$ to avoid dividing by zero. In any case, we would not consider $P(A|B)$ where $P(B) = 0$ because in a finite probability space $P(B) = 0$ means B is impossible.

EXAMPLE 4.10 _____

Do Example 4.9 using Theorem 4.3.

SOLUTION

a. $O = \{1, 3, 5\}$, $G = \{4, 5\}$, thus $O \cap G = \{5\}$. By Theorem 4.3,

$$P(O|G) = \frac{P(O \cap G)}{P(G)} = \frac{1}{6} \div \frac{2}{6} = \frac{1}{2}$$

b. $A = \{1, 3, 5\}$, $B = \{2, 3, 5\}$, thus $B \cap A = \{3, 5\}$. By Theorem 4.3,

$$P(B|A) = \frac{P(B \cap A)}{P(A)} = \frac{2}{8} \div \frac{6}{8} = \frac{2}{6} = \frac{1}{3}$$

EXAMPLE 4.11 _____

A bag contains six green, four red, and five blue balls. Three balls are selected at random without replacement. Let

$$C = \text{All three balls are the same color}$$
$$D = \text{All three balls are green}$$

Find

a. $P(D|C)$

b. $P(C|D)$

SOLUTION

a. By Theorem 4.3, $P(D|C) = P(D \cap C)/P(C)$. To calculate $P(C)$ note that C can occur in any of three mutually exclusive ways: all three balls may be red, all three green, or all three blue. Thus,

$$P(C) = P(\text{All three red}) + P(\text{All three green}) + P(\text{All three blue})$$

$$= \binom{4}{3} \bigg/ \binom{15}{3} + \binom{6}{3} \bigg/ \binom{15}{3} + \binom{5}{3} \bigg/ \binom{15}{3}$$

$$= \frac{4}{455} + \frac{20}{455} + \frac{10}{455}$$

$$= \frac{34}{455}$$

Note that $D \cap C = D$ so that $P(D \cap C) = P(D) = \binom{6}{3} \bigg/ \binom{15}{3} = 20/455$. Hence,

$$P(D|C) = \frac{20}{455} \div \frac{34}{455} = \frac{20}{34} = \frac{10}{17}$$

b. By Theorem 4.3, $P(C|D) = P(C \cap D)/P(D)$. As noted in part a, $C \cap D = D$, so

$$P(C/D) = \frac{P(C \cap D)}{P(D)} = \frac{P(D)}{P(D)} = 1$$

Thus, $P(C|D)$ is the certain event. We could have arrived at this conclusion without Theorem 4.3, however. If all three balls chosen are green, then they are certainly the same color.

Independent Events

It occasionally happens that the occurrence of one event has no effect on the probability of another. Two events A and B are **independent** if the occurrence (or nonoccurrence) of one has no effect on the probability of the other. If A and B are not independent, they are **dependent.** In terms of conditional probability, we have Note 4.3.

Note 4.3

A and B are independent if and only if $P(A|B) = P(A)$.

Putting Note 4.3 together with Theorem 4.3 yields Theorem 4.4.

Theorem 4.4

Events A and B are independent if and only if

$$P(A \cap B) = P(A) \cdot P(B)$$

EXAMPLE 4.12 _____

A card is selected at random from a well-shuffled deck of playing cards. Which of the following pairs of events are independent:

a. $A = $ A spade is chosen, $B = $ An ace is chosen
b. $A = $ A spade is chosen, $C = $ A black ace is chosen

SOLUTION

a. $A \cap B = $ The ace of spades, so $P(A \cap B) = 1/52$ ($P(A) = 13/52$ and $P(B) = 4/52$). Since $P(A)P(B) = (13/52)(4/52) = 1/52 = P(A \cap B)$, A and B are independent.
b. $A \cap C = $ The ace of spades, so $P(A|C) = P(A \cap C)/P(C) = 1/52 \div 2/52 = 1/2$. Thus, $P(A|C) \neq P(A)$, and A and C are dependent.

If we have more than two independent events, Theorem 4.4 can be extended.

Corollary 4.2   The Multiplication Law of Probability

If $A_1, A_2, \ldots, A_n$ are independent events, then

$$P(A_1 \cap A_2 \cap \cdots \cap A_n) = P(A_1) \cdot P(A_2) \cdots P(A_n)$$

Sometimes the very nature of events implies their independence. Consider Example 4.13

EXAMPLE 4.13 _____

A nickel, a dime, and a quarter are each flipped once. Assuming all three coins are balanced, what is the probability that all three coins turn up heads?

SOLUTION

The occurrence of a head on any one of the coins has no effect on the outcome of the other coins. The three events—head-on-nickel, HN; head-on-dime, HD; and

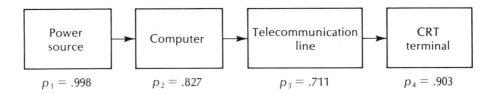

Figure 4.5

head-on-quarter, HQ—are independent. Thus,

$$P(HN \cap HD \cap HQ) = P(HN)P(HD)P(HQ) = \frac{1}{2} \times \frac{1}{2} \times \frac{1}{2} = \frac{1}{8}$$

EXAMPLE 4.14

Suppose a set of computer system components are linked in series, as depicted in Figure 4.5. Assume each component of the system depicted will operate without failing for one day with probabilities $p_1 = .998$ for the power source, $p_2 = .827$ for the computer, $p_3 = .711$ for the telecommunication line, and $p_4 = .903$ for the CRT terminal. If all four components operate independently, what is the probability of system failure? **System failure** is defined as the event of the failure of any one of the components of the system.

SOLUTION

Let $F =$ System failure. The occurrence of F means that one, two, three, or all four components fail. To calculate $P(F)$ it is easier to consider the complementary event $F' =$ No components fail. Since the components are independent, $P(F') = p_1p_2p_3p_4 = .530$ and $P(F) = 1 - P(F') = 1 - .530 = .470$. Thus, the system will fail about once every two days!

Remark If events A and B are mutually exclusive, they cannot occur together. If one of A or B occurs, the other cannot. Therefore, $P(A|B) = 0$ and $P(B|A) = 0$. Assuming A and B are mutually exclusive, can they be independent? If they were independent, we would have $P(A|B) = P(A)$. However, according to the preceding, this would imply $P(A) = 0$. Thus, two nonimpossible, mutually exclusive events must be dependent. This implies that two nonimpossible independent events cannot be mutually exclusive.

That A and B are dependent events does not imply a causal connection between the two. Dependence means that the occurrence of one event affects the *probability* of the other. It is possible that A causes B, or B causes A, or A and B are causally related to other events but not to each other. One of the great debates of our time has been whether or not smoking (event A) causes lung cancer (event B). It can be shown statistically that A and B are dependent, but this does not necessarily mean that A causes B. Events A and B may be caused by other factors, such as nervousness. Thus, there may be no causal connection between the two.

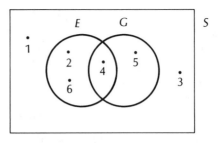

Figure 4.6

Probability of the Union of Two Events

As a final note, we consider how to compute $P(A \cup B)$ when A and B are not mutually exclusive. Toss a balanced die and let $E = \{2, 4, 6\}$ and $G = \{4, 5\}$. Thus, $E \cup G = \{2, 4, 5, 6\}$ and $P(E \cup G) = 4/6 = 2/3$. To see how this is related to $P(E)$ and $P(G)$, we refer to the Venn diagram of Figure 4.6.

Here $P(E) = 3/6$ and $P(G) = 2/6$. In this case $P(E \cup G) \neq P(E) + P(G)$ because in adding $P(E)$ and $P(G)$ we count the simple event $\{4\}$ twice—once for its occurrence in E and once for its occurrence in G. We must subtract from the sum $P(E) + P(G)$ the probability of the intersection $E \cap G$ so that simple events in the intersection are counted only once:

$$P(E \cup G) = P(E) + P(G) - P(E \cap G) = \frac{3}{6} + \frac{2}{6} - \frac{1}{6} = \frac{4}{6} = \frac{2}{3}$$

This argument is the same as the one that established the inclusion–exclusion principle in Chapter 3. Using the inclusion–exclusion principle on the events A and B viewed as subsets of the sample space S justifies Theorem 4.5.

Theorem 4.5

For any events A and B

$$P(A \cup B) = P(A) + P(B) - P(A \cap B)$$

Note that Theorem 4.5 generalizes Theorem 4.2(i) because if A and B are mutually exclusive, $P(A \cap B) = 0$.

EXAMPLE 4.15 _____

Select a card at random from a well-shuffled deck of playing cards. Let $A = A$ spade is chosen, $B = $ An ace is chosen, and $C = A$ black ace is chosen. Find

a. $P(A \cup B)$ b. $P(A \cup C)$

SOLUTION

a. By Theorem 4.5, $P(A \cup B) = P(A) + P(B) - P(A \cap B)$. Since $A \cap B =$ The ace of spades is chosen, $P(A) = 13/52$, $P(B) = 4/52$, and $P(A \cap B) = 1/52$. thus,

$$P(A \cup B) = \frac{13}{52} + \frac{4}{52} - \frac{1}{52} = \frac{16}{52} = \frac{4}{13}$$

b. $A \cap C =$ The ace of spades is chosen, so $P(A \cap C) = 1/52$. By Theorem 4.5,

$$P(A \cup C) = P(A) + P(C) - P(A \cap C) = \frac{13}{52} + \frac{2}{52} - \frac{1}{52} = \frac{14}{52}$$

$$= \frac{7}{26}$$

EXERCISES 4.2

1. Spin the spinner depicted in Figure 4.7 and record the number that comes up. Assume the spinner cannot stop on a dividing line. Let $A =$ A prime number comes up, and $B =$ A number greater than 3 comes up.
   a. Find $P(A|B)$ and $P(B|A)$.
   b. Are A and B independent?
2. A nickel, a dime, and a quarter are each flipped once. Let $C =$ Exactly two heads come up, and $D =$ At least two heads come up.
   a. Find $P(C|D)$ and $P(D|C)$.
   b. Are C and D independent?
3. Three cards are selected at random without replacement from a well-shuffled standard deck of playing cards. Let $A =$ Three hearts are chosen, $B =$ Exactly two red cards are chosen, $D =$ Three face cards are chosen, and $E =$ Three aces or three kings are chosen.

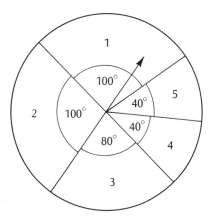

Figure 4.7

     a. Are E and D independent?

     b. Are D and A independent?

     c. Find $P(A \cup D)$ and $P(B \cup E)$.

4. A bag contains 7 white and 13 black balls. Four balls are selected at random without replacement. Let $C =$ Two white balls and two black balls are chosen, $D =$ At least one white ball is chosen, and $E =$ The number of black balls chosen exceeds the number of white balls chosen.

     a. Are E and D independent?

     b. Are C and D independent?

     c. Find $P(D \cup E)$ and $P(C \cup D)$.

5. In Example 4.14 we calculated the probability of system failure as .470, which is rather high. To decrease the probability of system failure we redesign the system to include redundancy. The weakest link in the computer system is the telecommunication line, whose probability that it will operate without failure for one day is $p_3 = .711$. We therefore connect a second telecommunication line from the computer to our CRT. If the first line should break down, the second will automatically be put into use. We assume the probability of operating for one day without failure on the second line is also .711. The new system configuration is given in Figure 4.8 with the associated probabilities. Find the probability of system failure for the newly configured system.

6. The manager of a programming project has divided the program into three modules, which she gives to programmers A, B, and C to write. From previous experience, the manager knows her programmers will produce a correct module in the alloted time with probabilities .6, .5, and .3, respectively. Assume the three programmers work independently of each other. What are the probabilities of the following events?

     a. Programmer A or programmer B produce correct modules.

     b. Programmer B or programmer C produce correct modules.

     c. All three programmers produce correct modules.

     d. At least one programmer produces a correct module.

     e. No programmer produces a correct module.

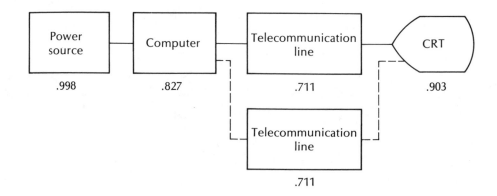

Figure 4.8

7. In your computer mathematics class, 45% of the students know Pascal, 25% know FORTRAN, and 10% know both Pascal and FORTRAN. If a student is selected at random from your class, find the following probabilities:
   a. If he knows Pascal, he knows FORTRAN.
   b. If he knows FORTRAN, he knows Pascal.
   c. He knows either FORTRAN or Pascal.

Problems and Projects

8. Prove Corollary 4.2 by mathematical induction.
9. **(The Birthday Paradox)** Suppose there are 24 people in your computer math class. What is the probability of the event B that at least two people in the class have the same birthday? By the same birthday we mean the same month and day, not necessarily the same year. To simplify matters, we will ignore leap years and the possibility of twins (or triplets, and so on) being in the class. It would be difficult to calculate $P(B)$ directly, so we consider instead $P(B')$ and use Theorem 4.2(ii).

   B' is the event that no two people in the class have the same birthday. Under the assumptions of the previous paragraph, the birthday of any given person in the class is independent of the birthday of any other person in the class. Suppose we select one person from the class and record her birthday. The probability that a second person chosen has a birthday different from the first person is 364/365. The probability that the third person chosen has a birthday different from the first two is 363/365, and so on. Hence, the probability that all 24 people have different birthdays is:

   $$P(B') = \frac{364}{365} \times \frac{363}{365} \times \cdots \times \frac{342}{365}$$

   and, therefore, $P(B) = 1 - P(B')$. Note that there are 23 factors in this product.
   a. Write a computer program to compute $P(B)$.
   b. Generalize the program written in part a. Write a program that reads an integer n and prints the probability that in a group of n people at least two have the same birthday. Run your program for $n = 30$, $n = 40$, and $n = 50$. You may be surprised at the answers!
10. Find a formula for $P(A \cup B \cup C)$ similar to the formula of Theorem 4.5. (A Venn diagram will be helpful.)

4.3 COMPOUND EXPERIMENTS AND REPEATED TRIALS

Compound Experiments

It is frequently convenient to consider a given experiment as composed of two or more simpler experiments. Such an experiment is called a **compound experiment.** As an example, consider the compound experiment consisting of flipping a balanced coin and rolling a balanced die. The sample space of this experiment is the set of all pairs in which the first element is an H or T, and the second element is an

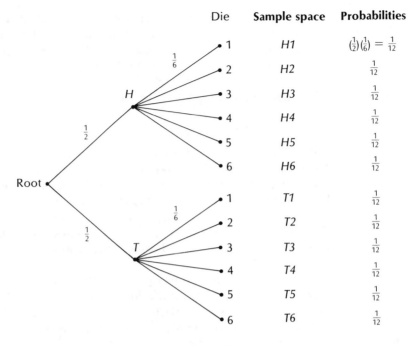

Figure 4.9

integer from 1 to 6.† The simple event *H4* indicates a head was flipped on the coin and a 4 tossed on the die. Since there are two possible outcomes on the coin and six on the die, by the multiplication principle, there are $2 \times 6 = 12$ simple events in the sample space S of this experiment.

A simple way to generate the elements of the sample space of a compound experiment is by a tree diagram. Tree diagrams also aid in computing the probabilities of the simple events. To draw a tree diagram of the coin–die experiment, we start with the root of the tree, a dot (●), and draw one branch for each possible outcome of flipping the coin (the first simple experiment of which the compound experiment is composed). We then label each branch accordingly. Each of these branches represents what could happen on a flip of the coin.

Now, from each branch just drawn, we draw one branch for each possible outcome of tossing the die (the second simple experiment in the compound experiment) and label each branch accordingly. See Figure 4.9. We shall explain the numbers on the branches of the tree presently.

Each path from the root of the tree to the end of a branch corresponds to a simple event in the sample space of the compound experiment. These simple events appear in the column headed **sample space** in Figure 4.9.

† The choice of putting the result of the coin flip first is arbitrary. We could have let the sample space be the set of all pairs in which the first element is an integer from 1 to 6, and the second element an *H* or *T*.

We now use the tree diagram of the sample space to obtain the probabilities of the simple events. Since the coin is balanced, the probability of flipping a T or H is $1/2$. We write these probabilities on their respective branches.

Since the die is balanced, the probabilities of tossing a 1, 2, 3, 4, 5, or 6 are each $1/6$. The outcome of tossing the die is independent of the outcome of flipping the coin, so the probability of each branch in the second part of the tree is $1/6$.

In Note 4.4 we state how to find the probability of a simple event in the sample space of a compound experiment.

Note 4.4

To find the probability of any simple event in the sample space, multiply the probabilities on the path from the root to the end of the branch corresponding to that simple event.

For example, $P(H1) = 1/2 \times 1/6 = 1/12$. The probabilities of all the simple events are given in the column headed probabilities in Figure 4.9.

EXAMPLE 4.16 _____

A bag contains eight white and two black balls. Three balls are selected at random from the bag, one at a time without replacement. Use a tree diagram to list the elements of the sample space and their associated probabilities. Also find the probabilities of the following events:

a. A = Selecting exactly two black balls

b. B = Selecting exactly two white balls

SOLUTION

We consider this a compound experiment. It is composed of selecting the first ball, then selecting the second, and finally selecting the third. Unlike the coin–die experiment, however, the probabilities associated with choosing the second ball are affected by the color of the first ball chosen. Likewise, the probabilities associated with choosing the third ball are affected by the colors of the balls chosen on the first two picks. Hence, the three simple experiments of which the compound experiment is composed are dependent. The tree diagram of the experiment is shown in Figure 4.10.

Note how to construct the second and third stages of the tree. If the first ball chosen from the bag is white, for example, the bag will contain seven white and two black balls. The probability of selecting a white ball (or black ball) in the second pick is, then, $7/9$ (or $2/9$ for a black ball). The tree is labeled accordingly.

The probabilities of the simple events are still computed by multiplying the probabilities on the path from the root to the end of the corresponding branch. The lowest branch of the tree is slightly different from the rest. If the first two balls chosen are black, there are only white balls remaining in the bag. Choosing a white

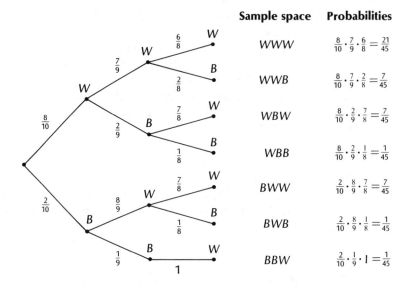

Figure 4.10

ball is then certain, giving a probability of 1 that the third ball is white. To complete the solution:

a. $P(A) = P(WBB, BWB, BBW) = \dfrac{1}{45} + \dfrac{1}{45} + \dfrac{1}{45} = \dfrac{3}{45} = \dfrac{1}{15}$

b. $P(B) = P(WWB, WBW, BWW) = \dfrac{7}{45} + \dfrac{7}{45} + \dfrac{7}{45} = \dfrac{21}{45} = \dfrac{7}{15}$

Repeated Trials

The most important type of compound experiment is that in which all the simple experiments are the same and independent. If there are n identical and independent simple experiments, the compound experiment is called **n independent, or repeated, trials.** As an example, consider the spinner depicted in Figure 4.11. The experiment consists of spinning the spinner three times. There are three repetitions of the same simple experiment (each simple experiment is called a "trial") and they are all independent. We thus have three repeated trials.

A simple event in an experiment consisting of n repeated trials is an ordered n-tuple. For example, a simple event in the spinner experiment could be denoted $(2, 1, 3)$. This means the first spin resulted in 2, the second in 1, and the third in 3. Since the individual trials are independent, the probability of a simple event is the product of the probabilities of its components. Thus, $P(2, 1, 3) = 1/4 \times 1/2 \times 1/4 = 1/32$.

The type of repeated trial occurring most often in applications is that in which each trial results in exactly one of two outcomes. One of these outcomes is arbitrarily called **success** and the other **failure**. A **binomial experiment** is a compound

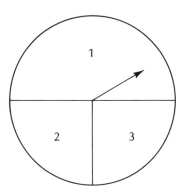

Figure 4.11

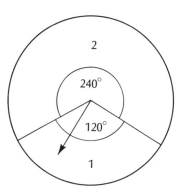

Figure 4.12

experiment consisting of n repeated trials in which each trial has two possible outcomes. Suppose, for example, we spin three times the spinner depicted in Figure 4.12.

We define a result of 1 as a success and a result of 2 as a failure. What is the probability of A = Exactly two successes? We first draw the tree diagram of the experiment (Figure 4.13). From this figure we see that

$$P(A) = P((1,1,2), (1,2,1), (2,1,1)) = \frac{2}{27} + \frac{2}{27} + \frac{2}{27}$$

$$= \frac{6}{27} = \frac{2}{9}$$

We could have obtained this probability in another way. We know that each simple event in A consists of two successes and one failure. Since the trials are independent, the probability of each simple event is the product of the probabilities of its components. No matter what their order in the products, these probabilities are 1/3, 1/3, and 2/3. The probability of any simple event in A is, therefore, $1/3 \times 1/3 \times 2/3 = (1/3)^2(2/3) = 2/27$. Now all we need do is count the number of

Sample space   Probabilities

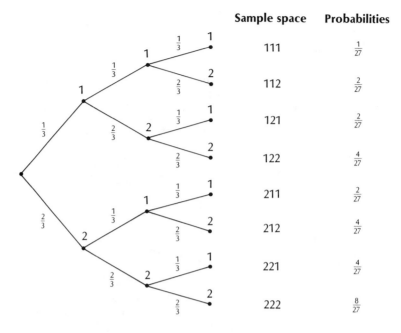

Sample space	Probabilities
111	$\frac{1}{27}$
112	$\frac{2}{27}$
121	$\frac{2}{27}$
122	$\frac{4}{27}$
211	$\frac{2}{27}$
212	$\frac{4}{27}$
221	$\frac{4}{27}$
222	$\frac{8}{27}$

Figure 4.13

simple events in A. This is an easy combinatorial problem. We want two successes in three trials. This can be done in $\binom{3}{2} = 3$ ways. Since each of the three simple events has the probability 2/27, then

$$P(A) = \binom{3}{2}\left(\frac{1}{3}\right)^2\left(\frac{2}{3}\right) = 3 \times \left(\frac{2}{27}\right) = \frac{6}{27} = \frac{2}{9}$$

This reasoning can be generalized. Consider a binomial experiment of n trials in which the probability of success on a single trial is p. The probability of failure is, thus, $1 - p = q$. To calculate the probability of exactly k successes, note first that the k successes can occur in $\binom{n}{k}$ ways. The probability of any one of these simple events is the product $p^k q^{n-k}$ since the probability of each of the k successes is p, and the probability of each of the $n - k$ failures is q. We thus have Theorem 4.6.

Theorem 4.6

The probability of exactly k successes in a binomial experiment of n trials is

$$\binom{n}{k}p^k q^{n-k}.$$

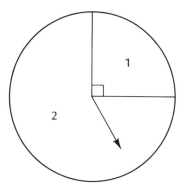

Figure 4.14

EXAMPLE 4.17 _____

Spin five times the spinner depicted in Figure 4.14. What are the probabilities of:

a. A = Exactly three 1s come up
b. B = At least four 1s come up
c. C = At least one 1 comes up.

SOLUTION

If we define a 1 coming up as a success, this is a binomial experiment of $n = 5$ trials. The probability of success is $p = 1/4$ and of failure $q = 1 - p = 3/4$.

a. By Theorem 4.6, with $k = 3$,

$$P(A) = \binom{5}{3}\left(\frac{1}{4}\right)^3\left(\frac{3}{4}\right)^{5-3} = 10 \cdot \left(\frac{1}{4}\right)^3\left(\frac{3}{4}\right)^2 = \frac{90}{1024}$$

b. We apply Theorem 4.6 twice because at least four successes means four or five successes:

$$P(B) = \binom{5}{4}\left(\frac{1}{4}\right)^4\left(\frac{3}{4}\right)^1 + \binom{5}{5}\left(\frac{1}{4}\right)^5\left(\frac{3}{4}\right)^0 = \frac{15}{1024} + \frac{1}{1024}$$

$$= \frac{16}{1024}$$

c. At least one success means one, two, three, four, or five successes. Rather than apply Theorem 4.6 five times, it is easier to compute the probability of C' = No successes. Theorem 4.6 with $k = 0$ gives:

$$P(C') = \binom{5}{0}\left(\frac{1}{4}\right)^0\left(\frac{3}{4}\right)^5 = \frac{243}{1024}$$

It follows that

$$P(C) = 1 - P(C') = 1 - \left(\frac{243}{1024} \right) = \frac{781}{1024}$$

EXERCISES 4.3

For each compound experiment in exercises 1 to 4, use a tree diagram to list the simple events in the sample space and calculate their probabilities.

1. A bag contains four red, three green, and two blue balls. Two balls are selected from the bag at random without replacement. Find the probability of
A = Selecting two balls of the same color.

2. Use the same bag as in exercise 1, and select two balls with replacement. Find the probability of the event A defined in exercise 1.

3. A balanced penny, nickel, and dime are each flipped once. What are the probabilities of
   a. A = Two heads coming up
   b. B = A head on the nickel
   c. C = A head on the penny or the nickel

4. A balanced red and a balanced green die are each tossed once.
   a. The sum of the numbers occurring on each die is recorded. Fill in the following table of probabilities:

Sum of dice	2	3	4	5	6	7	8	9	10	11	12
Probability											

   b. What is the probability that the larger of the two numbers appearing is greater than 4?
   c. What is the probability that the two numbers appearing are the same?

5. A quarter is flipped ten times. Find the probabilities of the following events:
   a. A = Exactly three heads occur
   b. B = Exactly eight heads occur
   c. C = At least eight heads occur
   d. D = At least three heads occur

6. A baseball player has a 0.250 batting average. Assume each at-bat is independent of the others. If he has four at-bats in a game, find the probability that he will have
   a. No hits   b. One hit   c. Two hits
   d. Three hits   e. Four hits

7. Each of the first three players in the lineup of a baseball team has a batting average of 0.300. Assume: (1) each player's performance at the plate is independent of the others, (2) no one hits into double or triple plays, and (3) no one gets on base other than by getting a hit. What is the probability the fourth batter in the lineup comes to bat in the first inning of a game?

8. Consider the hypothetical computer system shown in Figure 4.15. The probability of non-failure of each component is shown in the figure. If one computer breaks down, the other is automatically put to use. If one telecommunication

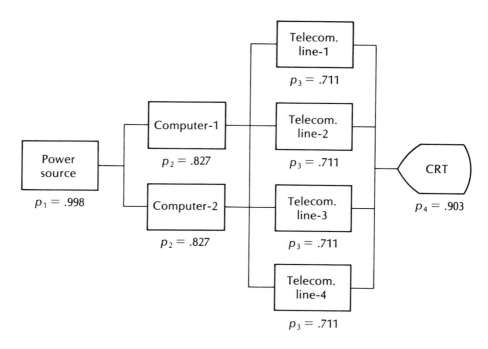

Figure 4.15

line breaks down, another is automatically put to use. What is the probability of system failure? (*Hint:* Consider the two computers as a unit and the four telecommunications lines as a unit, and find the probability of nonfailure for each unit.)

9. The spinner of Figure 4.16 is spun five times. Find the following probabilities: (*Hint:* In each case consider this a binomial experiment.)
   a. Exactly three 1s come up.
   b. A 2 comes up at least twice.
   c. No 2s come up.

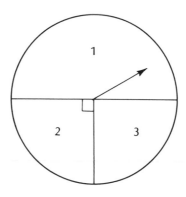

Figure 4.16

4.4   RANDOM VARIABLES AND EXPECTED VALUE

It is frequently useful to associate a number with each outcome of a random experiment. Consider, for example, the experiment of tossing a balanced coin three times. Associate with each outcome of this binomial experiment the number X of heads that appear. If the outcome of the experiment is THH, then $X = 2$. As another example, consider the experiment of betting \$1 on the number 22 on a roulette wheel. The number Y we associate with each spin of the wheel is the amount of money we win on that spin. If the number 22 comes up, $Y = 36$ (the payoff on a single number bet in roulette is 36 to 1). If the number 22 does not come up, $Y = -1$. The minus indicates a negative win, that is, a loss.

Random Variables

A **random variable** is a rule that associates a number with each outcome of a random experiment. In the preceding discussion, X and Y were random variables. Associated with each value of a random variable is the probability of its occurrence. For example, in the coin-tossing experiment, what is the probability of the event $\{X = 2\}$? We write this probability as $P(X = 2)$. Tossing three coins is a binomial experiment of three trials in which a success is a head and $p = 1/2$. Therefore, we require the probability of exactly two heads.

$$P(X = 2) = \binom{3}{2}\left(\frac{1}{2}\right)^2\left(\frac{1}{2}\right)^1 = 3\left(\frac{1}{4}\right)\left(\frac{1}{2}\right) = \frac{3}{8}$$

In the roulette example, $P(Y = 22) = 1/38$ since there are 38 numbers on a roulette wheel.

A table listing all possible values of a random variable and their associated probabilities is called the **distribution** of the random variable.

EXAMPLE 4.18

What are the distributions of the random variables X and Y just discussed?

SOLUTION

a. For the random variable X,

X	0	1	2	3
P	$\dfrac{1}{8}$	$\dfrac{3}{8}$	$\dfrac{3}{8}$	$\dfrac{1}{8}$

b. For the random variable Y,

Y	-1	$+36$
P	$\dfrac{37}{38}$	$\dfrac{1}{38}$

Since all possible values of the random variable appear in the table, the probabilities in the table must sum to 1.

EXAMPLE 4.19

Three balls are selected without replacement from a bag containing eight white and two black balls. Let the random variable X = The number of black balls chosen. What is the distribution of X?

SOLUTION

In Example 4.16 we computed the probability of each simple event as follows:

Simple event	WWW	WWB	WBW	WBB	BWW	BWB	BBW
Probability	$\dfrac{21}{45}$	$\dfrac{7}{45}$	$\dfrac{7}{45}$	$\dfrac{1}{45}$	$\dfrac{7}{45}$	$\dfrac{1}{45}$	$\dfrac{1}{45}$

The event $\{X = 0\} = \{WWW\}$

$\{X = 1\} = \{WWB, WBW, BWW\}$

$\{X = 2\} = \{WBB, BWB, BBW\}$

Hence, the distribution of X is

X	0	1	2
P	$\dfrac{21}{45}$	$\dfrac{21}{45}$	$\dfrac{3}{45}$

The distribution of a random variable may be depicted graphically.

EXAMPLE 4.20

Toss a balanced die once. Let the random variable X = The number that comes up. Draw a graph of the distribution of X.

SOLUTION

Since the die is balanced, we have the following distribution:

X	1	2	3	4	5	6
P	$\dfrac{1}{6}$	$\dfrac{1}{6}$	$\dfrac{1}{6}$	$\dfrac{1}{6}$	$\dfrac{1}{6}$	$\dfrac{1}{6}$

Draw a set of axes and label the values of X on the horizontal axis. Above each value place a point whose height above the horizontal axis represents the probability of X. See Figure 4.17.

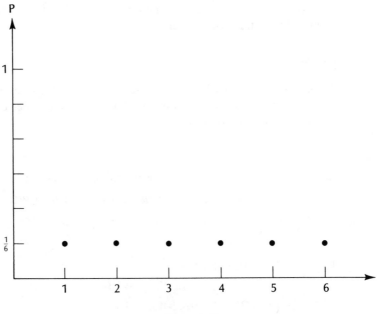

Figure 4.17

Since the points of the distribution of Example 4.20 lie on a horizontal line, the probabilities are evenly, or uniformly, distributed. For this reason, the equiprobable distribution of X is called a **uniform distribution;** that is, X is a **uniformly distributed random variable.**

EXAMPLE 4.21

Let the random variable $X =$ The number of heads that come up in five tosses of a balanced coin. Graph the distribution of X.

SOLUTION

Using Theorem 4.6 with $p = q = 1/2$, $n = 5$, and k successively 0, 1, 2, 3, 4, and 5, we have the distribution:

X	0	1	2	3	4	5
P	$\dfrac{1}{32}$	$\dfrac{5}{32}$	$\dfrac{10}{32}$	$\dfrac{10}{32}$	$\dfrac{5}{32}$	$\dfrac{1}{32}$

The graph of the distribution is shown in Figure 4.18.

Distributions similar to that of Example 4.21 arise frequently. We, therefore, make the following generalization. Consider a binomial experiment of n trials in which the probability of success on any one trial is p. Let the random variable

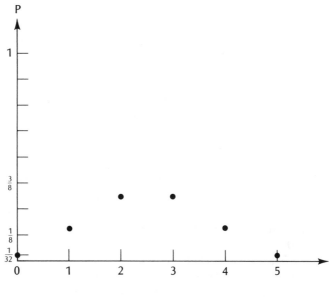

Figure 4.18

$X = $ The number of successes. As we saw in Section 4.3,

$$P(X = k) = \binom{n}{k} p^k q^{n-k}$$

The distribution of X is called the **binomial distribution.** The distribution of the random variable X of Example 4.21 is, therefore, a binomial distribution.

Expected Value

Writing the distribution of a random variable in the form of a table and drawing its graph are two ways in which a distribution may be depicted. However, it is useful to have simple numeric measures that describe certain properties of distributions. We now describe several such measures.

The **average,** or **mean,** $\bar{x}$ of a set of numbers $x_1, x_2, \ldots, x_n$ is obtained by adding the numbers and dividing the sum by the number of numbers, n. Using the summation notation introduced in Chapter 3, we have

$$\bar{x} = \frac{1}{n} \sum_{i=1}^{n} x_i$$

EXAMPLE 4.22 _____

Find the mean of 3, 7, 4, 9, 8, 4, 2, 9, 3, 1.

SOLUTION

$$\bar{x} = \left(\frac{1}{10}\right)(3 + 7 + 4 + 9 + 8 + 4 + 2 + 9 + 3 + 1) = \left(\frac{1}{10}\right)(50) = 5$$

The average, or expected value, of a random variable is a generalization of the concept of the mean. Consider the example introduced earlier of betting \$1 on the number 22 in a roulette game. The random variable Y = The amount of money won on one spin of the wheel. Suppose we play the game a large number of times, each time betting \$1 on the number 22. How much money would we expect to be ahead or behind at the end of play? If we play the game 380 times, we expect to win about 10 times. Recall $P(Y = 22) = 1/38$; that is, we win about 1/38 of the time. Consequently, we expect to lose about 370 times. Since each win pays \$36, our total winnings would be 10(36) = \$360, and our total losses would be 370(1) = \$370. Therefore, we would lose about \$10.

The expected value of Y, $E(Y)$, is computed using similar reasoning. Y takes on two possible values: -1 and $+36$. From the distribution of Y, Y takes on the value -1 about 37/38 of the time and the value $+36$ about 1/38 of the time. The expected value of Y is then

$$E(Y) = (-1)\left(\frac{37}{38}\right) + (+36)\left(\frac{1}{38}\right) = -\frac{37}{38} + \frac{36}{38} = -\frac{1}{38}$$

We interpret $E(Y) = -1/38$ as meaning that on average we lose (we lose because of the minus sign) 1/38 of each dollar we bet. Consequently, if we play 380 games, our total winnings are $380(-1/38) = -10$. That is, we lose \$10, as we already saw.

We now define expected value for an arbitrary random variable. Let X be a random variable with distribution

X	x_1	x_2	$\cdots$	x_n
P	p_1	p_2	$\cdots$	p_n

> The **expected value** $E(X)$ of X is defined as
>
> $$E(X) = \sum_{i=1}^{n} x_i p_i$$

EXAMPLE 4.23 ───

Flip a balanced coin ten times. Let X = The number of times heads comes up. Find $E(X)$.

SOLUTION

The distribution is as follows. Note that it is convenient to write the product $x_i p_i$ under each column of the distribution.

X	0	1	2	3	4	5	6	7	8	9	10	
P	$\dfrac{1}{1024}$	$\dfrac{10}{1024}$	$\dfrac{45}{1024}$	$\dfrac{120}{1024}$	$\dfrac{210}{1024}$	$\dfrac{252}{1024}$	$\dfrac{210}{1024}$	$\dfrac{120}{1024}$	$\dfrac{45}{1024}$	$\dfrac{10}{1024}$	$\dfrac{1}{1024}$	SUM
$x_i p_i$	0	$\dfrac{10}{1024}$	$\dfrac{90}{1024}$	$\dfrac{360}{1024}$	$\dfrac{840}{1024}$	$\dfrac{1260}{1024}$	$\dfrac{1260}{1024}$	$\dfrac{840}{1024}$	$\dfrac{360}{1024}$	$\dfrac{90}{1024}$	$\dfrac{10}{1024}$	5

Hence, $E(X) = 5$.

There is a simple formula for the expected value of a binomially distributed random variable.

Theorem 4.7

If X = The number of successes in a binomial experiment of n trials in which the probability of success on one trial is p, then $E(X) = np$.

The random variable of Example 4.23 is binomially distributed. Applying Theorem 4.7, we obtain $E(X) = 10(1/2) = 5$.

The idea of expected value may be extended to the individual values a random variable assumes. If we toss a die 6000 times, we expect each of the numbers 1, 2, 3, 4, 5, 6 to appear about 1000 times because the probability of any one of these numbers is 1/6. We would expect, therefore, about 1/6 of the 6000 tosses to result in any particular number. Thus, the expected number of occurrences of a given number is $6000(1/6) = 1000$.

Generalizing, let an experiment be repeated n times and let X be a random variable associated with the experiment. If x_i is a value of X having probability p_i, then the **expected number of occurrences** of x_i is np_i.

EXAMPLE 4.24

Three balls are selected without replacement from a bag containing eight white and two black balls. Let X = The number of black balls chosen. If this experiment is repeated 900 times, what would be the expected number of occurrences of $X = 1$?

SOLUTION

We are asked for the expected number of occurrences of x_i = One black ball is chosen, when the experiment is repeated 900 times. The distribution of X, as calculated in Example 4.19, is

X	0	1	2
P	$\dfrac{21}{45}$	$\dfrac{21}{45}$	$\dfrac{3}{45}$

Since $p_i = P(X = 1) = 21/45$, the expected number of occurrences is $np_i = 900(21/45) = 420$.

Another measure associated with a random variable is its standard deviation. The standard deviation measures how the values of the random variable are dispersed, or spread out, from the mean (that is, from the expected value) of the distribution. Let X be a random variable with the distribution

X	x_1	x_2	$\cdots$	x_n
P	p_1	p_2	$\cdots$	p_n

and let $\bar{X} = E(X)$. The **variance** of X is

$$\sigma^2(X) = \sum_{i=1}^{n} (x_i - \bar{X})^2 p_i$$

The **standard deviation** of X is $\sigma(X) = \sqrt{\sigma^2(X)}$.

EXAMPLE 4.25

Find the variance and the standard deviation of the random variable X of Example 4.24.

SOLUTION

For ease of calculation, it is convenient to arrange the distribution as in Table 4.1.

X	p_i	$x_i p_i$	$x_i - \bar{X}$	$(x_i - \bar{X})^2$	$(x_i - \bar{X})^2 p_i$
0	$\dfrac{21}{45}$	0	$-\dfrac{3}{5}$	$\dfrac{9}{25}$	$\left(\dfrac{9}{25}\right)\left(\dfrac{21}{45}\right) = \dfrac{189}{1125}$
1	$\dfrac{21}{45}$	$\dfrac{21}{45}$	$\dfrac{2}{5}$	$\dfrac{4}{25}$	$\left(\dfrac{4}{25}\right)\left(\dfrac{21}{45}\right) = \dfrac{84}{1125}$
2	$\dfrac{3}{45}$	$\dfrac{6}{45}$	$\dfrac{7}{5}$	$\dfrac{49}{25}$	$\left(\dfrac{49}{25}\right)\left(\dfrac{3}{45}\right) = \dfrac{147}{1125}$
	SUM $= \dfrac{27}{45}$				SUM $= \dfrac{420}{1125}$

Table 4.1

Thus, $\bar{X} = E(X) = 27/45 = 3/5$.

$$\sigma^2(X) = \frac{420}{1125} = .373$$

Thus, $\sigma(X) = \sqrt{0.373} = .611$.

It is sometimes more convenient to use the formula in the following theorem.

Theorem 4.8

The variance of a random variable X with distribution

X	x_1	x_2	$\cdots$	x_n
P	p_1	p_2	$\cdots$	p_n

and expected value $\bar{X} = E(X)$ is

$$\sigma^2(X) = \sum_{i=1}^{n} x_i^2 p_i - \bar{X}^2$$

EXAMPLE 4.26

Do Example 4.25 using the formula of Theorem 4.8.

SOLUTION

We can arrange the distribution as in Table 4.2.

X	p_i	$x_i p_i$	x_i^2	$x_i^2 p_i$
0	$\dfrac{21}{45}$	0	0	0
1	$\dfrac{21}{45}$	$\dfrac{21}{45}$	1	$\dfrac{21}{45}$
2	$\dfrac{3}{45}$	$\dfrac{6}{45}$	4	$\dfrac{12}{45}$
SUMS		$\dfrac{27}{45}$		$\dfrac{33}{45}$

Table 4.2

Thus, $\bar{X} = 27/45 = 3/5$.

$$\sum_{i=1}^{n} x_i^2 p_i = \frac{33}{45} = \frac{11}{15}$$

$$\sigma^2(X) = \frac{11}{15} - \left(\frac{3}{5}\right)^2 = \frac{11}{15} - \frac{9}{25} = \frac{28}{75} = .373$$

Hence, $\sigma(X) = \sqrt{0.373} = .611$.

Just as there is a formula for the expected value of a binomially distributed random variable, there is a formula for its standard deviation.

Theorem 4.9

If X = The number of successes in a binomial experiment of n trials, in which the probability of success on one trial is p and the probability of failure is $q = 1 - p$, then $\sigma^2(X) = npq$ and $\sigma(X) = \sqrt{npq}$.

EXAMPLE 4.27

Flip a balanced coin ten times. Let X = The number of heads that come up. Find $E(X)$, $\sigma^2(X)$, and $\sigma(X)$.

SOLUTION

We have $n = 10$, $p = q = 1/2$. Using Theorems 4.7 and 4.9, $E(X) = np = 10(1/2) = 5$:

$$\sigma^2(X) = npq = 10\left(\frac{1}{2}\right)\left(\frac{1}{2}\right) = 2.5$$

$$\sigma(X) = \sqrt{2.5} = 1.58$$

It is important to understand the meaning of the standard deviation. Consider this example. Two people take two different IQ tests, test A and test B. Each scores 120 on his or her respective test. Each test has an average score of 100. Who is more intelligent (as measured by IQ)? With the information given so far we cannot tell. Suppose, however, the standard deviation on test A is $\sigma_A = 5$ and the standard deviation on test B is $\sigma_B = 10$. This means the scores on test A are more tightly packed about the average of 100 than the scores on test B because $\sigma_A < \sigma_B$. Since there is less variation in the scores of test A, it should be harder to achieve a score of 120 on test A than to achieve the same score on test B. Therefore, we judge the person who took test A to be more intelligent (as measured by IQ).

Sums of Random Variables

Random variables can be combined arithmetically. Suppose we have two bags, each containing white and black balls. Bag A contains three white and two black balls. Bag B contains two white and three black balls. One ball is selected at random from each bag. Let X = The number of white balls selected from bag A, and let Y = The number of white balls selected from bag B. We can define the random variable $Z = X + Y$ = The total number of white balls selected. Being a random variable, Z has a distribution, an expected value, and a variance.

EXAMPLE 4.28

Find the distribution, expected value, and variance of Z.

SOLUTION

Since the values of X can be either 0 or 1, and similarly for Y, the variable Z can assume the values 0, 1, or 2. We can view this experiment as a 2-stage compound experiment and use a tree diagram to compute the probabilities of column 2 in Table 4.3.

z_i	p_i	$z_i p_i$	$z_i - \bar{Z}$	$(z_i - \bar{Z})^2$	$p_i(z_i - \bar{Z})^2$
0	$\dfrac{6}{25}$	0	-1	1	$\dfrac{6}{25}$
1	$\dfrac{13}{25}$	$\dfrac{13}{25}$	0	0	0
2	$\dfrac{6}{25}$	$\dfrac{12}{25}$	1	1	$\dfrac{6}{25}$
		$\bar{Z} = 1$			$\sigma^2(Z) = \dfrac{12}{25}$

Table 4.3

Thus, $E(X) = 1$ and $\sigma^2(X) = 12/25$.

The expected value and variance of $X + Y$ are related to the expected values and variances of X and Y. Before stating the theorem giving those relations, we need a definition. The random variables X and Y are **independent random variables** if the occurrence of a value of one random variable does not affect the probability of occurrence of any value of the other random variable. Since the number of white balls chosen from bag A does not affect the number of white balls chosen from bag B, the random variables X and Y are independent.

> **Theorem 4.10**
>
> i. If X and Y are any random variables, $E(X + Y) = E(X) + E(Y)$.
> ii. If X and Y are independent random variables, $\sigma^2(X + Y) = \sigma^2(X) + \sigma^2(Y)$.

EXAMPLE 4.29

Verify Theorem 4.10 for the random variables X and Y of Example 4.28.

SOLUTION

The calculations for X are contained in Table 4.4 and those for Y in Table 4.5.

x_i	p_i	$x_i p_i$	x_i^2	$x_i^2 p_i$
0	$\dfrac{2}{5}$	0	0	0
1	$\dfrac{3}{5}$	$\dfrac{3}{5}$	1	$\dfrac{3}{5}$

$E(X) = \dfrac{3}{5}$

$\sigma^2(X) = \dfrac{3}{5} - \left(\dfrac{3}{5}\right)^2 = \dfrac{6}{25}$

Table 4.4

y_i	p_i	$y_i p_i$	y_i^2	$y_i^2 p_i$
0	$\dfrac{3}{5}$	0	0	0
1	$\dfrac{2}{5}$	$\dfrac{2}{5}$	1	$\dfrac{2}{5}$

$E(Y) = \dfrac{2}{5}$

$\sigma^2(Y) = \dfrac{2}{5} - \left(\dfrac{2}{5}\right)^2 = \dfrac{6}{25}$

Table 4.5

Hence, $E(Z) = 1 = E(X) + E(Y) = 3/5 + 2/5.$

$$\sigma^2(Z) = 12/25 = \sigma^2(X) + \sigma^2(Y) = \frac{6}{25} + \frac{6}{25} = \frac{12}{25}$$

EXERCISES 4.4

1. Find the mean for each of the following sets of numbers:
   a. 42, 87, 91, 53, 49, 31, 1, 20, 92, 18, 10, 59
   b. 150, 488, 600, 125, 179, 315, 208, 82, 263, 859

2. The standard deviation of a set of numbers $x_1, x_2, \ldots, x_n$ is defined as the following

$$\sigma = \sqrt{(1/n) \sum_{i=1}^{n} (x_i - \bar{x})^2}$$

where $\bar{x}$ is the mean of the set of numbers. Find the standard deviation for each of the sets of numbers in exercise 1.

3. Two balanced dice, one red and one green, are rolled. Let X = The sum of the numbers that turn up.
   a. Write the distribution of X in tabular form and graph the distribution. Consider this a compound experiment in which one simple experiment is rolling the red die and the other is rolling the green die.
   b. Find $E(X)$ and $\sigma^2(X)$.

4. A bag contains ten balls, three white and seven black. Three balls are selected at random from the bag without replacement. Let X = The number of white balls chosen.
   a. Write the distribution of X in tabular form and graph the distribution.
   b. Find $E(X)$ and $\sigma^2(X)$.

5. Spin the spinner of Figure 4.19. If a prime number comes up, you win that number of dollars. If a nonprime comes up, you lose that number of dollars. Let X = Your winnings.
   a. Write the distribution of X in tabular form and graph the distribution.
   b. Find $E(X)$.
   c. If you play the game 1000 times, what would be your expected winnings?

6. A manufacturer of integrated circuits knows from experience that 5% of the components he produces are defective. The quality control procedure is to reject an assembly-line run if three or more components out of a random sample of ten components are defective.
   a. What is the probability of rejecting an assembly-line run?
   b. If X = The number of defective components in a sample of size 10, what are $E(X)$ and $\sigma^2(X)$?

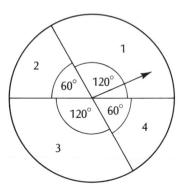

Figure 4.19

7. If a pair of balanced dice are rolled 3600 times, what are the expected number of occurrences of each of the possible sums: 2, 3, . . . , 12? See exercise 3.

8. The game of Over-and-Under is played with a pair of dice. There are three possible bets: (i) the sum of the dice is under 7 (paying even money), (ii) the sum of the dice is over 7 (paying even money), or (iii) the sum of the dice is 7 (paying 5-to-1). The dice are rolled. If the sum of the dice agrees with your bet, you win. Otherwise, you lose. Let the random variable X = Your winnings.
   a. Find $E(X)$ if you always bet under 7.
   b. Find $E(X)$ if you always bet over 7.
   c. Find $E(X)$ if you always bet equal to 7.
   d. Are any of these betting schemes advantageous to you?

9. A balanced dime is tossed 20 times and a balanced quarter is tossed 40 times. Let X = The total number of heads that come up. What are $E(X)$ and $\sigma^2(X)$?

10. Two bags each contain white and black balls. Bag A contains three white and two black balls. Bag B contains two white and two black balls. One ball is selected at random from bag A and its color noted; then it is placed in bag B. Then one ball is selected at random from bag B. Let X = The number of white balls selected from bag A, and Y = The number of white balls selected from bag B.
   a. Find the distributions of X and Y. (A tree diagram will be useful.)
   b. Are X and Y independent random variables?
   c. Find $E(X)$ and $\sigma^2(X)$.
   d. Find $E(Y)$ and $\sigma^2(Y)$.
   e. Find $E(X + Y)$ and $\sigma^2(X + Y)$.

Problems and Projects

11. Write a program to compute the mean of a set of numbers. Do not assume the number of numbers input to the program is known. Run the program with each set of numbers in exercise 1.

12. Write a program to compute the standard deviation of a set of numbers. Do not assume the number of numbers input to the program is known. Run the program with each set of numbers in exercise 1.

13. Write a program that inputs the distribution of a random variable X, and calculates and prints the mean and standard deviation of X. Test your program by running it with the distributions of exercises 3, 4, and 5.

CASE STUDY 4A   RANDOM NUMBERS AND SIMULATION

Random Experiments

If we roll a balanced die a number of times, we obtain a sequence of the digits 1 to 6 in a "mixed up" or random order. For example, we actually rolled a die 20 times and obtained the following sequence:

5, 6, 3, 3, 4, 4, 6, 6, 3, 2, 6, 5, 2, 4, 1, 5, 4, 5, 2, 4

We believe this sequence to be random because the numbers result from a random experiment. Suppose we need, for whatever reason, the results of 6000 rolls of a balanced die. If we had the time and the fortitude, we could roll our die 6000 times and count the number of times 1 appears, the number of times 2 appears, and so on. Since the die is balanced, we expect each number to appear about 1000 times. Suppose, however, we have neither the time nor the courage. Can we simulate 6000 rolls of a balanced die on a computer? Can we make a computer generate a sequence of 6000 digits from 1 to 6 in some "mixed up" or random order?

Certain random experiments, like tossing a die, can be simulated on a computer. Such **probabilistic** (or **nondeterministic**) **simulations** are important in many applications such as: the flow of traffic through an intersection, how an epidemic spreads through a population, the arrival of telephone calls at an exchange, the behavior of a nation's economy, how the space shuttle behaves under various wind conditions, and how a configuration of computer hardware will behave. In a simulation, a computer model of a system is developed. By studying the behavior of the model under varying conditions, we hope to learn how the real system will perform. It is preferable to study a computer model rather than the real system because it increases our understanding of the real system, and it is easier and less costly to manipulate the computer model.

Nondeterministic simulations are based on sequences of random numbers. Experimental methods of obtaining random sequences, like tossing a die 6000 times, are generally time-consuming and costly. They also have the undesirable property that the sequence of numbers is not reproducible. If some anomaly should appear in a simulation based on an experimentally generated sequence of random numbers, it would be almost impossible to duplicate the sequence of steps leading to the irregularity. It is, therefore, difficult to deduce the source of the error.

In this case study, we show how a computer can generate random sequences. In doing so, we can take advantage of the speed of the computer and, if need be, run exactly the same experiment several times.

Defining the notion of a sequence of random numbers is not easy because the concept means different things to different people. We define a sequence of random numbers as follows: A sequence S of integers between 1 and N is a sequence of **random numbers** if, when choosing a number at random from S, each number has an equal probability of being selected. For example, the sequence of numbers generated by tossing a die qualifies as a sequence of random numbers between 1 and 6 provided the die is balanced and the sequence is long.

Pseudorandom Numbers

Computer methods that produce sequences of random numbers are unlike the experimental method. A computer produces a sequence of random numbers in a well-determined order. What is random about such sequences is not their method of generation but rather their statistical properties. Computer-generated sequences of random numbers are, therefore, referred to as **pseudorandom numbers** to distinguish them from sequences generated by random experiments. We shall investigate some statistical properties of pseudorandom numbers in the exercises.

Many pseudorandom number generators are based on the following scheme. To generate the sequence $r_1, r_2, \ldots,$ let $r_1 = s$. The remainder of the sequence is computed by using the following congruence:

$$r_{i+1} = (kr_i + c) \bmod m \qquad (1)$$

Obviously, to use (1) we need to know the values of s, k, c, and m. How best to choose these parameters is a difficult question whose answer depends on considerations from number theory and the architecture of the computer on which the scheme is implemented. We will be content to experiment with several choices of these parameters.

First, consider $s = 1$, $k = 1$, $c = 3$, and $m = 7$. Then $r_1 = 1$ and (1) becomes

$$r_{i+1} = (r_i + 3) \bmod 7$$

Hence, $r_2 = (r_1 + 3) \bmod 7 = 4 \bmod 7 = 4$

$r_3 = (r_2 + 3) \bmod 7 = (4 + 3) \bmod 7 = 7 \bmod 7 = 0$

$r_4 = (r_3 + 3) \bmod 7 = (0 + 3) \bmod 7 = 3$

$r_5 = (r_4 + 3) \bmod 7 = (3 + 3) \bmod 7 = 6$

$r_6 = (r_5 + 3) \bmod 7 = (6 + 3) \bmod 7 = 2$

$r_7 = (r_6 + 3) \bmod 7 = (2 + 3) \bmod 7 = 5$

$r_8 = (r_7 + 3) \bmod 7 = (5 + 3) \bmod 7 = 1$

From r_8 onward the sequence repeats the pattern 1, 4, 0, 3, 6, 2, 5 infinitely often. All sequences of pseudorandom numbers generated by (1) have this property: After at most m numbers a sequence of numbers repeats. The length of the sequence that repeats is called the **cycle length** of the pseudorandom number generator.

As another example, choose $s = 0$, $k = 1$, $c = 6$, and $m = 8$. The congruence (1) becomes

$$r_{i+1} = (r_i + 6) \bmod 8$$

The sequence of numbers generated is

$$0, \quad 6, \quad 4, \quad 2, \quad 0, \quad 6, \quad 4, \quad 2, \quad 0 \ldots$$

This pseudorandom number generator has a cycle length of four.

A sequence that repeats itself is hardly what we can call random. The only way to avoid this, however, is to choose a very large value for m. A common value of m on computers with 32-bit registers is $2^{31} = 2,147,483,648$. With appropriate choices of s, k, and c the cycle length of the resulting pseudorandom number generator can be made equal to 2^{31}.

Another objection to sequences produced by (1) might be that they always start with the same number. This can be overcome by varying the value of s each time (1) is used. If k, c, and m are held constant, the way in which the sequence is generated is determined by the choice of s. Thus, s is frequently called the **seed** of the sequence. If, for example, we take $s = 3$ and $k = 1$, $c = 3$, and $m = 7$, as in our first example, the resulting sequence is

$$3, \quad 6, \quad 2, \quad 5, \quad 1, \quad 4, \quad 0, \quad 3 \ldots$$

This is the original sequence, but starting in a different place.

To this point we have seen how a long sequence of distinct pseudorandom numbers can be generated by making m in congruence (1) large. We have also seen how the beginning of the sequence may be varied by choosing different values for the seed s. We still, however, have a problem. The numbers generated might not be in the range of numbers required. For example, if we wish to simulate tosses of a die, our random numbers must be integers from 1 to 6, inclusive. As the first step around this range problem, we must normalize the random numbers produced by (1).

Normalizing a random number means converting that number to a number between 0 and 1. This can be done for numbers generated by (1) by dividing each number by m. The first pseudorandom number sequence we generated, 1, 4, 0, 3, 6, 2, 5, 1, . . . , would, if normalized, become

$$0.1429, \quad 0.5714, \quad 0.0000, \quad 0.4286, \quad 0.8571, \quad 0.2857, \quad 0.7143, \quad 0.1429, \ldots$$

We now must convert a normalized pseudorandom number r, $0 \le r \le 1$, to a number in the required range. In the die-tossing simulation, r must be converted to an integer between 1 and 6, inclusive. We illustrate the general procedure using this case: (1) multiply the pseudorandom number r by 6, obtaining a number between 0 and 5; (2) truncate the result, obtaining an integer between 0 and 5; (3) add 1, giving an integer between 1 and 6.

The normalized sequence of pseudorandom numbers thus obtained would yield:

	Truncate	Add 1
$6 \times 0.1429 = 0.8574$	0	1
$6 \times 0.5714 = 3.4284$	3	4
$6 \times 0.0000 = 0.0000$	0	1
$6 \times 0.4286 = 2.5716$	2	3
$6 \times 0.8571 = 5.1426$	5	6
$6 \times 0.2857 = 1.7142$	1	2
$6 \times 0.7143 = 4.2858$	4	5

Thus, this pseudorandom number generator simulates the sequence of rolls: 1, 4, 1, 3, 6, 2, 5, . . .

The random number generator (1) with $k = 25{,}173$, $c = 13{,}849$ and $m = 65{,}536$ (Scheid, 1982) provides an acceptable sequence of pseudorandom numbers.

$$r_{i+1} = (25{,}173r_i + 13{,}849) \bmod 65{,}536 \tag{2}$$

We now illustrate how the generator (2) can be used to simulate 6000 tosses of a die.

EXAMPLE 4.30

Write a Pascal program to simulate 6000 tosses of a balanced die, using the pseudo-random number generator (2). The program should print the number of times each number comes up in the simulation.

SOLUTION

```
program dieroll(input, output);
var i, seed, randnum, roll: integer;
    results: array [1..6] of integer;
    normalizedrandnum: real;
begin (*dieroll*)
   for i := 1 to 6 do
      results [i] := 0;
   readln (seed);
   for i:= 1 to 6000 do
      begin
         randnum := (25173 * seed + 13849) mod 65536 ;
         normalizedrandnum := randnum/65536;
         roll := trunc(6*normalizedrandnum) + 1;
         results[roll] := results[roll] + 1;
         seed := randnum
      end;
   for i := 1 to 6 do
      writeln (results[i]);
end
```

The program of Example 4.30 simulates rolls of a balanced die because the normalized pseudorandom numbers generated (the values of NORMALIZEDRANDNUM) are uniformly (or evenly) distributed between 0 and 1. The digits produced by the simulation program will, therefore, be uniformly distributed. That is, each digit will appear about 1/6 of the time. Recall this is precisely what we meant by a sequence of random numbers.

Remark   The preceding discussion gives the key to what we mean by a satisfactory pseudorandom number generator. Define a random variable R to be the value of the number generated by a given pseudorandom number generator. The random number generator is satisfactory if R has a uniform distribution.

The technique used in Example 4.30, that is, multiplying by 6 and adding 1 to obtain a number in the desired range, may be used only when the probability space of the experiment is equiprobable. How can we use a normalized pseudorandom number

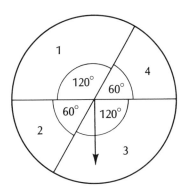

Figure 4.20

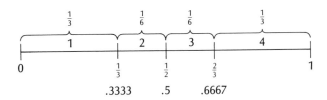

Figure 4.21

generator, which produces numbers uniformly in the interval from 0 to 1, in a simulation of a non–uniformly distributed random variable? We illustrate the method by considering how to simulate spins of the spinner depicted in Figure 4.20.

Let the random variable X = The number that comes up. The distribution of X is

X	1	2	3	4
P	$\dfrac{1}{3}$	$\dfrac{1}{6}$	$\dfrac{1}{3}$	$\dfrac{1}{6}$

To make a pseudorandom number generator simulate this disturbution, divide the
interval from 0 to 1 in the same ratio as given by the distribution of X and note the values
bounding the subintervals. See Figure 4.21.

As indicated in Figure 4.21, the first interval represents 1, the second 2, and so on.
If a number generated by the pseudorandom number generator lies in the interval i, the
number i has occurred in the simulation. If, for example, 0.4613 is generated, a 2 has
occurred.

EXAMPLE 4.31

Write a Pascal program to simulate 60 spins of the spinner in Figure 4.20. Use the pseudorandom number generator (2).

SOLUTION

```
Program spinner (input, output);
var i, seed, randnum :integer;
    normalizedrandnum: real;
begin (* spinner *)
   readln (seed);
   for i := 1 to 60 do
      begin
         randnum := (25173 * seed + 13849) mod 65536 ;
         normalizedrandnum := randnum/65536;
         if normalizedrandnum < 0.3333
            then writeln('1')
         else if normalizedrandnum < 0.5
            then writeln('2')
         else if normalizedrandnum < 0.6667
            then writeln('3')
         else
            writeln('4');
         seed := randnum;
         end;
end.
```

EXERCISES: CASE STUDY 4A

1. Using the pseudorandom number generator (1), write the first ten numbers produced for each of the following choices of s, k, c, and m.
   a. $s = 41$, $k = 1$, $c = 0$, $m = 100$
   b. $s = 73$, $k = 1$, $c = 5$, $m = 100$
   c. $s = 61$, $k = 91$, $c = 0$, $m = 100,000$

2. A scheme for generating a large number of pseudorandom numbers is suggested by F. Gruenberger and G. Jaffray (1965). Define

$$r_{i+1} = 2r_i \bmod 999999893$$

   and

$$s_{i+1} = 2s_i \bmod 999999883$$

   Let r_1 and s_1 be integers larger than 500,000,000. To obtain the pseudorandom number sequence $t_1, t_2, \ldots$ define $t_i = r_i + s_i$.
   a. Write a program to generate the sequence t_i.
   b. Using the generator of part a, simulate 6000 tosses of a balanced die (see exercise 7).

3. An old and unsatisfactory method of generating a pseudorandom number sequence is the mid-square, or inner-product method. We illustrate the method with 4-digit numbers. Start with an arbitrary 4-digit number, say $r_1 = 2222$. To obtain r_2, square r_1 and take the middle four digits of the answer as r_2. To illustrate:

$$r_1 = 2222 \qquad r_1^2 = 49\widetilde{3728}4$$
$$r_2 = 9372 \qquad r_2^2 = 87\widetilde{8343}84$$
$$r_3 = 8343 \qquad r_3^2 = 69\widetilde{6056}49$$
$$r_4 = 6056 \qquad \cdots\cdots$$

a. Write a program that generates 1000 numbers using the mid-square method and 4-digit numbers. The program should input the starting number r_1.

b. Why is the mid-square method unsatisfactory? To answer this question, it would help to run the program of part a several times with different r_1 values and investigate its outputs.

4. Using the pseudorandom number generator (2), write programs to simulate:

a. 1000 tosses of a balanced coin. The total number of heads and tails should be printed. What is the expected number of heads? How many heads actually occur in your simulation?

b. 6000 tosses of a balanced die (see Example 4.30). The total number of occurrences of each face should be printed. What is the expected number of 1s? How many 1s occur in your simulation?

c. 380 spins of a roulette wheel. The total number of occurrences of each number should be printed. What is the expected number of 22s? How many 22s occur in your simulation?

5. Write and run a program that simulates 360 spins of the spinner of Figure 4.22. The program should output only the total number of times each section comes up in the simulation. What is the expected number of occurrences of section 3 on the spinner? How many 3s occur in your simulation?

6. Consider the following game based on the spinner of exercise 5. Spin the spinner. If a prime number comes up, you win that number of dollars. If a nonprime comes up, you lose that number of dollars.

a. Write a program that simulates 360 plays of the game. The program should output your total earnings at the end of the 360 plays.

b. How does the output of the simulation compare to your expected winnings?

7. By each of the following methods, write a program that simulates rolling a pair of dice, one red and one green, 3600 times. Each program should calculate the sum of the dice on each simulated roll and make a tally of how many times each sum comes up. At the end of the simulation, the programs should print the total number of times each sum appears.

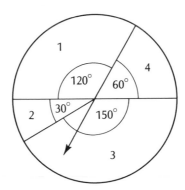

Figure 4.22

    a. The experiment can be thought of as a compound experiment—roll one die, then the other, and note their sum. Let $Z =$ The sum of the numbers that come up. Use a tree diagram to find the distribution of Z. Note that it is a nonuniform distribution. Base your program on the distribution of Z.

    b. Again note that the experiment is a compound experiment. This time, however, write a program that does the following: A roll of the red die is simulated; a roll of the green die is simulated; the sum of the two numbers is then computed.

    c. How do the outputs of parts a and b compare? How should they compare?

Problems and Projects

Exercises 9–12 involve several simple statistical tests that a sequence of pseudorandom *digits* should pass before being judged satisfactory. We assume the sequence of digits produced by the pseudorandom digit generator of exercise 8 is used as input to test each of the programs.

  8. Using the pseudorandom number generator (2), write a program fragment that produces a sequence of 1000 random digits (that is, a random sequence of digits 0, 1, . . . , 9).

  9. (The Frequency Test) If a pseudorandom digit generator is truly random, the expected frequency of occurrence of each digit should be 1/10. Thus, the expected number of occurrences of each digit in a sequence of 1000 random digits is 100.

    a. Write a program to count the number of occurrences of each digit produced by a pseudorandom digit generator. Test your program.

    b. How do the results of part a compare to the expected number of occurrences of each digit?

 10. (The Serial Test) The serial test counts the appearance of each combination of two digits, ranging from 00 to 99, in a sequence of random digits. For example, in the sequence 12312567, the pair 12 appears twice; 23, 31, 25, 56, and 67 each appear once.

    a. In a sequence of 1000 random digits, what is the expected number of occurrences of each pair of digits from 00 to 99?

    b. Write a program to count the number of occurrences of each combination of two digits in a sequence of pseudorandom digits. Test your program.

    c. Compare the results of parts a and b.

 11. (The Maximum Test) For three consecutive digits a *maximum* occurs when the center digit is greater than either of the two outside digits. For example, 271 is a maximum.

    a. In a sequence of 1000 random digits, what is the expected number of maximums?

    b. Write a program to count the number of maximums in a sequence of pseudorandom digits. Test your program.

    c. Compare the results of parts a and b.

 12. On the basis of exercises 9b, 10c, and 11c, how good is the pseudorandom number generator (2)? Note that a quantitative statistical test, called the chi-square (χ^2) test, may be used to answer 9b, 10c, and 11c. The reader is referred to any standard book in statistics for a discussion of the χ^2 test.

REFERENCES

Feller, W. *An Introduction to Probability Theory and Its Applications.* 3d ed. New York: Wiley and Sons, 1968.

Goldberg, A. *Probability: An Introduction.* Englewood Cliffs, N.J.: Prentice-Hall, 1964.

Gruenberger, F. and Jaffray, G. *Problems for Computer Solution.* New York: Wiley and Sons, 1965.

Rotando, L. M. *Finite Mathematics for Business, Social Sciences and Liberal Arts.* New York: D. Van Nostrand, 1980.

Scheid, F. *Computers and Programming.* New York: McGraw-Hill, 1982.

A quick glance at the table of contents of almost any book in computer programming will show a chapter on functions. What are functions and how can they be defined in a computer program? Many programming languages have the facility to define recursive functions. What is a recursive function?

If you have available several algorithms for solving a problem, how can you decide which is the most efficient one? Is there a way to quantitatively compare the efficiencies of algorithms?

When is a problem considered unsolvable, even on a large computer?

A code is a correspondence between letters of one alphabet and letters of another alphabet. When is a code well defined? When can an encoding be decoded? Can you decode this message:

AVKZ FEMVZ ZVHGIMV SGXLUH AEXLGA THUKWQ

In this chapter we shall study functions and learn the mathematics necessary to answer all these questions. We will discuss functions in general, as well as several special classes of functions. We also discuss graphs and limits of functions, and describe how to combine functions in various ways to produce new functions.

5.1 RELATIONS

Product Sets

An **ordered pair** consists of two elements, one of which is designated the first element and the other the second element. We write such a pair as (a, b). The element a is the first element and b is the second. Two ordered pairs (a, b) and (c, d) are equal, and we write $(a, b) = (c, d)$ if and only if $a = c$ and $b = d$, that is, if they have the same first element and the same second element. The ordered pair (b, a) is, therefore, different from the pair (a, b) because the first and second elements of each are different.

Let A and B be sets. The **product** of A and B, $A \times B$ (read "A cross B"), is the set of all ordered pairs with the first element from A and the second from B. Using set notation,

$$A \times B = \{(a, b) : a \in A, b \in B\}$$

RELATIONS AND FUNCTIONS

EXAMPLE 5.1

Find the products of the following sets:

a. $A \times B$ where $A = \{a, b\}$, $B = \{1, 2, 3\}$.
b. $R \times R$ where $R = \{$All real numbers$\}$.

SOLUTION

a. $A \times B = \{(a, 1), (a, 2), (a, 3), (b, 1), (b, 2), (b, 3)\}$
b. $R \times R$, sometimes written as R^2, is the set of all ordered pairs of real numbers. Recall from elementary algebra that points in the plane can be coordinatized by the familiar Cartesian coordinate system. See Figure 5.1.

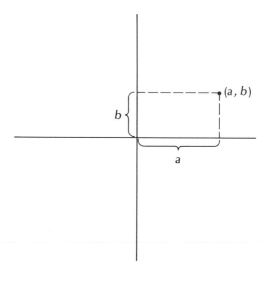

Figure 5.1

In such a coordinate system, each point in the plane is identified by an ordered pair of real numbers, called the "coordinates of the point." The first coordinate is the distance of the point from the vertical axis. The second coordinate is the distance of the point from the horizontal axis. Thus, geometrically, the set of all ordered pairs of real numbers, that is, $R \times R$, represents the entire plane.

We can plot ordered pairs even when their elements are not numbers. We illustrate the construction of such a **coordinate diagram** in Example 5.2.

EXAMPLE 5.2

Construct the coordinate diagram of $A \times B$ where $A = \{a, b\}$ and $B = \{1, 2, 3\}$.

SOLUTION

The diagram is given in Figure 5.2.

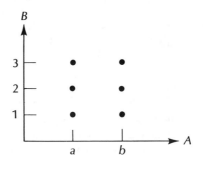

Figure 5.2

Assuming A and B are finite sets, there is a relationship between the number of elements in $A \times B$ and the number of elements in A and B as stated in Theorem 5.1. For a finite set S, $|S|$ denotes the number of elements in S.

Theorem 5.1

Let A and B be finite sets. Then

$$|A \times B| = |A| \cdot |B|$$

The proof is a simple application of the multiplication principle. Since there are $|A|$ ways of choosing a first element in an ordered pair and $|B|$ ways of choosing a second element, there are $|A| \cdot |B|$ ways of choosing ordered pairs.

Relations

A **relation** R from set A to set B is a subset of $A \times B$. If $(a, b) \in R$, we say a is related to b and we write $a \, R \, b$. If $(a, b) \notin R$, we say a is not related to b and we write $a \, \cancel{R} \, b$. To consider an example, let A be the set of all women and B the set of all men. $A \times B$ is the set of all possible women-men pairs. Let M be the relation defined by

$$M = \{(a, b) : a \text{ is married to } b\}$$

Thus, $a \, M \, b$ if and only if a is a woman married to man b. We read $a \, M \, b$ as "a is married to b."

As another example, let Z be the set of all integers. Define the relation F on $Z \times Z$ by

$$F = \{(a, b) : a \equiv b \bmod 5\}$$

Thus, $a \, F \, b$ if and only if $a \equiv b \bmod 5$. Hence, $8 \, F \, 3$, but $9 \, \cancel{F} \, 6$.

The **domain** of a relation R is the set of all first elements of the ordered pairs belonging to R. The **range** of a relation R is the set of all second elements of the ordered pairs belonging to R. The domain of the relation M (just defined) is the set of all married women, and the range is the set of all married men.

When the domain and range of a relation are finite sets, there are two ways of depicting the relation graphically. The **arrow diagram** of a relation is formed by enclosing the elements of the domain in an oval on the left and the elements of the range in an oval on the right. When an element of the domain is related to an element of the range, we draw an arrow from the element of the domain to the corresponding element of the range. We can draw the coordinate diagram or **graph** of a relation by drawing a diagram as we did in Example 5.2. However, we plot only those points belonging to the relation.

EXAMPLE 5.3 _____

Draw the arrow diagram and graph of the relation R from A to B where $A = \{1, 2, 3, 4, 5\}$, $B = \{a, b, c, d\}$, and $R = \{(3, b), (2, a), (2, b), (1, a), (4, c)\}$.

SOLUTION

The arrow diagram is shown in Figure 5.3(a) and the graph is shown in Figure 5.3(b).

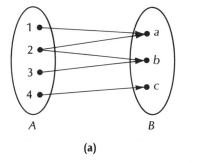

(a)

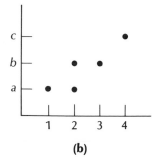

(b)

Figure 5.3

Equivalence Relations

A relation R from A to A is an **equivalence relation** on A if it has the following properties:

  i. $a R a$ for all a in the domain of R **(Reflexive Property)**

  ii. If $a R b$, then $b R a$ **(Symmetric Property)**

  iii. If $a R b$ and $b R c$, then $a R c$ **(Transitive Property)**

EXAMPLE 5.4

Which of the following are equivalence relations:

a. Let U be any set and let A and B be subsets of U. Define S by $A S B$ if and only if $A \subset B$.

b. Let Z be the set of integers. Define F by $a F b$ if and only if $a \equiv b$ mod 5.

SOLUTION

a. S is not a equivalence relation on U. S is reflexive because every set is a subset of itself. S is transitive: If $A \subset B$ and $B \subset C$, then $A \subset C$. However, S is not symmetric because if $A \subset B$, then $B \subset A$ only if $A = B$.

b. For the relation F on Z, the reflexive, symmetric, and transitive properties become:

  i. $a \equiv a$ mod 5 for any integer a

  ii. If $a \equiv b$ mod 5, then $b \equiv a$ mod 5

  iii. If $a \equiv b$ mod 5 and $b \equiv c$ mod 5, then $a \equiv c$ mod 5

These properties were established in Theorem 1.4 for any modulus m, not only $m = 5$. Theorem 1.4 can, therefore, be restated as follows: "Congruence modulo m is an equivalence relation."

An equivalence relation R on a set A is called an *equivalence* relation because it imposes a sameness or equivalence on the elements of A. An equivalence relation R on A partitions A into disjoint subsets A_i called **equivalence classes** such that

$$A = \bigcup_{i=1}^{n} A_i$$

These equivalence classes A_i have the property that $a R b$ if and only if a and b belong to the same equivalence class.

For example, the equivalence relation F of Example 5.4 partitions the set of integers Z into five disjoint classes Z_0, Z_1, Z_2, Z_3, Z_4, where Z_i is the set of integers with last residue i. Thus, Z_0 through Z_4 are the residue classes modulo 5. Two integers in the same residue class are equivalent in that they are congruent modulo 5.

If R is an equivalence relation on the set A, each element $a \in A$ determines an equivalence class, denoted by $[a]$, defined as follows: $[a] = \{x : x \in A, a R x\}$. $[a]$

is, therefore, the set of all elements in A equivalent to a. The set of all equivalence classes $[a]$ is called the **quotient set** of A by R and is denoted by A/R. Thus, if $\mathscr{Z}$ is the set of integers and F is the equivalence relation of Example 5.4, $[i] = \mathscr{Z}_i$ and $Z/F = \{Z_0, Z_1, Z_2, Z_3, Z_4\}$.

Theorem 5.2   The Partition Theorem

Let A be any set and R an equivalence relation on A. Then the quotient set of A by R is a partition of A; that is, every element of A belongs to an equivalence class and if $[a] \neq [b]$, then $[a]$ and $[b]$ are disjoint.

We now prove Theorem 5.2. To show that every element of A is in an equivalence class, we have $a \in [a]$ for all $a \in A$ because of the reflexive property of equivalence relations. Now, suppose $[a] \neq [b]$. If $[a]$ and $[b]$ are not disjoint, there is an element c such that $c \in [a]$ and $c \in [b]$. Since $c \in [a]$, $c\,R\,a$. By the symmetric property of equivalence relations, $a\,R\,c$. Thus, we have $a\,R\,c$ and, because $c \in [b]$, $c\,R\,b$. By the transitive property of equivalence relations, it follows that $a\,R\,b$. But, if $a\,R\,b$, then $[a] = [b]$. This contradicts our assumption that $[a] \neq [b]$, establishing our theorem.

As a last example, let S be the set of all fractions a/b, where a and b are integers, $b \neq 0$. We define the relation R on S by $(a/b)\,R\,(c/d)$ if and only if $ad = bc$. We leave it as an exercise to show that R is an equivalence relation. If $(a/b)\,R\,(c/d)$, then a/b and c/d are said to be **equivalent fractions**. Since R is an equivalence relation, it partitions the set S of all fractions into equivalence classes. Each fraction in a given equivalence class represents the same rational number. For example, the fractions 1/2, 2/4, and 9/18 are all in the same equivalence class. Each such fraction represents the rational number one-half. Thus, a rational number can be thought of as an equivalence class of the relation R.

EXERCISES 5.1

1. Let $A = \{a\}$, $B = \{1, 2, 3, 4\}$, $C = \{x, y\}$. List the elements of each of the following product sets and draw their coordinate diagrams.
   a. $A \times B$ **b.** $B \times B$ **c.** $B \times C$ **d.** $C \times B$
2. Which of the following relations are equivalence relations?
   a. Two triangles are related if they are congruent.
   b. Two triangles are related if they are similar.
   c. Two people are related if they have the same parents.
   d. Two straight lines are related if they are perpendicular.
3. Which of the following relations defined on the integers are equivalence relations?
   a. $a\,R\,b$ if and only if $a = 2^k \cdot b$ for some integer k.
   b. $a\,S\,b$ if and only if $a|b$ or $b|a$.
   c. $a\,T\,b$ if and only if $a + b$ is divisible by 3.

4. Describe the partition of the integers—that is, the equivalence classes—determined by each of the equivalence relations of exercise 3.

5. It is stated in the text that if R is an equivalence relation on A, then $a R b$ if and only if $[a] = [b]$. Write a detailed proof of this statement.

6. The concept of the product of sets can be extended to products of three or more sets. Let $A_1, A_2, \ldots, A_n$ be sets. Then $A_1 \times A_2 \times \cdots \times A_n = \prod_{i=1}^{n} A_i$ consists of all ordered n-tuples of elements, the first from A_1, the second from $A_2, \ldots,$ the nth from A_n.

   a. Using the sets A, B, C defined in exercise 1, find $A \times B \times C$.

   b. How many elements are in $A_1 \times A_2 \times \cdots \times A_n$? That is, generalize Theorem 5.1 to the product of an arbitrary number of sets.

   c. An ***n*-ary relation** is a subset of a product of n sets. Relations, as we defined them in this section, are, therefore, 2-ary or binary relations. Let Z be the set of integers. Define the ternary relation S on $Z \times Z \times Z$ as follows: $S = \{(a, b, c) : c = a + b\}$. List five elements from S.

Problems and Projects

7. A relation R on a set A is **antisymmetric** if $a R b$ and $b R a$ imply $a = b$. A relation R on a set A is called a **partial ordering** on A if it is reflexive, antisymmetric, and transitive. A partial ordering is usually denoted by $\preceq$ and is read as "a precedes b." Thus $\preceq$ is a partial ordering of A if

   i. $a \preceq a$ for all a in A.

   ii. $a \preceq b$ and $b \preceq a$ implies $a = b$.

   iii. $a \preceq b$ and $b \preceq c$ implies $a \preceq c$.

   a. Show that the relation "less or equal" ($\leq$) on the set of real numbers is a partial ordering.

   b. Let U be any set. Show that the relation "is a subset of," $\subseteq$, on $\mathscr{P}(U)$ is a partial ordering.

   c. Let N be the set of natural numbers. Show that the relation "a divides b," $|$, is a partial ordering on N.

8. A partial ordering $\preceq$ on a set A is called *partial* because it is possible for two elements of A to be incomparable. That is, it is possible that there are two elements a and b such that neither $a \preceq b$ nor $b \preceq a$.

   a. Find such incomparable pairs for the partial orderings of exercise 7(b) and 7(c).

      A partial ordering $\preceq$ on a set A is called a **total ordering** on A if every two elements of A are comparable; that is, if a, $b \in A$, then either $a \preceq b$ or $b \preceq a$. The elements of a set A on which a total ordering is defined can be arranged in a linear sequence. For this reason, a total ordering is sometimes called a "linear ordering."

   b. Show that the relation of Example 7(a) is a total ordering.

   c. Let Z be the set of integers. Define the relation "is a multiple of" on Z. Is this relation a linear ordering? A partial ordering?

   d. Let W be the set of all words in the English language. Define the relation L on W by: $w_1 L w_2$ if and only if w_1 precedes w_2 alphabetically. Is L a partial ordering? A linear ordering?

5.2   FUNCTIONS

One of the most important concepts in all of computer science and mathematics is that of a function. In this section we define the concept of a function, introduce function notation, and consider several examples.

A **function** (or **mapping**) from A into B is a relation from A to B in which no two distinct ordered pairs have the same first element. A is called the **domain** of the function and B is called the **codomain** of the function. For a relation to be a function it *cannot* contain ordered pairs like $(1, a)$ and $(1, c)$, because they have the same first element. An alternate and useful way of defining a function is the following: A **function** is a rule that assigns to each element of a set (the domain) a unique element in another set (the codomain).

The domain and codomain of a function can be sets of numbers or sets of other types of objects, such as character strings (a **character string** is a sequence of letters, digits, and special characters such as the period, comma, and the blank). Functions are usually given names like f, g, h, SQR, MAX, and so on.

EXAMPLE 5.5

Let the domain of a function f be $V = \{1, 2, 3, 4, 5\}$ and the codomain be $W = \{a, b, c, d\}$. We can specify the assignment rule as shown in Figure 5.4(a). Note that the element c of the codomain is not assigned to any element of the domain.

There are several ways of stating that f assigns to an element of the domain an element of the codomain. The following statements are all equivalent:

a. f **maps** 1 into b.

b. The **value** of f at 1 is b.

c. The **image** of 1 under f is b.

Since a function f is a relation, it has a **range**—the set of all second elements of the ordered pairs of f. Using the terminology of (c), however, we can describe the range of f as the set of all elements in the codomain that are images of elements in the domain.

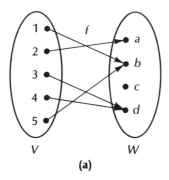
(a)

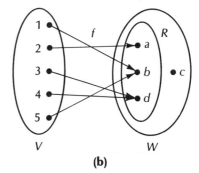
(b)

Figure 5.4

In Example 5.5, the domain is $\{1, 2, 3, 4, 5\}$, the codomain is $\{a, b, c, d\}$, and the range is $\{a, b, d\}$. The relationship between the codomain and range is illustrated in Figure 5.4(b).

Finally, when focusing on the whole domain rather than a single element, we say that f **maps** V **into** W. This is denoted by $f : V \rightarrow W$.

Note 5.1 gives the three important points to note about the function f of Example 5.5 and about functions in general.

Note 5.1

1. The assignment rule of a function must assign to each element of the domain a *unique* element of the codomain. Thus, an assignment such as that shown in Figure 5.5 is *not* allowed.

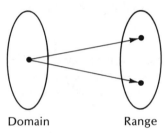

Domain   Range

Figure 5.5

2. Two different elements in the domain *can* be assigned to the same element of the codomain. For example, 1 and 5 both get mapped to b under the function f of Example 5.5.

3. There *may* be some elements in the codomain that have *no* element mapped to them. The element c in the codomain of the function f of Example 5.5 has no elements mapped to it.

The assignment rule for the function f of Example 5.5 can be given using function notation. For example, that f maps 1 to b is written $f(1) = b$ and is read "f of 1 equals b." The other assignments may be written as $f(2) = a$, $f(3) = d$, $f(4) = d$, $f(5) = b$. Function notation is most useful for a function whose domain is an infinite set. Consider the function g that maps each real number into its square. Thus, for example, $g(-2) = 4$, $g(3) = 9$, and so on. Because the domain of g is infinite, we cannot draw a simple diagram, nor can we list all the assignments as we did for the function f. We can, however, state the assignment rule more compactly in the form of an equation using function notation as follows: $g(x) = x^2$. This equation states that the function g assigns to the number x the number x^2. The letter x is referred to as the **variable** or **argument** of the function.

Recalling that a function is a relation, we can write the function g as follows:

$$g = \{(x, y) : y = x^2\}$$

or

$$g = \{(x, x^2) : x \text{ is a real number}\}$$

EXAMPLE 5.6

If $h(x) = 2x^2 - x + 3$, find $h(-3)$, $h(0)$, and $h(4)$.

SOLUTION

To calculate the value of a function at a given value of x, substitute the value of x into the equation that defines the function:

$$h(-3) = 2(-3)^2 - (-3) + 3 = 18 + 3 + 3 = 24$$
$$h(0) = 2(0)^2 - (0) + 3 = 0 - 0 + 3 = 3$$
$$h(4) = 2(4)^2 - (4) + 3 = 32 - 4 + 3 = 31$$

Note 5.2 discusses the domain and codomain of a function.

Note 5.2

Whenever the domain of a function is not specified, we shall assume the domain is the largest set of real numbers for which the function "makes sense." The function $f(x) = \sqrt{x}$, for example, has domain the set of nonnegative real numbers. The function $g(x) = 1/(x - 1)$ has domain the set of all real numbers except $x = 1$ because division by zero is undefined.

We also assume that the codomain of a function is the set of real numbers whenever the codomain of the function is not explicitly stated. Determining the range of a function is often a difficult problem and requires techniques we shall not discuss in this text. We will, however, specify the range when it is easy to find.

Some functions have "split" definitions like the one in Example 5.7.

EXAMPLE 5.7

If

$$f(x) = \begin{cases} x^2 + 1 & \text{if } x < 0 \\ 0 & \text{if } x = 0 \\ 1 & \text{if } 0 < x \le 1 \\ -x & \text{if } x > 1 \end{cases}$$

find $f(-2)$, $f(0)$, $f(1/2)$, $f(1)$, and $f(2)$.

SOLUTION

The rule for computing the function values of f varies depending on the value of x. To find $f(-2)$, we note that $x = -2 < 0$, so we must use the first part of the definition of f, $f(x) = x^2 + 1$. Therefore, $f(-2) = (-2)^2 + 1 = 4 + 1 = 5$.

To find $f(0)$ we use the part of the definition of f that is valid for $x = 0$, namely $f(0) = 0$. Since $0 < 1/2 \leq 1$, $f(1/2) = 1$. Since $0 < 1 \leq 1$, $f(1) = 1$. Since $2 > 1$, $f(2) = -2$.

We now consider several examples of functions in which either the domain or range (or both) are not sets of numbers.

EXAMPLE 5.8

Let the function LEN assign to a character string the number of characters in the string. We shall enclose character strings in quotes. The quotes, however, are not part of the character string.

a. Describe the domain and range of LEN.

b. Find LEN("COMPUTER MATHEMATICS") and LEN("ALFRED E. NEWMAN").

SOLUTION

a. The domain of LEN is the set of all possible character strings. The range of LEN is the set of nonnegative integers.

b. LEN("COMPUTER MATHEMATICS") = 20 and LEN("ALFRED E. NEWMAN") = 16. Remember that a blank space is a character.

EXAMPLE 5.9

Let the function ALPH assign to an integer between 1 and 26 the corresponding lowercase letter of the alphabet.

a. Describe the domain and range of ALPH.

b. Find ALPH(6) and ALPH(22).

SOLUTION

a. The domain of ALPH is the set $\{1, 2, \ldots, 26\}$. The range of ALPH is the alphabet $\{a, b, \ldots, z\}$.

b. ALPH(6) = "f", and ALPH(22) = "v".

EXAMPLE 5.10 ───────────────────────────────────

Let the function FIRSTCHR assign to a character string its first character.

a. Describe the domain and range of FIRSTCHR.

b. Find FIRSTCHR("NEWMAN"), FIRSTCHR("L"), FIRSTCHR("1983").

SOLUTION

a. The domain of FIRSTCHR is the set of all character strings. The range of FIRSTCHR is the set of all alphanumeric characters.

b. FIRSTCHR("NEWMAN") = "N", FIRSTCHR("L") = "L", FIRSTCHR-("1983") = "1".

───

The domain of a function can be any set, even a product set. If a function $f: A \times B \to C$, then f assigns to each ordered pair $(a, b) \in A \times B$ an element of C. Such a function is called a "function of two variables." We now consider three examples.

EXAMPLE 5.11 ───────────────────────────────────

Consider the function $k(x, y) = 2x + y$. Find $k(0,0)$, $k(3, -1)$, and $k(5,9)$.

SOLUTION

The function k maps a pair of real numbers to a real number. Thus, $k: R \times R \to R$. To find the value of the function for a given pair of values x and y, replace x and y in the equation defining k by their given values.

$$k(0,0) = 2(0) + 0 = 0$$
$$k(3,1) = 2(3) + (-1) = 5$$
$$k(5,9) = 2(5) + 9 = 19$$

───

EXAMPLE 5.12 ───────────────────────────────────

Consider the function $MAX(x, y)$ whose value is the larger of the real numbers x and y. Find $MAX(0,0)$, $MAX(3, -1)$, and $MAX(5,9)$.

SOLUTION

Like the function k of Example 5.11, $MAX: R \times R \to R$. $MAX(0,0) = 0$, $MAX(3, -1) = 3$, and $MAX(5,9) = 9$.

───

EXAMPLE 5.13 _____

Consider the function CONCAT(S, T), which concatenates two character strings S and T. For example, if $S =$ "HEL" and $T =$ "LO", CONCAT(S, T) = "HELLO".

a. Describe the domain and range of CONCAT.

b. If $S =$ "COMP", $T =$ "U",   and $U =$ "TER",   find   CONCAT(S, T), CONCAT(T, S), and CONCAT(S, U).

SOLUTION

a. The domain of CONCAT is the set of all ordered pairs of character strings. The range is the set of all character strings. Thus, if $\mathscr{S}$ is the set of all character strings,

$$\text{CONCAT}: \mathscr{S} \times \mathscr{S} \to \mathscr{S}$$

b. We find CONCAT(S, T) = "COMPU", CONCAT(T, S) = "UCOMP", and CONCAT(S, U) = "COMPTER"

_____

EXERCISES 5.2

1. State whether or not each diagram in Figure 5.6 defines a function from $D = \{1, 2, 3, 4\}$ into $C = \{5, 6, 7, 8, 9\}$. If the diagram is not that of a function, explain why.

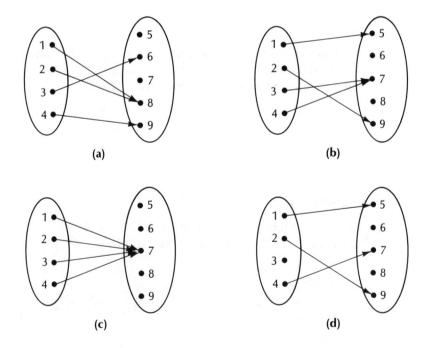

(a)   (b)

(c)   (d)

Figure 5.6

2. Write each of the following function rules as an equation using function notation:
   a. f assigns to each real number 1 more than twice the number.
   b. g assigns to each negative real number the number 1 and to each nonnegative real number 2 more than the number.
   c. A is the total amount of an investment of $1000 earning 6% simple interest per year for x years.
3. Let $f(x) = x^3$ and $g(x) = -x^2 + 3x - 2$. Find $f(-4)$, $f(0)$, $f(2)$, $g(-3)$, $g(0)$.
4. Consider the function $\text{ABS}(x) = |x|$, the absolute value of x.
   a. What are the domain and range of $\text{ABS}(x)$?
   b. Find $\text{ABS}(-6)$, $\text{ABS}(-2)$, $\text{ABS}(0)$, $\text{ABS}(7)$.
5. Consider the function $\text{SGN}(x)$, defined as follows:

$$\text{SGN}(x) = \begin{cases} +1 & \text{if } x > 0 \\ 0 & \text{if } x = 0 \\ -1 & \text{if } x < 0 \end{cases}$$

   a. What are the domain and range of $\text{SGN}(x)$?
   b. Find $\text{SGN}(-11)$ and $\text{SGN}(14)$.
6. Define the function $\text{WHAT}(x)$ as follows: $\text{WHAT}(x) = x \cdot \text{SGN}(x)$.
   a. Find $\text{WHAT}(-6)$, $\text{WHAT}(-2)$, $\text{WHAT}(0)$, and $\text{WHAT}(7)$.
   b. What familiar function is $\text{WHAT}(x)$?
7. Define the function of two variables $\text{INDEX}(C, S)$ as follows: The variable C can be any single character; S can be any character string. $\text{INDEX}(C, S)$ is the position in string S of the first occurrence of the character C. If S does not contain the character C, the value of INDEX is 0. Let $S =$ "MISSISSIPPI" and $T =$ "BULLETIN". Find $\text{INDEX}(\text{"I"}, S)$, $\text{INDEX}(\text{"S"}, S)$, $\text{INDEX}(\text{"I"}, T)$, $\text{INDEX}(\text{"U"}, S)$, and $\text{INDEX}(\text{"U"}, T)$.
8. Let A be any finite set. The function $\text{CARD}(A)$, the cardinality of A, is defined as the number of elements of A.
   a. Find $\text{CARD}(D)$, where $D = \{\text{All cards in a standard deck}\}$.
   b. Find $\text{CARD}(S)$, where $S = \{x : x \text{ is an integer}, 10 \le x \le 20\}$.
   c. If $\text{CARD}(B) = 0$, what is B?
9. The range of the function LEN, defined in Example 5.8, is the set of non-negative integers. For what character string S is $\text{LEN}(S) = 0$?
10. Let

$$f(x) = \frac{2x + 1}{x - 1}$$

   a. What is the domain of f?
   b. Find $f(-3)$, $f(0)$, $f(4)$, $f(100)$, $f(1000)$, and $f(10000)$.
   c. Based on the results of part b, what do you think is happening to the values of $f(x)$ as x gets larger and larger?
11. Let $f(x) = \sqrt{4x - 1}$.
   a. What is the domain of f?
   b. Find $f(1/2)$, $f(3)$, and $f(9)$.

5.3   GRAPHS OF FUNCTIONS

In Section 5.1 we introduced the idea of the coordinate diagram or graph of a relation, and hence a function, when the domain and range are finite sets. If the domain and range of a function are infinite subsets of the real numbers, the simple technique described in Section 5.1 to construct the graph of a function is not adequate. In this section we describe more general techniques for drawing the graph of a function whose domain and range are subsets of R.

Since a function is a relation, it determines a set of ordered pairs of real numbers. The function $f(x) = 3x + 2$ is the set of ordered pairs

$$f = \{(x, y): y = 3x + 2\}$$

Each ordered pair of real numbers in a function can be considered as the coordinates of a point in a Cartesian coordinate system and, therefore, can be plotted as a point. Figure 5.7 shows three ordered pairs of the function f just defined, plotted on a coordinate system.

The set of all ordered pairs of the function f plotted in a Cartesian coordinate system is called the **graph of f.**

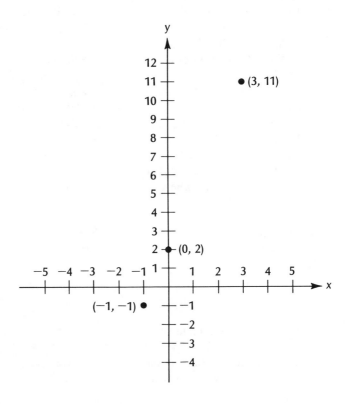

Figure 5.7

The graph of a function f is equivalent to the graph of the equation $y = f(x)$ as described in elementary algebra. (An *equation* is a statement that two expressions are equal. As we shall soon see, an equation does not necessarily determine a function.) Therefore, the graph of $f(x) = 3x + 2$ is the same as the graph of $y = 3x + 2$. This is the graph of the equation of the straight line having slope 3 and y-intercept 2. The graph of $f(x) = 3x + 2$ is, therefore, the straight line depicted in Figure 5.8.

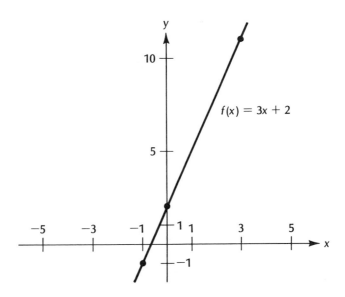

Figure 5.8

EXAMPLE 5.14 _____

Draw the graph of $f(x) = x^2$.

SOLUTION

The graph of $f(x) = x^2$ is the same as the graph of the equation $y = x^2$. To draw the graph of a function, calculate the coordinates of several points that lie on the graph by using the equation of the function. Plot the resulting points. Then, draw a smooth curve connecting the plotted points.

The calculation can best be arranged in tabular form. It is usually desirable to use a good mix of values for x. Therefore, we choose some negative as well as some positive values for x. The resulting graph is depicted in Figure 5.9.

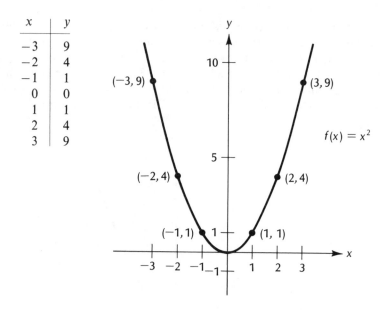

x	y
-3	9
-2	4
-1	1
0	0
1	1
2	4
3	9

Figure 5.9

It is important to note that not all equations in x and y determine functions. For example, $y^2 = x$ does not define a function. The relation defined by $y^2 = x$, $\{(x, y): y^2 = x\}$, contains the ordered pairs $(9, -3)$ and $(9, 3)$ and so is not a function.

Some graphs can be recognized as the graphs of functions by using the simple geometric test given in Note 5.3.

Note 5.3

For a graph to be the graph of a function, any given vertical line can intersect the graph in at most one point.

If, therefore, the number b is not in the domain of the function, the vertical line $x = b$ will not intersect the graph (why?). Also, if the number b is in the domain of the function, the vertical line $x = b$ intersects the graph exactly once (why?).

Figure 5.10 shows the graph of $y^2 = x$. Note that some vertical lines intersect the graph twice. Therefore, the graph is not the graph of a function.

EXAMPLE 5.15 _____

Which of the graphs of Figure 5.11 are graphs of functions?

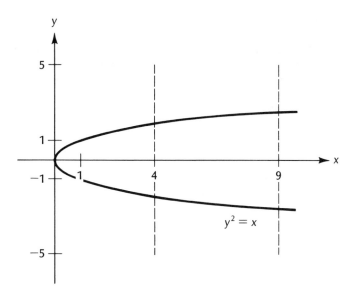

Figure 5.10

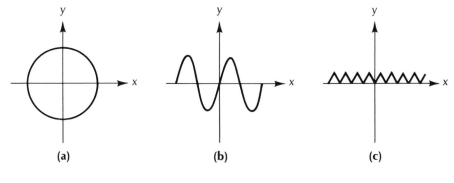

Figure 5.11

SOLUTION

The graph in (*a*) is not the graph of a function because there are vertical lines (for example, the *y*-axis) that intersect the graph twice. Graphs (b) and (c) are graphs of functions.

In Example 5.16 we introduce the so-called closed-dot and open-dot convention for drawing graphs.

EXAMPLE 5.16

Graph the function FLOOR(x), where FLOOR(x) is defined as the largest integer not exceeding x. Thus, FLOOR(7.34) = 7, FLOOR(-2.9) = -3, and FLOOR(5) = 5.

SOLUTION

If $n \leq x < (n + 1)$, where n is an integer, then

$$\text{FLOOR}(x) = n$$

Thus, the function FLOOR equals the constant n in the interval between the integers n and $n + 1$. Then, the value of FLOOR jumps to $n + 1$ when $x = n + 1$. The graph of $y = \text{FLOOR}(x)$ is shown in Figure 5.12.

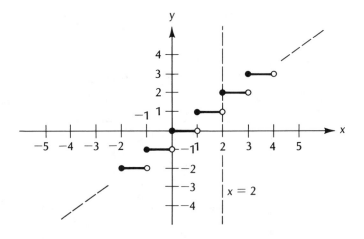

Figure 5.12

Note that the left endpoint of each line segment has a closed dot and the right endpoint an open dot. The closed dot means that the left endpoint of the segment is on the graph. The open dot means that the right endpoint of the segment is not on the graph. It follows that the vertical line at $x = 2$, shown by the dashed line in Figure 5.12, intersects the graph in only one point, namely $(2, 2)$.

EXAMPLE 5.17

Graph the function

$$f(x) = \begin{cases} x^2 + 1 & \text{if } x < 0 \\ 0 & \text{if } x = 0 \\ 1 & \text{if } 0 < x \leq 1 \\ -x & \text{if } x > 1 \end{cases}$$

SOLUTION

Using the open and closed dot notation of the previous example, we obtain the graph of Figure 5.13.

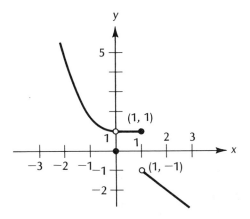

Figure 5.13

Important in many applications are the points where the graph of a function intersects the axes, if any such points exist. The **y-intercept** of a function f is the value of y where the graph of the function $y = f(x)$ intersects the y-axis. Note 5.4 explains how to find the y-intercept.

Note 5.4

To find the y-intercept of the function $y = f(x)$, let $x = 0$ and solve for y: $y = f(0)$.

The **x-intercepts,** or **zeros,** of a function f are those x values where the graph of $y = f(x)$ intersects the x-axis.

Note 5.5

To find the x-intercepts of a function $y = f(x)$, set $y = 0$ and solve the resulting equation: $f(x) = 0$.

Finding the zeros of a function is, in general, a very difficult problem requiring sophisticated mathematical techniques. In this book we shall discuss only those functions whose zeros may be found using the techniques of elementary algebra. Finding the zeros of $y = f(x)$ by solving $f(x) = 0$ is frequently called "finding the roots of the equation $f(x) = 0$."

EXAMPLE 5.18

Find the y-intercept and zeros of:

a. $f(x) = 3$ b. $g(x) = 2x - 1$ c. $h(x) = x^2 - 3x + 2$

SOLUTION

a. $f(x) = 3$ has y-intercept 3 and no zeros.
b. The y-intercept of $g(x) = 2x - 1$ is $g(0) = -1$. To find the zeros of $g(x)$, we solve $2x - 1 = 0$, obtaining $x = 1/2$.
c. The y-intercept of $h(x) = x^2 - 3x + 2$ is $h(0) = 2$. To find the x-intercepts, we solve $x^2 - 3x + 2 = 0$. Factoring, we obtain $(x - 2)(x - 1) = 0$. Thus, either $x = 2$ or $x = 1$, and $h(x)$ has zeros 2 and 1.

As we draw the graphs of functions, it becomes obvious that the y values of certain functions sometimes decrease as x gets larger, and sometimes increase as x gets larger. For example, as illustrated in Figure 5.14, the function $f(x) = x^2$ decreases as x gets larger through negative values of x, becomes 0 at $x = 0$, and then increases as x gets larger through positive values of x.

Of particular importance are those functions that always increase or always decrease. We assume a and b are in the domain of f. A function $y = f(x)$ is **increasing** if whenever $a < b$, $f(a) < f(b)$. The function $y = f(x)$ is **nondecreasing** if

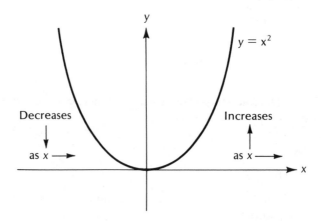

Figure 5.14

whenever $a < b$, $f(a) \leq f(b)$. Similarly, $y = f(x)$ is **decreasing** if whenever $a < b$, $f(a) > f(b)$. Finally, $y = f(x)$ is **nonincreasing** if whenever $a < b$, $f(a) \geq f(b)$.

EXAMPLE 5.19

Draw the graph of $f(x) = x^3$ and prove it is an increasing function.

SOLUTION

The graph is shown in Figure 5.15. It should be clear from this that $f(x) = x^3$ is an increasing function. To *prove* it, however, requires a bit of algebraic manipulation. Assume a and b are real numbers with $a < b$. We must show that $f(a) < f(b)$ or $a^3 < b^3$. If a and b have opposite signs, b must be positive and a negative; so b^3 is positive and a^3 negative. In this case, therefore, $a^3 < b^3$. Now assume a and b have the same sign and consider $b^3 - a^3 = (b - a)(b^2 + ab + a^2)$. Multiply out the right side to verify the factorization. The second factor on the right must be positive (why?). Also, the first factor on the right is positive because $b > a$. Thus, $b^3 - a^3 > 0$ or $a^3 < b^3$.

Besides increasing or decreasing, another property you may notice from the graphs of certain functions is symmetry. There are many kinds of symmetry, but we shall discuss only the two simplest types. A function $y = f(x)$ is an **even function**

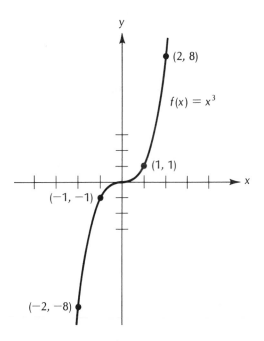

Figure 5.15

if for every value a in its domain $f(-a) = f(a)$. Therefore, for an even function f, whenever the point $(a, f(a))$ is on the graph, so is the point $(-a, f(a))$. This means that the graph of an even function is symmetric with respect to the y-axis. A function $y = f(x)$ is an **odd function** if for every value a in its domain $f(-a) = -f(a)$. Therefore, for an odd function f, whenever the point $(a, f(a))$ is on the graph, so is the point $(-a, -f(a))$. This means that the graph of an odd function is symmetric with respect to the origin.

EXAMPLE 5.20

a. Show that $f(x) = x^2 + 1$ is an even function.

b. Show that $g(x) = x^3 - x$ is an odd function.

SOLUTION

a. To show $f(x) = x^2 + 1$ is even, we must show that $f(-a) = f(a)$ for any real number a. To see this, note that $f(-a) = (-a)^2 + 1 = a^2 + 1 = f(a)$, because $(-a)^2 = a^2$. The symmetry of the graph of f about the y-axis is evident from Figure 5.16(a).

b. To show that $g(x) = x^3 - x$ is an odd function, we must show $g(-a) = -g(a)$ for any real number a. We have

$$g(-a) = (-a)^3 - (-a) = -a^3 + a = -(a^3 - a) = -g(a)$$

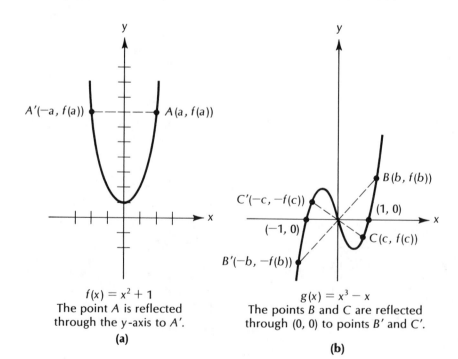

$f(x) = x^2 + 1$
The point A is reflected through the y-axis to A'.
(a)

$g(x) = x^3 - x$
The points B and C are reflected through $(0, 0)$ to points B' and C'.
(b)

Figure 5.16

by the usual sign rules of algebra. The graph of $g(x)$, shown in Figure 5.16(b), illustrates its symmetry through the origin.

EXERCISES 5.3

1. Draw the graph of the function $f(x) = x^4$.
2. Draw the graph of the function $g(x) = -x^2 + 3x - 2$.
3. Draw the graph of the function $\text{ABS}(x) = |x|$. Is this function even or odd?
4. Draw the graph of the function

$$\text{SGN}(x) = \begin{cases} +1 & \text{if } x > 0 \\ 0 & \text{if } x = 0 \\ -1 & \text{if } x < 0 \end{cases}$$

   Is this function even or odd?
5. Draw the graph of the function

$$h(x) = \begin{cases} -x & \text{if } x \leq -1 \\ 1 & \text{if } -1 < x < 1 \\ -x + 1 & \text{if } x \geq 1 \end{cases}$$

6. Which of the graphs in Figure 5.17 are the graphs of functions?
7. Graph the function $\text{CEIL}(x)$, where $\text{CEIL}(x) = n + 1$ if $n < x \leq (n + 1)$.
8. **a.** Show that the function $f(x) = 5x - 9$ is increasing.
   b. Show that the function $g(x) = -3x + 4$ is decreasing.
9. What is true of a function that is both nonincreasing and nondecreasing for all values of x?
10. Give an example of a function that is neither even nor odd.
11. **a.** Can a function have two y-intercepts? Justify your answer.
    b. Can there be such a thing as symmetry with respect to the x-axis for a function? Justify your answer.

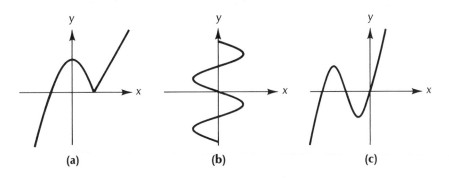

Figure 5.17

Problems and Projects

12. Write a program that inputs a real number X and outputs the values of FLOOR(X) and CEIL(X) (see Example 5.16 and exercise 7). The Pascal TRUNC function would be helpful.

13. Write a program that inputs a real number X and outputs the value of the function $h(X)$ defined in exercise 5.

5.4 THREE CLASSES OF FUNCTIONS

In this section we consider polynomial, rational, and exponential functions, along with some of their more important properties.

Polynomials

A function in the form

$$f(x) = a_n x^n + a_{n-1} x^{n-1} + \cdots + a_1 x + a_0 \qquad a_n \neq 0$$

where n is a nonnegative integer is called a **polynomial function of degree n.** The **coefficients** of the polynomial are the real numbers $a_n, a_{n-1}, \ldots, a_1, a_0$. Although not necessary, we shall always write polynomials with the exponents of the terms decreasing, going from left to right. This is sometimes called **standard form.** Thus, $5x^3 - 4x^2 + 7x + 3$ is in standard form, but $x - 4x^2 + 8$ is not. If a polynomial is in standard form, its **degree** is the exponent of the left-most, or **leading,** term. Note that some of the coefficients of a polynomial may be 0; that is, some terms may be missing. For example, the coefficients of x and x^2 in the polynomial $x^3 + 1$ are both 0. Finally, we note that the domain of any polynomial function equals the set of all real numbers.

Some polynomials have special names depending on their degree. A polynomial of degree 0 ($n = 0$) is called a **constant function.** A polynomial of degree 1 ($n = 1$) is called a **linear function.** A polynomial of degree 2 ($n = 2$) is called a **quadratic function.** Thus the general forms for constant, linear, and quadratic functions are

$$f(x) = a_0 \qquad \text{(Constant Function)}$$
$$f(x) = a_1 x + a_0 \qquad \text{(Linear Function)}$$
$$f(x) = a_2 x^2 + a_1 x + a_0 \qquad \text{(Quadratic Function)}$$

EXAMPLE 5.21 _____

Draw the graphs of the following functions:

a. $f(x) = 3$ b. $g(x) = 2x - 1$ c. $h(x) = x^2 - 3x + 2$

SOLUTION

The graphs are shown in Figure 5.18. $f(x)$ is a constant function whose value is 3. f maps every value of x to 3. The linear function $g(x)$ has a graph that is a straight line. The graph of the quadratic function $h(x)$ is called a parabola.

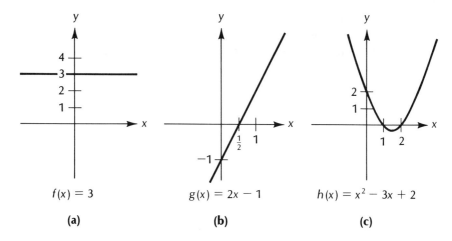

$$f(x) = 3$$

(a)

$$g(x) = 2x - 1$$

(b)

$$h(x) = x^2 - 3x + 2$$

(c)

Figure 5.18

The y-intercept of a polynomial is always its constant term. The x-intercept of the linear function $f(x) = ax + b$ is $-b/a$. The x-intercepts, if any, of the quadratic polynomial $f(x) = ax^2 + bx + c$ are given by the quadratic formula:

$$x = \frac{-b \pm \sqrt{b^2 - 4ac}}{2a}$$

One x-intercept is obtained by using the $+$ sign in the formula, the other by using· the $-$ sign. If the expression under the square root sign in the quadratic formula (this expression is called the "discriminant") is negative, the quadratic polynomial has no x-intercepts.

If the degree of *every* term of a polynomial is even, then the polynomial is an even function and its graph is symmetric with respect to the y-axis. If the degree of *every* term of a polynomial is odd, then the polynomial function is an odd function and its graph is symmetric with respect to the origin.

In general, polynomials are neither increasing nor decreasing. They may increase for some values of x and decrease for others. To study this behavior properly requires the techniques of calculus and, therefore, will not be discussed in this book.

Rational Functions

The second class of functions we consider in this section is the class of rational functions. A function of the form

$$f(x) = \frac{P(x)}{Q(x)}$$

where $P(x)$ and $Q(x)$ are polynomials, is called a **rational function.**

Since a fraction can become zero only when the numerator is zero, the zeros of the rational function $f(x)$ are the same as the zeros of $P(x)$ (the polynomial in its numerator) provided these zeros are in the domain of f. Since we are never allowed to divide by 0, the zeros of $Q(x)$ (the polynomial in its denominator) are the only real numbers *not* in the domain of f. It is important in many applications, however, to know how $f(x)$ behaves *near* the zeros of $Q(x)$ and how $f(x)$ behaves for very large values of x.

EXAMPLE 5.22 _____

Draw the graph of

$$f(x) = \frac{2x - 1}{x - 1}$$

by constructing a table of values for the function.

SOLUTION

First we find the intercepts. Let $x = 0$, and obtain $f(0) = 1$ as the y-intercept. The x-intercept is the zero of $2x - 1$; that is, $x = 1/2$. Since $x = 1$ is a zero of the denominator, 1 is not in the domain of f. We do, however, calculate the values of f for several values of x close to 1. We also calculate the values of f for several large values of x, both positive and negative.

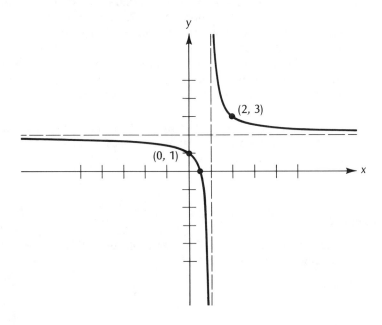

Figure 5.19

x	-1000	-100	-10	0	0.5	0.9	0.95	0.99	1.01	1.05
$f(x)$	1.999	1.990	1.909	1	0	-8	-18	-98	102	22

x	2	10	100	1000
$f(x)$	3	2.111	2.010	2.001

The graph of $f(x)$ is shown in Figure 5.19.

Observe the behavior of $f(x)$ near $x = 1$, the zero of the denominator of $f(x)$. As x gets close to 1 through values less than 1 (for example, 0.5, 0.9, 0.95, 0.99), the values of the function become arbitrarily large in the negative direction. As x gets close to 1 through values greater than 1 (such as 2, 1.05, 1.01), the values of the function become arbitrarily large in the positive direction. The line $x = 1$ is called a **vertical asymptote** of the graph. This behavior is typical. See Note 5.6.

Note 5.6

If $x = a$ is a zero of $Q(x)$ and $P(a) \neq 0$, the line $x = a$ is a vertical asymptote of

$$f(x) = \frac{P(x)}{Q(x)}$$

We also see from the graph of Figure 5.19 that as x gets arbitrarily large through positive values (such as 10, 100, 1000), the values of the function seem to be getting arbitrarily close to 2 from values greater than 2. Also, as x gets arbitrarily large through negative values (such as -10, -100, -1000), the values of the function seem to be getting arbitrarily close to 2 from values less than 2. The line $y = 2$ is called a **horizontal asymptote** of the graph.

A rational function can have several vertical asymptotes, one for each zero of $Q(x)$, but, as we shall see, only one horizontal asymptote. To study horizontal asymptotes we need the concept of the limit of a function.

Limits

A very important concept in studying the behavior of functions for large values of x is that of the limit of a function as x approaches infinity.

Remark   We study only limits as x approaches infinity because of their usefulness in comparing the rates of growth of functions. These comparisons are used to measure the efficiency of computational algorithms. For a more general and rigorous treatment of limits, the reader is referred to any standard text in calculus.

The function $f(x)$ **approaches the limit L as x approaches $+\infty$,** written $\lim_{x \to +\infty} f(x) = L$, if the values of $f(x)$ get arbitrarily close to L as x gets arbitrarily large.

It is easy to see using a calculator that $\lim_{x \to +\infty} (1/x) = 0$. Likewise, it is easy to see that $\lim_{x \to +\infty} (1/x^2) = 0$. In fact, we have the general result given in Note 5.7, which is used in the remainder of this section.

Note 5.7

If $n > 0$, $\lim_{x \to +\infty} (c/x^n) = 0$ for any constant c.

EXAMPLE 5.23

Show that

$$\lim_{x \to +\infty} \frac{2x - 1}{x - 1} = 2$$

SOLUTION

The technique for finding the limit as x approaches $+\infty$ of a rational function is to first divide the numerator and denominator by the highest power of x appearing in either place and then let x approach $+\infty$:

$$\frac{2x - 1}{x - 1} = \frac{2 - (1/x)}{1 - (1/x)}$$

Since $\lim_{x \to +\infty} (1/x) = 0$, we have

$$\lim_{x \to +\infty} \frac{2x - 1}{x - 1} = \lim_{x \to +\infty} \frac{2 - (1/x)}{1 - (1/x)} = \frac{2}{1} = 2$$

EXAMPLE 5.24

Find the following limits:

a. $\lim\limits_{x \to +\infty} \dfrac{5x^2 + 3x - 4}{x^2 + 1}$ b. $\lim\limits_{x \to +\infty} \dfrac{3x + 4}{x^2 + x}$ c. $\lim\limits_{x \to +\infty} \dfrac{x^3 + 3x - 1}{x^2 + 7x + 3}$

SOLUTION

a. Dividing numerator and denominator by x^2, we have

$$\lim_{x \to +\infty} \frac{5x^2 + 3x - 4}{x^2 + 1} = \lim_{x \to +\infty} \frac{5 + (3/x) - (4/x^2)}{1 + (1/x^2)} = \frac{5 + 0 + 0}{1 + 0} = 5$$

b. Dividing numerator and denominator by x^2, we have

$$\lim_{x \to +\infty} \frac{3x + 4}{x^2 + x} = \lim_{x \to +\infty} \frac{(3/x) + (4/x^2)}{1 + (1/x)} = \frac{0 + 0}{1 + 0} = 0$$

c. Dividing numerator and denominator by x^3, we have

$$\lim_{x \to +\infty} \frac{x^3 + 3x - 1}{x^2 + 7x + 3} = \lim_{x \to +\infty} \frac{1 + (3/x^2) - (1/x^3)}{(1/x) + 7(1/x^2) + (3/x^3)} = \frac{1 + 0 + 0}{0 + 0 + 0} = \frac{1}{0}$$

Because we cannot divide by zero, how can we interpret this result? The numerator in the limit approaches 1. The denominator approaches 0 and is always positive. If we divide 1 by smaller and smaller numbers, the quotients will become larger and larger. Thus, we write

$$\lim_{x \to +\infty} \frac{x^3 + 3x - 1}{x^2 + 7x + 3} = +\infty$$

That is, the value of the fraction becomes arbitrarily large as x becomes arbitrarily large.

Another way of viewing limits of rational functions as x approaches $+\infty$ is to consider only the leading (that is, highest degree) terms in the numerator and denominator. For very large values of x, the leading term of a polynomial is the numerically dominant term. For example, if $x = 100$, x^3 is much larger than $5x + 1$ (1,000,000 compared to 501). We can, therefore, reason as follows: In the fraction $(x^3 + 3x - 1)/(x^2 + 7x + 3)$ the numerator is dominated by x^3 and the denominator by x^2. Thus, for very large values of x the fraction should be approximately $x^3/x^2 = x$. As $x \to +\infty$, the fraction should, therefore, behave like x and approach $+\infty$.

We can also do parts a and b of Example 5.24 this way: $(5x^2 + 3x - 4)/(x^2 + 1)$ behaves like $5x^2/x^2 = 5$ for large values of x. The fraction $(3x + 4)/(x^2 + x)$ behaves like $3x/x^2 = 3/x$ for large values of x and, thus, approaches 0.

Limits as $x \to -\infty$ are handled in basically the same way as limits with $x \to +\infty$. The difference is, naturally, that x becomes large (that is, large in absolute value) through negative values. The following result enables us to treat limits of rational functions as $x \to -\infty$ in the same way as limits with $x \to +\infty$.

Note 5.8

If $n > 0$, $\lim_{x \to -\infty} (c/x^n) = 0$ for any constant c.

EXAMPLE 5.25

Find the following limits:

a. $\displaystyle \lim_{x \to -\infty} \frac{3x^2 - 2x + 1}{2x^2 + 3x - 5}$ b. $\displaystyle \lim_{x \to -\infty} \frac{5x^3 + 3x}{9x^4 - 4x^3 + 2}$ c. $\displaystyle \lim_{x \to -\infty} \frac{x^5 + 1}{x^2 + 1}$

SOLUTION

a. Using the dominant term technique described earlier, $(3x^2 - 2x + 1)/(2x^2 + 3x - 5)$ behaves like $3x^2/2x^2 = 3/2$ for negative x with large absolute value. The limit is, therefore, 3/2.

b. Similarly, for negative x with very large absolute value $(5x^3 + 3x)/(9x^4 - 4x^3 + 2)$ behaves like $5x^3/9x^4 = 5/9x$. Since $\lim_{x \to -\infty} (5/9x) = 0$,

$$\lim_{x \to -\infty} \frac{5x^3 + 3x}{9x^4 - 4x^3 + 2} = 0$$

c. Using the dominant term technique, $(x^5 + 1)/(x^2 + 1)$ behaves like $x^5/x^2 = x^3$. As x becomes arbitrarily large through negative values, x^3 will also become arbitrarily large through *negative values*. Therefore,

$$\lim_{x \to -\infty} \frac{x^5 + 1}{x^2 + 1} = -\infty$$

Returning to our discussion of horizontal asymptotes of rational functions, we make the following note (Note 5.9).

Note 5.9

The rational function $f(x) = P(x)/Q(x)$ has the horizontal asymptote $y = b$ if $\lim_{x \to +\infty} f(x) = b$.

From our discussion of limits as $x \to -\infty$ and the limits of Example 5.25, we can conclude that if $f(x)$ is a rational function and $\lim_{x \to +\infty} f(x) = b$, where b is finite, then $\lim_{x \to -\infty} f(x) = b$. Thus, a rational function can have only one horizontal asymptote.

Exponential Functions

The third class of functions we consider in this section is the class of exponential functions. We shall assume the reader is familiar with the meaning of exponents (for example, $x^5 = x \cdot x \cdot x \cdot x \cdot x$) and the rules for manipulating exponents, given in Note 5.10.

Note 5.10   The Rules of Exponents

E1. $x^n \cdot x^m = x^{n+m}$ E4. $x^0 = 1, \quad x \neq 0$ E7. $\left(\dfrac{x}{y}\right)^n = \dfrac{x^n}{y^n}$

E2. $\dfrac{x^n}{x^m} = x^{n-m}$ E5. $x^{-n} = \dfrac{1}{x^n}$

E3. $(x^n)^m = x^{nm}$ E6. $(xy)^n = x^n \cdot y^n$

An **exponential function** is any function of the form $f(x) = a \cdot b^{kx}$, where a, b, and k are constants and $b > 1$. The number b is called the **base** of the exponential function. Why require $b > 1$? We exclude negative values of b to avoid complex numbers. If we allow b to be 1, the function will reduce to $f(x) = a$, the constant function. Finally, we exclude values of b between 0 and 1 because functions using such values can be rewritten using reciprocals. Suppose, for example, we are given the function $f(x) = (2/3)^x$. The base for the function in this form is $b = 2/3$. We can, however, write $f(x) = (3/2)^{-x}$ using property E5. The base for the function is now $b = 3/2$, the reciprocal of the original base.

Before stating the important properties of exponential functions, we consider some examples.

EXAMPLE 5.26

Graph the following exponential functions:

a. $h(x) = 2 \cdot 3^{2x}$ b. $g(x) = 2^{-x}$

SOLUTION

Following are tables of values for the functions; their graphs are given in Figure 5.20.

x	-4	-3	-2	-1	0	1	2	3	4
$h(x) = 2 \cdot 3^{2x}$	0.0003	0.0027	0.0247	0.2222	2	18	162	1458	13122

x	-4	-3	-2	-1	0	1	2	3	4
$g(x) = 2^{-x}$	16	8	4	2	1	$\frac{1}{2}$	$\frac{1}{4}$	$\frac{1}{8}$	$\frac{1}{16}$

Recall that when evaluating $h(x)$ we must raise 3 to the power $2x$ before multiplying by 2.

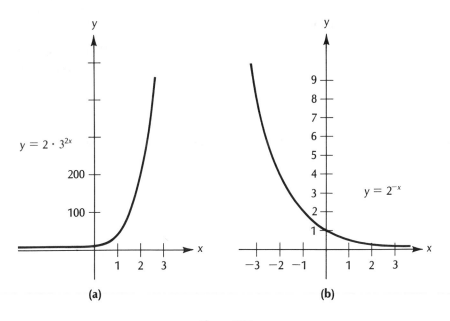

(a)   (b)

Figure 5.20

Exponential functions have the properties given in Note 5.11.

Note 5.11   Properties of $f(x) = a \cdot b^{kx}$

1. The y-intercept of f is a.
2. If $k > 0$, the left half of the x-axis is a horizontal asymptote. This is so because if $k > 0$, $\lim_{x \to -\infty} b^{kx} = 0$.
3. If $k < 0$, the right half of the x-axis is a horizontal asymptote. This is so because if $k < 0$, $\lim_{x \to +\infty} b^{kx} = 0$.
4. If a and k have the same sign, f is an increasing function.
5. If a and k have opposite signs, f is a decreasing function.

An important characteristic of exponential functions is their rate of growth or decay. The term *growth* is used for increasing functions and *decay* is used for decreasing functions. As evidenced by the graphs of exponential functions, increasing exponential functions like $h(x) = 2 \cdot 3^{2x}$ grow *very* rapidly. For example, $h(4) = 13{,}122$, which is an increase of nearly 12,000 over the value of $h(3)$. Likewise, decreasing exponential functions decay very rapidly. That is, they approach their horizontal asymptote very rapidly.

An exponential function that approaches $+\infty$ as $x \to +\infty$, in fact, grows faster than *any* polynomial function. We state without proof the following theorem.

Theorem 5.3

If $P(x)$ is a polynomial such that $\lim_{x \to +\infty} P(x) = +\infty$ and $f(x) = a \cdot b^{kx}$ where a and k are both positive, then $\lim_{x \to +\infty} P(x)/f(x) = 0$.

It follows from Theorem 5.3 that

$$\lim_{x \to +\infty} \frac{10x^7 + 4x^5 - 3x + 9}{10^x} = 0$$

Although for small values of x the numerator in this limit is larger than the denominator, for large values of x the denominator is much larger than the numerator. Try evaluating numerator and denominator for $x = 10$. Also, see Case Study 5A.

The most important base for an exponential function in advanced mathematics is an irrational number denoted by the letter e. The number e is defined by the following limit:

$$e = \lim_{x \to +\infty} \left(1 + \frac{1}{x}\right)^x$$

The following table should convince you that the value of e is approximately 2.7182818 . . . .

x	$\left(1+\dfrac{1}{x}\right)^x$
1	2
10	2.59374
10^2	2.70481
10^3	2.71692
10^4	2.71815
10^5	2.71825
10^6	2.71828
10^7	2.7182818

The exponential function $f(x) = e^x$ is so important it is sometimes called *the* exponential function.

EXAMPLE 5.27

Construct a table of values for $f(x) = e^x$ and draw its graph.

SOLUTION

Using a scientific calculator we obtain the following table and the graph in Figure 5.21.

x	-3	-2	-1	-0.5	0	0.5	1	2	3
$f(x) = e^x$	0.0498	0.1353	0.3679	0.6065	1	1.6487	2.7183	7.3891	20.0855

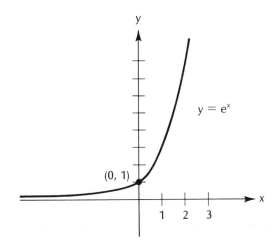

Figure 5.21

EXERCISES 5.4

1. Draw the graphs of the following polynomial functions. Find all the intercepts and note where the functions are increasing and decreasing:
   a. $f(x) = x^2 - 4$ **b.** $g(x) = x^2 - 2x + 1$ **c.** $h(x) = x^3 - 2x^2 - x + 2$

2. State and prove a general theorem that states for which values of a and b the linear function $h(x) = ax + b$ is increasing.

3. A power function is a function of the form $f(x) = x^n$. Use the laws of exponents to prove that the power function $f(x) = x^n$ is an even function if n is even and is an odd function if n is odd.

4. Draw the graphs of the following rational functions. Draw all asymptotes with dashed lines:

   a. $f(x) = \dfrac{x^2 + 1}{x + 2}$ **b.** $g(x) = \dfrac{x^2 + 1}{x^2 - 1}$

5. Find the following limits:

   a. $\lim\limits_{x \to +\infty} \dfrac{5x^2 + 3}{7x^2 - 9}$ **b.** $\lim\limits_{x \to +\infty} \dfrac{6x^2 - 3}{2x + 8}$

   c. $\lim\limits_{x \to +\infty} \dfrac{x^2 + (1/x)}{x^2 + (1/x)}$ **d.** $\lim\limits_{x \to +\infty} \dfrac{2^x}{5^x}$ (*Hint:* Use E7.)

6. If n and m are constants, what is $\lim_{x \to +\infty} x^n/x^m$? Consider all cases.

7. Graph the exponential functions:
   a. $f(x) = 3 \cdot 5^x$ **b.** $g(x) = 4^{-x}$ **c.** $h(x) = e^{2x}$

Problems and Projects

8. We noted after our definition of a rational function $f(x) = P(x)/Q(x)$ that a zero of f is a zero of P provided the zero is in the domain of f—that is, provided it is not also a zero of Q. Thus, values of x that make both P and Q zero must be avoided. (We also alluded to this fact in Note 5.6.) How does a rational function behave near such values? How does $f(x) = (x^2 - 1)/(x - 1)$ behave near $x = 1$?
   a. Investigate the question computationally. Evaluate f for values of x near 1 (both less and greater than 1).
   b. Investigate algebraically. Factor the numerator of f. Reduce the fraction to lowest terms and let x get arbitrarily close to 1 in the resulting expression.

9. An interesting limit is $\lim_{x \to 0} x^x$. Investigate this limit computationally using a calculator or computer. Use only positive values of x. You might try 0.01, 0.001, 0.0001, and so on.

10. Find $\lim_{x \to +\infty} [x - x^2/(x + 1)]$. (*Hint:* The answer is *not* zero. Find a common denominator first.)

5.5 OPERATIONS ON FUNCTIONS: INVERSES

In this section we discuss how to combine functions to yield new functions by using the four basic arithmetic operations of addition, subtraction, multiplication, and division, and a new operation called "composition."

We first define **equality of functions.** Let f and g have the same domain D. Then $f = g$ if $f(x) = g(x)$ for all $x \in D$. Thus, if we define $f(x) = 2x + 4$ and $g(x) = 2(x + 2)$, $f = g$ because $2x + 4 = 2(x + 2)$ for all x.

Let $f(x)$ and $g(x)$ have a common domain D, which we assume is a subset of the real numbers. Let the ranges of f and g be subsets of the real numbers. We can now define the following:

1. The **sum of f and g, $f + g$,** is defined by

$$(f + g)(x) = f(x) + g(x)$$

2. The **difference of f and g, $f - g$,** is defined by

$$(f - g)(x) = f(x) - g(x)$$

3. The **product of f and g, fg,** is defined by

$$(fg)(x) = f(x) \cdot g(x)$$

4. The **quotient of f and g, f/g,** is defined by

$$(f/g)(x) = \frac{f(x)}{g(x)}$$

for those values of x where $g(x) \neq 0$.

Note that the domain of each of $f + g$, $f - g$, and fg is the set D. However, the domain of f/g is D *excluding* the zeros of $g(x)$.

EXAMPLE 5.28 _____

Find the sum, product, difference, and quotient of $f(x) = 3x + 5$ and $g(x) = 4x - 3$.

SOLUTION

Let

$$s(x) = (f + g)(x) = f(x) + g(x) = (3x + 5) + (4x - 3) = 7x + 2$$
$$p(x) = (fg)(x) = f(x)g(x) = (3x + 5)(4x - 3) = 12x^2 + 11x - 15$$
$$d(x) = (f - g)(x) = f(x) - g(x) = (3x + 5) - (4x - 3) = -x + 8$$
$$q(x) = (f/g)(x) = f(x)/g(x) = (3x + 5)/(4x - 3)$$

The domain of f and g is the set of real numbers. The domain of each of s, p, and d is, therefore, the set of real numbers. The domain of $q(x)$ is, however, the set of real numbers excluding 3/4 because $4x - 3 = 0$ when $x = 3/4$.

A function $f:X \rightarrow Y$ is said to be a **one-to-one** (written **1-1**) **function,** or simply **one-to-one,** if different elements of the domain have different images. The function f in Figure 5.22 is 1-1, but the function g is not because $g(1) = g(3) = 1$.

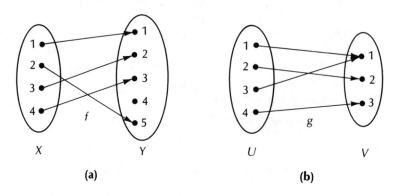

Figure 5.22

Equivalent to the definition of a 1-1 function is the statement in Note 5.12.

Note 5.12

$f:X \rightarrow Y$ is 1-1 if $f(a) = f(b)$ implies $a = b$.

EXAMPLE 5.29

Prove that $f(x) = 3x + 4$ is 1-1.

SOLUTION

To show f is 1-1 we must show that $f(a) = f(b)$ implies $a = b$. Therefore, we start with

$$f(a) = 3a + 4 = 3b + 4 = f(b)$$

Subtracting 4 from both sides yields $3a = 3b$. Dividing both sides by 3 yields $a = b$. Thus, f is 1-1.

A function $f:X \rightarrow Y$ is **onto** if every element of the codomain Y is the image of some element in the domain X. Thus, f is onto if the codomain and range of f are the same set. The function f of Figure 5.22 is not onto because 4 in Y is not the image of an element in X. The function g of Figure 5.22 is onto. Note from the functions depicted in Figure 5.22 that the concepts of 1-1 and onto are independent: A function can be 1-1 without being onto and a function can be onto without being 1-1. Finally, note that, by definition, a function always maps its domain onto its range.

Suppose a function maps the real numbers to the real numbers. We can sometimes determine whether the function is 1-1 or onto using the rules in Note 5.13.

Note 5.13

If *no* horizontal line intersects the graph of f more than once, f is 1-1.
If for every value b in the codomain of f the horizontal line $y = b$ intersects the graph of f at least once, f is onto.

EXAMPLE 5.30 _____

Determine graphically whether the following functions are 1-1 or onto. Consider the domain and codomain for each to be the real numbers R:

a. $f(x) = x^2$ b. $g(x) = 2x + 1$ c. $h(x) = x^3 - x$

SOLUTION

The graphs are shown in Figure 5.23.

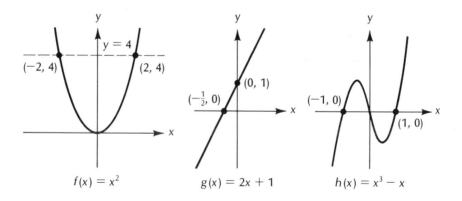

Figure 5.23

a. $f(x)$ is neither 1-1 nor onto. The dashed horizontal line $y = 4$ intersects the graph at two points $(-2, 4)$ and $(2, 4)$. Since $f(2) = f(-2) = 4$, two values of x are mapped to the same value of y and the function is not 1-1. No horizontal line below the x-axis intersects the graph; that is, there is no value of x with a negative square. The function f is, therefore, not onto.

b. $g(x)$ is both 1-1 and onto. The graph of g is a straight line that is not parallel to the x-axis. Because nonparallel lines must intersect exactly once, (1) *every* horizontal line, such as $y = -2$, must intersect the graph, making g onto, and (2) every horizontal line intersects the graph in exactly one point making g 1-1.

c. $h(x)$ is not 1-1. Many lines, the x-axis in particular, intersect the graph in more than one point. $h(x)$ is, however, onto because every horizontal line intersects the graph at least once.

Remark   If we consider the codomain of $f(x) = x^2$ to be the nonnegative reals, so that the codomain of f equals its range, f becomes an onto function. Can you show this from the graph? Also, some functions that are not 1-1 may be made 1-1 if considered on a different domain. $f(x) = x^2$ is not 1-1 when the domain is the set R of real numbers. However, if we restrict the domain of f to R^*, the set of nonnegative real numbers, then f is 1-1. Can you show this from the graph?

Composition of Functions

Consider two functions f and g where the range of f is contained in the domain of g

$$f : X \to Y \qquad \text{and} \qquad g : Y \to Z$$

We can take an element $a \in X$ and find its image $f(a)$ under f. Since the range of f is contained in the domain of g, we can find the image of $f(a)$ under g—namely, $g(f(a))$. This procedure is called the **composition** of functions and is defined as follows: $(g \circ f)(x) = g(f(x))$.

EXAMPLE 5.31 _____

Find $g \circ f$ for the functions depicted in Figure 5.24.

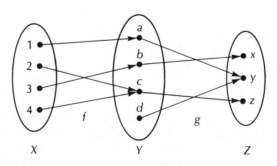

Figure 5.24

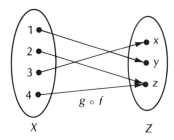

Figure 5.25

SOLUTION

$$(g \circ f)(1) = g(f(1)) = g(a) = y$$
$$(g \circ f)(2) = g(f(2)) = g(c) = z$$
$$(g \circ f)(3) = g(f(3)) = g(b) = x$$
$$(g \circ f)(4) = g(f(4)) = g(c) = z$$

The diagram of $g \circ f$ is given in Figure 5.25.

EXAMPLE 5.32

Let $f(x) = 2x - 5$ and $g(x) = 5x^2$. Find:

a. $(f \circ g)(x)$ b. $(g \circ f)(x)$ c. $(f \circ f)(x)$

SOLUTION

The equation for a function is a way of expressing in algebraic form the rule that defines the function. Thus, $f(x) = 3x - 5$ will take any number or expression, multiply it by 3, and subtract 5 from the resulting product.

a. $(f \circ g)(x) = f(g(x)) = f(5x^2) = 3(5x^2) - 5 = 15x^2 - 5$
b. $(g \circ f)(x) = g(f(x)) = g(3x - 5) = 5(3x - 5)^2$
$$= 5(9x^2 - 30x + 25)$$
$$= 45x^2 - 150x + 125$$

Note that in general $f \circ g \neq g \circ f$.

c. $(f \circ f)(x) = f(f(x)) = f(3x - 5) = 3(3x - 5) - 5 = 9x - 20$

Inverse Functions

One of the most fruitful general ideas in mathematics is that of an inverse. Addition and multiplication have inverse operations—namely, subtraction and division. Subtraction undoes addition. Division undoes multiplication. The operations of squaring a number and taking the square root of a number are also inverse operations. For example, $9^2 = 81$. We can undo this squaring by taking the square root, $\sqrt{81} = 9$. Relations also have inverses, and some functions have inverses that are also functions. We now explain how.

Suppose R is a relation from A to B. The **inverse R^{-1}** of R is the relation from B to A defined as

$$R^{-1} = \{(b, a) : (a, b) \in R\}$$

Thus, if the relation R is given as a set of ordered pairs, then to find R^{-1} simply invert each ordered pair. R^{-1} undoes the relation R because the following identity holds: $(R^{-1})^{-1} = R$. To get R^{-1} we invert the ordered pairs in R. To get the inverse relation of R^{-1}, namely $(R^{-1})^{-1}$, we invert the ordered pairs of R^{-1}. This, however, just gives back to us the ordered pairs in R.

EXAMPLE 5.33 ───────────────────────────────

Find the inverse relation of

$$R = \{(3, b), (2, a), (5, b), (1, a), (4, c)\}$$

SOLUTION

$$R^{-1} = \{(b, 3), (a, 2), (b, 5), (a, 1), (c, 4)\}$$

───

Since a function $f : X \to Y$ is a relation, f has an inverse relation. But, the inverse relation may not be a function. The relation R of Example 5.33 is a function, but R^{-1} is not a function. We now investigate those conditions on a function f that will guarantee that f^{-1} is also a function. In fact, we shall say that a function has an inverse only if its inverse relation is also a function.

Suppose $f : X \to Y$. The function f will map an element $a \in X$ to $f(a) \in Y$. We want the inverse relation f^{-1} to map $f(a)$ to a, and we want f^{-1} to be a function. First, for f^{-1} to be a function, the domain of f^{-1} must be Y. This means that f must be onto. Also, for f^{-1} to be a function, each element of Y must be assigned by f^{-1} to *exactly* one element of X. This means each element of Y can be the image of only one element in X. Thus, f must be 1-1. We now state the definition of the inverse of $f : X \to Y$ be 1-1 and onto. Then the **inverse of f, $f^{-1} : Y \to X$** is a function and $(f^{-1} \circ f)(x) = x$ for all $x \in X$ and $(f \circ f^{-1})(y) = y$ for all $y \in Y$. Thus the reader should keep Note 5.14 in mind.

┌───┐

Note 5.14

A function has an inverse if and only if it is both 1-1 and onto.

└───┘

EXAMPLE 5.34

Find the inverse of the function depicted in Figure 5.26.

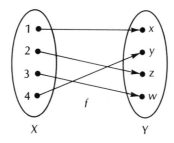

Figure 5.26

SOLUTION

The function f is 1-1 and onto so the inverse exists. f^{-1} is shown in Figure 5.27. Note, for example, that f^{-1} maps y to 4 because f maps 4 to y.

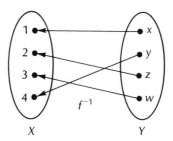

Figure 5.27

Consider the function $f(x) = x^3$. Since f is 1-1 and onto, f^{-1} exists. What is $f^{-1}(8)$? We must answer the question: What number x gets mapped to 8 by f? Or, for what x is $f(x) = 8$? Or, for what x is $x^3 = 8$? The answer is the cube root of 8, which is 2. Thus, the inverse of $f(x) = x^3$ is the function that takes the cube root of a number $f^{-1}(x) = \sqrt[3]{x}$.

EXAMPLE 5.35

Show that $f(x) = x^3$ and $g(x) = \sqrt[3]{x}$ are inverse functions.

SOLUTION

It is easy to see graphically that f is both 1-1 and onto. To show that f and g are inverses of one another, however, we must show that $(f \circ g)(x) = x$ and that $(g \circ f)(x) = x$:

$$(f \circ g)(x) = f(\sqrt[3]{x}) = (\sqrt[3]{x})^3 = x$$
$$(g \circ f)(x) = g(x^3) = \sqrt[3]{(x^3)} = x$$

We can sometimes find the inverse of a function by the technique in Note 5.15.

Note 5.15

To find the inverse of a function $y = f(x)$:

1. Solve the equation $y = f(x)$ for x in terms of y.
2. In the resulting equation, replace x by y and y by x.
3. f^{-1} equals the right side of the equation found in step 2.

EXAMPLE 5.36 _____

Use the technique of Note 5.15 to find the inverse of

a. $f(x) = 4x - 1$ b. $f(x) = x^3$

SOLUTION

a. First we solve $y = 4x - 1$ for x in terms of y:

$$y + 1 = 4x \quad \text{or} \quad 4x = y + 1 \quad \text{or} \quad x = \frac{y + 1}{4}$$

Now replace x by y and y by x, obtaining $y = (x + 1)/4$. Therefore, $f^{-1}(x) = (x + 1)/4$.

To verify this is actually the inverse function,

$$(f \circ f^{-1})(x) = f(f^{-1}(x)) = f\left(\frac{x + 1}{4}\right) = 4 \cdot \frac{x + 1}{4} - 1$$

$$= (x + 1) - 1$$
$$= x$$

and

$$(f^{-1} \circ f)(x) = f^{-1}(f(x)) = f^{-1}(4x - 1) = \frac{(4x - 1) + 1}{4}$$

$$= \frac{4x}{4} = x$$

b. Solve $y = x^3$ for x in terms of y: $x = \sqrt[3]{y}$. Interchange x and y: $y = \sqrt[3]{x}$. Hence, $f^{-1}(x) = \sqrt[3]{x}$. We verified these functions are inverses in Example 5.35.

Logarithmic Functions

One of the most important inverse functions is the inverse of an exponential function. We now restrict the codomain of our exponential functions to make them 1-1 onto functions. For example, we will restrict the codomain of $f(x) = 2^x$ to the set of positive real numbers. So restricted, it is easy to see from its graph that $f(x) = 2^x$ is both 1-1 and onto. From Note 5.14 it follows that f^{-1} exists. What is $f^{-1}(x)$? To be more specific, what is $f^{-1}(64)$? That is, what number x gets mapped by f to 64? Or, what value of x is such that $f(x) = 2^x = 64$? A little computation shows that $x = 6$. Thus, $f^{-1}(64) = 6$.

The inverse of an exponential function is called a **logarithmic function** (or **logarithm**). Given the exponential function $f(x) = b^x$, the inverse of f is called the **logarithm base b** and is denoted by $f^{-1}(x) = \log_b x$. This is read "f inverse of x equals the log base b of x." Thus, instead of writing $f^{-1}(64) = 6$, we write $\log_2 64 = 6$. In what follows, keep in mind Note 5.16.

Note 5.16

$\log_b x = p$ is equivalent to $b^p = x$.

EXAMPLE 5.37

Find the following logarithms:

a. $\log_3 9$ b. $\log_2(1/8)$ c. $\log_e 1$ d. $\log_{10} 1000$ e. $\log_{16} 4$

SOLUTION

a. Let $\log_3 9 = p$. This means $3^p = 9$, so $p = 2$. Hence, $\log_3 9 = 2$.
b. Let $\log_2(1/8) = p$. This means $2^p = 1/8$, so $p = -3$. Hence, $\log_2(1/8) = -3$.
c. Let $\log_e 1 = p$. This means $e^p = 1$, so $p = 0$. Hence, $\log_e 1 = 0$.

d. Let $\log_{10}1000 = p$. This means $10^p = 1000$, so $p = 3$. Hence, $\log_{10}1000 = 3$.

e. Let $\log_{16}4 = p$. This means $16^p = 4$, so $p = 1/2$. Hence, $\log_{16}4 = 1/2$.

The logarithm, or log function, has the properties given in Note 5.17.

Note 5.17   Properties of Logarithmic Functions

1. Since $\log_b x$ and b^x are inverse functions, we have

$$\log_b b^x = x \qquad \text{and} \qquad b^{\log_b x} = x$$

2. $\log_b 1 = 0$ and $\log_b b = 1$

3. The domain of $f(x) = \log_b x$ is the set of positive real numbers.

4. The function $f(x) = \log_b x$ is increasing, 1-1, and onto. The graph of the function is shown in Figure 5.28.

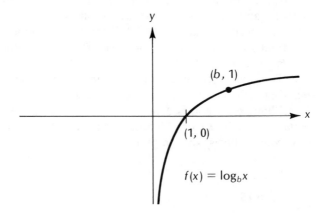

Figure 5.28

5. As $x \rightarrow +\infty$, $\log_b x$ approaches $+\infty$. The log function, however, grows at a slower rate than *any* polynomial. To be more precise, if $P(x)$ is a polynomial that approaches $+\infty$ as $x \rightarrow +\infty$,

$$\lim_{x \rightarrow +\infty} \frac{\log_b x}{P(x)} = 0$$

6. The negative y-axis is a vertical asymptote of the graph of $f(x) = \log_b x$.

Remark   Some books, particularly high school algebra texts, write $\log_{10}x$ as $\log x$, the base being understood. In some calculus texts, $\log x$ is understood as meaning $\log_e x$, although most calculus texts use ln x for $\log_e x$. Many computer science texts use $\log x$ to mean $\log_2 x$. In this book we shall always write the base explicitly.

Remark   Pascal does not have an exponentiation operator as do other programming languages. In BASIC, for example, to code B^X we would write B ↑ X. To code B^X in Pascal, we can use the build-in functions EXP(X) and LN(X). If B is any positive number, there is a real number C such that $B = e^C$. This real number $C = \log_e B$, by the definition of the logarithm. Hence, $B^X = (e^C)^X = e^{CX} = e^{X \cdot \log_e B}$. It follows that in Pascal to get $Y = B^X$, we code

$$C := LN(B);$$

$$Y := EXP(C * X);$$

To use composition of functions, we code

$$Y := EXP(LN(B) * X);$$

EXERCISES 5.5

1. Find the sum, difference, product, and quotient of each pair of the following functions and state the domain of each:
   a. $f(x) = 2x$, $g(x) = x - 3$
   b. $f(x) = 3x + 5$, $g(x) = 2 - x$
   c. $f(x) = x^2$, $g(x) = x^3$
2. Which of the following functions are 1-1? Which are onto? Assume the domain and codomain of each is the set of real numbers:
   a. $f(x) = 5x + 3$ **b.** $g(x) = x^3 + 2x$ **c.** $h(x) = |x|$.
3. In Examples 5.8 and 5.10 we described two functions with domain the set of all character strings. LEN maps each character string to the number of characters the string contains. FIRSTCHR maps a character string to its first character. Are either of these functions 1-1?
4. In Example 5.9 we defined the function ALPH, which maps each integer between 1 and 26 to the corresponding lowercase letter of the alphabet. Show that the function ALPH is 1-1.
5. All linear functions are 1-1 and onto with one class of exceptions. What is this class of exceptions?
6. For each of the following pairs of functions, find $f \circ g$, $g \circ f$, and $g \circ g$:
   a. $f(x) = 2x$, $g(x) = x - 3$ **b.** $f(x) = 3x + 5$, $g(x) = 2 - x$
   c. $f(x) = x^2$, $g(x) = x^3$ **d.** $f(x) = 2x + 7$, $g(x) = |x|$
7. The domain of a composite function $f \circ g$ must sometimes be restricted to a subset of the domain of g because of the definition and the domain of f.
   a. Let $f(x) = \sqrt{x}$ and $g(x) = x - 1$. The domain of g is the set R of all real numbers. The domain of f is the set of all nonnegative real numbers R^*. Find $f \circ g$. To what set of real numbers must you restrict the domain of $f \circ g$ for the function to be defined?

b. Referring to exercises 3 and 4, describe the function ALPH ∘ LEN; that is, what does the function do? To what set of character strings must you restrict the domain of the function for the composition to be defined?

8. Describe the function ALPH^{-1}. Describe the function $\text{ALPH}^{-1} \circ \text{FIRSTCHR}$.

9. Describe the function LEN ∘ FIRSTCHR. Consider all cases.

10. Show that the following pairs of functions are inverses of each other:

a. $f(x) = x + 1$, $g(x) = x - 1$

b. $f(x) = \dfrac{1}{x}$, $g(x) = \dfrac{1}{x}$, $x \neq 0$

11. Find the inverse of each of the following functions. Specify the domain of each inverse.

a. $f(x) = -5x + 1$ **b.** $f(x) = \dfrac{2x + 1}{x - 1}$

12. Find the following:

a. $\log_4 64$ **b.** $\log_{10} 0.5$ **c.** $\log_8 8^7$

d. $\log_e(1/e^6)$ **e.** $\log_3(1/9)$ **f.** $e^{\log_e 15}$

Problems and Projects

13. Assume f_1 and f_2 are even functions and g_1 and g_2 are odd functions. Prove the following:

a. $f_1 + f_2$ and $f_1 f_2$ are even functions.

b. $g_1 + g_2$ and $f_1 g_1$ are odd functions.

14. Let $f : X \to Y$ and $g : Y \to Z$.

a. Prove that if f and g are 1-1, so is $g \circ f$.

b. Prove that if f and g are onto, so is $g \circ f$.

c. If f and g are 1-1 and onto, prove that $g \circ f$ has an inverse and $(g \circ f)^{-1} = f^{-1} \circ g^{-1}$.

15. Assume $y = f(x)$ is a function mapping a subset of the real numbers into the real numbers and that it has an inverse function $f^{-1}(x)$. How are the graphs of $f(x)$ and $f^{-1}(x)$ related? Consider the graphs of the following:

a. $y = 2^x$ and $y = \log_2 x$

b. $y = 4x - 1$ and $y = (x + 1)/4$

(*Hint:* Consider what the procedure of Note 5.15 does geometrically.)

16. If $f(x) = a \cdot b^{kx}$, find $f^{-1}(x)$.

CASE STUDY 5A   COMPUTER FUNCTIONS

Functions are used in almost every computer program. In this case study, we discuss how functions can be used in Pascal.

Like most other programming languages, Pascal has **built-in** (or **library**) **functions**, which the programmer can use by simply using the function name in a program. Table 5.1 contains a list of some standard built-in functions of Pascal.

	Parameter Type	Result Type	Returns
ABS(X)	INTEGER or REAL	Same as parameter	Absolute value of X
ARCTAN(X)	INTEGER or REAL	REAL	Arctangent of X in radians
COS(X)	INTEGER or REAL	REAL	Cosine of X (X is in radians)
X DIV Y	INTEGER	INTEGER	Integer quotient of X by Y
EXP(X)	REAL or INTEGER	REAL	e to the X power
LN(X)	REAL or INTEGER	REAL	Natural logarithm of X
X MOD Y	INTEGER	INTEGER	Remainder when X is divided by Y
ROUND(X)	REAL	INTEGER	X rounded
SIN(X)	REAL or INTEGER	REAL	Sine of X (X is in radians)
SQR(X)	REAL or INTEGER	Same as parameter	Square of X
SQRT(X)	REAL or INTEGER	REAL	Square root of X
TRUNC(X)	REAL	INTEGER	X truncated

Table 5.1

In addition to built-in functions, Pascal allows programmers to define their own functions. These are called **user-defined functions.** For example, to define the equivalent of the function $f(x) = 2x + 1$, we would code

```
FUNCTION F(X:REAL): REAL;
    BEGIN
        F := 2 * X + 1;
    END;
```

The definition of the function must be placed in the program before the place or places where the function is used. The definition must begin with the word FUNCTION, followed by the function name, followed in parentheses by a list of the formal parameters and their types, followed by the type of the function itself. The value to be returned by the function is stored in the function name (F in our example) by using the function name on the left side of an assignment statement within the function procedure. To use the function in the program, use the function name (followed by a list of actual parameters) in an expression. For example, Y := F(10); .

To illustrate the use of both user-defined and built-in functions, we consider Example 5.38.

EXAMPLE 5.38 _____

Write a Pascal program that accepts as input two positive numbers, which represent the lengths of the legs of a right triangle, and prints the hypotenuse of the triangle.

SOLUTION

By the Pythagorean Theorem, if the legs of a right triangle are A and B, its hypotenuse is $\sqrt{A^2 + B^2}$. Thus, we have the following program:

```
program hypotenuse (input, output);
var a, b, c: real;
function hyp (x, y:real) : real;
begin (* hyp*)
   hyp := sqrt (x*x + y*y);
end (* hyp *);
begin (*hypotenuse*)
   writeln ('Enter the lengths of the two legs');
   readln (a,b);
   c := hyp(a,b);
   writeln ('The hypotenuse is ', c);
end.
```

The function HYP computes the hypotenuse of the right triangle. The assignment statement giving HYP its value uses the built-in function SQRT.

We can also use the composition of functions in a program by nesting function references. The following program does the same thing as the program of Example 5.38:

```
program hypotenuse2 (input, output);
var a, b, c: real;
function sumsquare (x, y:real) : real;
begin (* sumsquare*)
   sumsquare := (x*x + y*y);
end (* sumsquare *);
begin (*hypotenuse*)
   writeln ('Enter the lengths of the two legs');
   readln (a,b);
c := sqrt (sumsquare (a,b));
   writeln ('The hypotenuse is ', c);
end.
```

The assignment statement giving C a value is equivalent to the composition of functions SQRT ∘ SUMSQUARE.

Many of the limits we described in Section 5.4 can be verified by using built-in and/or user-defined functions.

EXAMPLE 5.39

Write a Pascal program to verify that

$$\lim_{x \to +\infty} \frac{\log_e x}{x} = 0$$

SOLUTION

We write a program that prints the values of $\log_e x / x$ for the powers of 10^n, $0 \le n \le 7$.

```
program limit1 (output);
var x, y: real;
    i: integer;
begin
    x := 1.0;
    for i := 1 to 8 do
        begin
            y := ln (x) / x;
            writeln (x,y);
            x := 10*x;
        end;
end.
```

A sample printout is

```
1.0000000000000E+00  0.0000000000000E+00
1.0000000000000E+01  2.3025850951676E-01
1.0000000000000E+02  4.6051701996525E-02
1.0000000000000E+03  6.9077552762382E-03
1.0000000000000E+04  9.2103403590497E-04
1.0000000000000E+05  1.1512925453187E-04
1.0000000000000E+06  1.3815510606014E-05
1.0000000000000E+07  1.6118096158157E-06
```

Both built-in and user-defined functions have domains and ranges. A built-in function usually checks input values to verify that they are in the domain of the function. The SQRT function, for example, has domain the set of nonnegative real numbers. If a negative number is input to this function, the program terminates and a message is displayed at the terminal. For example, if we have the statement

$$A := SQRT(-4)$$

in a Pascal program, the program will terminate execution when it reaches the statement.

Care must also be taken with the range of built-in functions. The Pascal function SIN(X) has domain the set of real numbers. Its range is the set of all numbers between -1 and $+1$ inclusive—that is, the interval $[-1, 1]$. Any computation using SIN(X) must take this range into account. Suppose we have the statement

$$B := SQRT(SIN(X) - 2)$$

in a Pascal program. The program will terminate when it reaches this line because SIN(X) $- 2$ is negative for every value of X in the domain of SIN and negative numbers are not in the domain of SQRT.

Function Procedures

Many functions used in computer programs cannot be defined in a one-line statement. Functions defined using several lines of computer code are sometimes called **function procedures.** Suppose we wish to define a function C that accepts as input the total weekly sales T of a salesperson and computes the commission, rounded to the nearer dollar, based on the following:

$$\text{Commission} = \begin{cases} 0.1T & \text{if } 0 \le T < 100 \\ 0.2T & \text{if } 100 \le T \end{cases}$$

There is, therefore, a 10% commission on sales less than $100 and a 20% commission on sales of $100 or more. The domain of the function C is the set of nonnegative real numbers. The range of C is the set of nonnegative integers. Our program, therefore, should verify that T is nonnegative and should compute C as a nonnegative integer.

EXAMPLE 5.40

Write a Pascal program that accepts the total weekly sales, WEEKLYSALES, and prints the commission as described.

SOLUTION

```
program weeklypay (input, output);
var weeklysales : real;
function commission (t:real): integer;
    begin
        if t < 100.0 then
            commission := round (0.1*t)
        else
            commission := round (0.2*t);
    end;
begin (* weeklypay *)
    writeln ('Enter weekly sales');
    readln (weeklysales);
    if weeklysales < 0.0 then
        begin
            writeln ('Invalid value entered');
            writeln ('Computation of commission terminated');
        end
    else
        writeln ('The commission is ', commission(weeklysales));
end.
```

The value of WEEKLYSALES is first checked by the program. If a negative value is entered, an appropriate message is printed. The built-in function ROUND is used to round the commission to the nearer dollar.

Recursive Functions

In some programming languages it is possible to have a function subroutine include a call to itself. Such a function is said to be defined **recursively**. Recursively defined functions are important in many applications as well as in theoretical computer science. Not all programming languages allow recursive function definitions. BASIC and FORTRAN do not allow recursion. Pascal, PL/I, ALGOL, and LOGO do allow recursion. Since a complete discussion of recursion would take us far beyond the scope of this book, we shall restrict ourselves to a few simple examples.

Recall our discussion of the factorial function from Chapter 3. In this section we shall denote $n!$ in function notation as FAC(n); that is, FAC(n) = $n!$ In Chapter 3 we noted, for example, that FAC(8) = $8 \cdot 7 \cdot 6 \cdot 5 \cdot 4 \cdot 3 \cdot 2 \cdot 1$ = 40,320. The factorial function can be defined recursively as follows:

$$\begin{cases} \text{FAC}(0) = 1 & (1) \\ \text{FAC}(n) = n \cdot \text{FAC}(n-1) \quad n \geq 1 & (2) \end{cases}$$

Equation (2), the **recurrence formula,** defines FAC(n) in terms of FAC($n-1$). To calculate FAC($n-1$), we use equation (2) again, which defines FAC($n-1$) in terms of FAC($n-2$), and so on. Equation (1) provides a way for the recursion (that is, the function calling itself) to stop. Once FAC(0) is reached, its value is known to be 1 by equation (1).

To take a concrete example, we use the recursive definition FAC(n) to compute FAC(4).

$$\begin{aligned} \text{FAC}(4) &= 4 \cdot \text{FAC}(3) & \text{(equation 2)} \\ \text{FAC}(3) &= 3 \cdot \text{FAC}(2) & \text{(equation 2)} \\ \text{FAC}(2) &= 2 \cdot \text{FAC}(1) & \text{(equation 2)} \\ \text{FAC}(1) &= 1 \cdot \text{FAC}(0) & \text{(equation 2)} \\ \text{FAC}(0) &= 1 \end{aligned}$$

from equation 1. Now, working backwards:

$$\begin{aligned} \text{FAC}(0) &= 1 \\ \text{FAC}(1) &= 1 \cdot \text{FAC}(0) = 1 \cdot 1 = 1 \\ \text{FAC}(2) &= 2 \cdot \text{FAC}(1) = 2 \cdot 1 = 2 \\ \text{FAC}(3) &= 3 \cdot \text{FAC}(2) = 3 \cdot 2 = 6 \\ \text{FAC}(4) &= 4 \cdot \text{FAC}(3) = 4 \cdot 6 = 24 \end{aligned}$$

EXAMPLE 5.41

Write a Pascal program that accepts a positive integer as input and prints the number and its factorial. Use a recursively defined function.

SOLUTION

```
program factorial (input,output);
var n: integer;
function fac(x:integer): integer;
    begin
        if x = 0 then
            fac := 1
        else
            fac := x*fac(x-1);
    end;
begin (* factorial *)
    writeln ('Enter a positive integer');
    readln (n);
    writeln ('Factorial',n,' is',fac(n));
end.
```

As a second example of a recursively defined function we consider the sequence of **Fibonacci numbers.** The sequence begins with the two integers 1, 1. The third and following numbers are defined by the rule: The nth number of the sequence equals the sum of the two immediately preceding numbers. Thus, the sequence is:

$$1, \quad 1, \quad 2, \quad 3, \quad 5, \quad 8, \quad 13, \quad 21, \quad 34, \quad 55, \quad 89, \ldots$$

We define the function FIB(n) as the nth Fibonacci number. So FIB(5) = 5 and FIB(8) = 21. The Fibonacci sequence may be defined recursively as follows:

$$\begin{cases} \text{FIB}(1) = 1 \\ \text{FIB}(2) = 1 \\ \text{FIB}(n) = \text{FIB}(n-1) + \text{FIB}(n-2) \qquad n \geq 3 \end{cases}$$

Since the recurrence formula defines FIB(n) in terms of two of its predecessors, we need two function values to terminate the recursion.

EXAMPLE 5.42

Write a Pascal program that accepts a positive integer N as input and outputs the Nth Fibonacci number. Use a recursively defined function subroutine.

SOLUTION

```
program fibonacci (input, output);
var n: integer;
function fib(x: integer): integer;
    begin
      if x = 1 then fib := 1
      else if x = 2 then fib := 1
      else   fib := fib(x-1) + fib(x-2);
    end;
begin (* fibonacci *)
    writeln ('Enter a positive integer');
    readln (n);
    writeln ('The ', n,'-th Fibonacci number is ', fib(n));
end.
```

Remark   Although programs that use recursion are frequently short and *look* efficient, there is considerable overhead in implementing recursion. Programmers frequently replace recursive algorithms with iterative ones whenever they are available.

EXERCISES: CASE STUDY 5A

1. Using the built-in function EXP(X) for e^x, write a Pascal program that prints a table of values of e^x as x ranges from -2 to $+2$ in increments of 0.1.
2. Write programs to verify the following limits:

   a. $\lim\limits_{x \to +\infty} \dfrac{3x^2 + 1}{2x^2 + x + 9} = \dfrac{3}{2}$ **b.** $\lim\limits_{x \to +\infty} \dfrac{5x + 9}{x^2} = 0$

   c. $\lim\limits_{x \to +\infty} \dfrac{x^6}{e^x} = 0$ **d.** $\lim\limits_{x \to +\infty} \dfrac{\log_e(\log_e x)}{\log_e x} = 0$

3. Write a user-defined function subroutine for the function:

$$S(x) = \begin{cases} -2 & \text{if } x < 0 \\ 0 & \text{if } x = 0 \\ 3x & \text{if } x > 0 \end{cases}$$

4. a. Write a user-defined function $M(X, Y)$ that returns the value of the larger of X and Y.

   b. Write a program using the function M of part a that accepts three real numbers as input and outputs the largest of the three input values.

5. The power function $P(X, N) = X^N$, N a positive integer, can be defined recursively as follows:

$$\begin{cases} P(X, 1) = X \\ P(X, N) = X \cdot P(X, N-1) & N > 1 \end{cases}$$

   a. Write a Pascal program that accepts as input a real number X and a positive integer N and outputs the Nth power of X. Use the recursive definition of the power function.

   b. Run the program for the values $X = 2$, $N = 5$.

Problems and Projects

6. Define the function $NI(X)$, where X is a nonnegative integer as follows:

$$\begin{cases} NI(X) = X - 10 & \text{if } X > 100 \\ NI(X) = NI(NI(X + 11)) & \text{if } 0 \leq X \leq 100 \end{cases}$$

   a. Find $NI(101)$, $NI(100)$, $NI(99)$, $NI(98)$, and $NI(97)$.

   b. Why is this function called the "91 function"? (See Stanat and McAllister, 1977.)

   c. Write a Pascal function subroutine that will compute $NI(X)$. Test the function by writing a program that prints out the values of NI listed in part a.

7. Write programs to verify the following limits:

   a. $\lim\limits_{n \to +\infty} \dfrac{e^n}{n!} = 0$ **b.** $\lim\limits_{n \to +\infty} \dfrac{n!}{n^n} = 0$

8. The binomial coefficients (see Chapter 3) can be computed using the recursion formula

$$\begin{cases} BIN(n, 0) = 1 \\ BIN(n, n) = 1 \\ BIN(n, m) = BIN(n-1, m) + BIN(n-1, m-1) \end{cases}$$

   a. Write a Pascal program that accepts as input two integers N and M and outputs the binomial coefficient $B(N, M)$. Use the recurrence formula.

   b. Test your program by making it compute $\binom{5}{5}$ and $\binom{7}{3}$.

9. Ackerman's function $A(m, n)$ is a function of two integer variables, which is defined recursively as follows:

$$\begin{cases} A(0, n) = n + 1 & n \geq 0 \\ A(m, 0) = A(m - 1, 1) & m > 0 \\ A(m, n) = A(m - 1, A(m, n - 1)) & m > 0, n > 0 \end{cases}$$

a. Compute $A(2, 3)$. As you compute values for Ackerman's function save the values in a table.

b. How many function values did you compute to obtain the answer to part a?

c. Write a Pascal function subroutine that computes Ackerman's function. Test the function by writing a program that inputs values for N and M and prints the value of $A(M, N)$. Input the values 2 and 3. (*Note:* For small values of M and N your computer's stack can overflow and cause the program to abort. For example, on a Honeywell-68/DPS main frame computer, an attempt to calculate $A(4, 5)$ caused stack overflow.)

CASE STUDY 5B   ALGORITHM ANALYSIS: ORDERS OF MAGNITUDE

One of the more important activities of a computer scientist is analyzing the algorithms he or she constructs. This analysis is undertaken for any of several reasons:

1. To estimate the runtime of an algorithm. These estimates can be obtained by carefully analyzing the number of operations used by the algorithm (additions, subtractions, multiplications, divisions, comparisons, procedure calls, and so on) and their relative costs.

2. To compare the efficiencies of several algorithms that solve the same class of problems. We would, of course, like to choose the most efficient algorithm—this saves time, money, and keeps the employer very happy!

3. To help decide if an algorithm is optimal. It is possible to *prove* mathematically that certain algorithms are optimal in their solution of a particular problem in the sense that no substantially more efficient algorithm *can* be found to solve the problem. If an optimal algorithm is found to solve a problem, then we should not bother to look for a more efficient one!

Estimating the runtime of an algorithm is largely a combinatorial problem and can be solved with the help of the techniques of Chapter 3.

Deciding if an algorithm is optimal involves comparing the efficiencies of algorithms. The efficiency of an algorithm is measured by its complexity function, which we define later. To help compare two algorithms, therefore, we need techniques for comparing two functions. A frequently used notation for this comparison is the so-called Landau notation, O and o, read "big-oh" and "little-oh."

We shall give an intuitive treatment of O and o, since a rigorous definition is possible only after a course in calculus. We assume in this case study that all functions have as their domain the set of positive integers. Consequently, the variable used in our function definitions will be n rather than x.

Let $\lim_{n \to +\infty} f(n) = \lim_{n \to +\infty} g(n) = +\infty$. To say that $\textbf{\textit{f}(n) = O(\textit{g}(n))}$, read "f is big-oh of g," means that $f(n)$ grows no faster than $g(n)$ as $n \to +\infty$. Thus, the growth of $f(n)$ is, in a sense, bounded by the growth of $g(n)$. If $f(n) = O(g(n))$ and $g(n) = O(f(n))$, then each of f and g grows no faster than the other. In this case, f and g are said to have the **same order of magnitude.** To say $\textbf{\textit{f}(n) = o(\textit{g}(n))}$, read "f is little-oh of g," means that $g(n)$ grows much faster than $f(n)$. Thus, g is said to have a **higher order of magnitude** than f. Note that if $f(n) = o(g(n))$, then $f(n) = O(g(n))$. That is, if f grows more slowly than g, it grows no faster than g.

The O and o relations are usually established by taking limits.

Theorem 5.4

Let $\lim_{n \to +\infty} f(n) = \lim_{n \to +\infty} g(n) = +\infty$ and let $\lim_{n \to +\infty} f(n)/g(n) = L$. Then,

i. If $0 < L < +\infty$, f and g are of the same order of magnitude.

ii. If $L = 0$, $f(n) = o(g(n))$.

iii. If $L = +\infty$, $g(n) = o(f(n))$.

We already established some O and o relations in Section 5.4. For example:

1. $\lim_{n \to +\infty} (5n^2 + 3n - 4)/(n^2 - 1) = 5$ implies by Theorem 5.4 that $5n^2 + 3n - 4$ and $n^2 - 1$ have the same order of magnitude. In fact, our discussion following Example 5.24 may be stated as in Note 5.18.

Note 5.18

A polynomial has the same order of magnitude as its leading term.

---

2. $\lim_{n \to +\infty} (3n + 4)/(n^2 + n) = 0$ implies by Theorem 5.4 that $3n + 4 = o(n^2 + n)$.

Using Theorem 5.4, we can restate Theorem 5.3 as in Note 5.19.

Note 5.19

If $P(n)$ is a polynomial such that $\lim_{x \to +\infty} P(n) = +\infty$ and $f(n) = a \cdot b^{kn}$, where a and b are both positive, then $P(n) = o(f(n))$.

More compactly, an exponential function has a higher order of magnitude than any polynomial function.

To begin comparing the efficiencies of algorithms, we need several definitions. Let the algorithm A solve a certain class of problems. For example, A could be an algorithm for testing whether a list of numbers is ordered. Let n be some measure of the size of the problem A solves. If A, for example, is an algorithm for testing whether a list of numbers is ordered, n could be the number of elements in the list. If A, on the other hand, is an algorithm for evaluating a polynomial, n could be the degree of the polynomial.

The **complexity function** $C_A(n)$ of an algorithm A of size n is defined to be the maximum number of basic operations that can be performed by A. Thus, $C_A(n)$ is a measure of the worst case of the algorithm, that is, the slowest the algorithm can run. Once the number of operations and the time required for each operation are known, the total time required by the algorithm can be computed. Thus, the complexity function is sometimes called the **time complexity** or, simply, the **time of the algorithm** A.

An algorithm A is a **polynomial time algorithm,** or simply a **polynomial algorithm,** if there exists a polynomial $p(n)$ such that $C_A(n) = O(p(n))$. An algorithm that is not a polynomial algorithm is called an **exponential algorithm.** The **order of magnitude of an algorithm** is the order of magnitude of its complexity function. Thus, for example, an algorithm A is of order n^3 if $C_A(n) = O(n^3)$.

EXAMPLE 5.43

What are the complexity functions for the algorithms SORT1 and SORT2 of Case Study 3B? What is the order of magnitude of each algorithm?

SOLUTION

The size n of the algorithms is the number of elements in the list $A(i)$. As derived in Case Study 3B, $C_{\text{SORT1}}(n) = 2n + 1 = O(n)$ and $C_{\text{SORT2}}(n) = n^2 + 3n + 1 = O(n^2)$. Thus, SORT1 has order n and SORT2 has order n^2.

As a simple example of an exponential algorithm, consider the following game. I pick an n-bit binary integer, and you must guess the integer. Suppose you choose for your algorithm the following procedure G: Start with the binary integer consisting of n 0s. If this is not the number I chose, add 1 and test again. This procedure is, in essence, a sequential search through the 2^n n-bit binary integers. In the worst possible case it will take 2^n questions to guess the correct number (if I choose the n-bit binary consisting of all 1s). Therefore, the complexity function of G is $C_G(n) = 2^n$ and the algorithm is exponential.

Can we find a better algorithm for our game? Let us try the following procedure H for guessing the binary number: Ask if the first bit is 0. If I answer no, the bit is 1. If I answer yes, the bit is 0. Now ask the same question for the second and following bits. Using this procedure we can guess the number by asking n questions, one question per bit. Hence, the complexity function of the algorithm H is $C_H(n) = n$.

To compare these two complexity functions consider Table 5.2. Obviously, algorithm H is far faster than algorithm G.

A COMPARISON OF COMPLEXITY FUNCTIONS

n	$C_G(n) = 2^n$	$C_H(n) = n$
10	1024	10
20	1048576	20
30	1073741824	30

Table 5.2

Computer scientists try to avoid exponential algorithms because the explosive growth of their complexity functions makes even moderately sized problems intractable, even with today's super-fast computers. To take a simple example, suppose our algorithm G with complexity function $C_G(n) = 2^n$ is implemented on a computer that can do one comparison per microsecond ($= 1/10^6$ sec). If the problem is to guess a 40-bit number ($n = 40$), it would take $2^{40} = 1,099,511,627,776$ comparisons to complete the algorithm in the worst case. It would, therefore, take about 1,099,512 sec or about 13 days to complete the computation.

However, if we use the polynomial algorithm H with complexity function $C_H(n) = n$, the same problem could be solved in 40 comparisons. On our hypothetical computer, this algorithm would take 0.00004 *sec* to run. The message is clear: Avoid exponential algorithms, if possible.

Common orders of magnitude are n, $\log_2 n$, $n \cdot \log_2 n$, n^2, n^3, and 2^n. Table 5.3, which lists some values for these functions, is useful in gaining insight into these orders of magnitude.

COMMON ORDERS OF MAGNITUDE

$\log_2 n$	n	$n \cdot \log_2 n$	n^2	n^3	2^n
0	1	0	1	1	2
1	2	2	4	8	4
2	4	8	16	64	16
3	8	24	64	512	256
4	16	64	512	4096	65536
5	32	160	1024	32768	4294967296

Table 5.3

Note how, for any particular row of the table (say row 5) each value is large compared to the value on its left.

Remark   If the size n of a problem is not too large, it may be more efficient to use the algorithm with a higher order of magnitude. (Recall that orders of magnitude are meaningful only for large values of n.) Thus, an algorithm of order $10n^2$ would be preferred over one of order $100n$ for a problem of size $n < 10$ because $10n^2 < 100n$ if $n < 10$.

Remark   There exist many problems for which there is no known polynomial time algorithm. In fact, there is a class of important problems, called **NP-Complete** problems, that are equivalent in the following sense: If any NP-Complete problem can be solved in polynomial time, then *all* NP-Complete problems can be solved in polynomial time. Since no polynomial time algorithm has yet been found for any NP-Complete problem, one is inclined to believe that no such algorithm *can* exist, that is, that these problems are intrinsically exponential.

EXERCISES: CASE STUDY 5B

1. Establish the following O and o relations using Theorem 5.4:
   a. $n^{100} = O(3^n)$ b. $(8n^2 - 3n + 1) = O(5n^2)$
2. For each pair of complexity functions, determine the values of n for which A is preferred over B, and conversely.

	Algorithm A	Algorithm B
Problem 1	$5n^3$	2^n
Problem 2	$n(\log_2 n)^2$	$n^2/\log_2 n$
Problem 3	$n^{2.9}$	$n^3/2$

3. The order of magnitude of an expression involving a root can sometimes be easily established. For example, consider $\sqrt{3n^4 - 5n^2 + 9}$. The polynomial under the radical has order of magnitude n^4. Its square root, therefore, has order of magnitude $\sqrt{n^4} = n^2$. Using similar reasoning, establish the following:

   a. $\sqrt{5n^5 + 3n - 4} = O(n^2\sqrt{n})$ b. $\dfrac{\sqrt{9n^6 + 3n^2 - 7}}{\sqrt{n^2 + 3n + 6}} = O(n^2)$

4. What is true of $f(n)$ if we know $f(n) = O(k)$ for some constant k?
5. If $n^a = o(n^b)$, what is the relationship of a to b?
6. Find the complexity functions and orders of magnitude for the standard algorithm for evaluating a polynomial and Horner's algorithm for evaluating a polynomial. (See Case Study 3B.)
7. In Case Study 2A, we discussed the sequential search and binary search algorithms, which we denote by S and B. Let the size of a search algorithm be the number of elements in the list to be searched. The basic operation will be comparison.
   a. Show that the complexity function for the sequential search algorithm is $C_S(n) = n$. Remember to consider the worst case.

b. Show that the complexity function for the binary search algorithm is $C_B(n) = \log_2 n$. Remember to consider the worst case. (*Hint:* Each iteration in the binary search algorithm cuts the remainder of the list in half. The number of comparisons necessary is the number of times the list must be cut in half to be left with one element. If the number of comparisons is denoted by m, we must have $n/2^m = 1$ or $2^m = n$.)

c. Which algorithm is more efficient? Prove your answer.

CASE STUDY 5C   CRYPTOGRAPHY

Let $\mathcal{A}$ be the standard 26 letter English alphabet. An **encoding** of $\mathcal{A}$ is a 1-1 function of $\mathcal{A}$ onto itself

$$f: \mathcal{A} \to \mathcal{A}$$

An encoding maps a letter of the alphabet to another letter of the alphabet. An encoding must be 1-1, because if $f(a_1) = f(a_2) = b$, then we would not know how to **decode** b. Would b be decoded as a_1 or a_2? The **decoding** of f is defined to be the inverse function f^{-1}. (Why are we sure f^{-1} exists?)

The encoding f can be extended to arbitrary sequences of letters (words) in a natural way:

$$f(a_1 a_2 \ldots a_n) = f(a_1) f(a_2) \ldots f(a_n)$$

To encode a word, therefore, encode it one letter at a time.

EXAMPLE 5.44 _____

Consider the encoding defined by

```
A B C D E F G H I J K L M N O P Q R S T U V W X Y Z
↓ ↓ ↓ ↓ ↓ ↓ ↓ ↓ ↓ ↓ ↓ ↓ ↓ ↓ ↓ ↓ ↓ ↓ ↓ ↓ ↓ ↓ ↓ ↓ ↓ ↓
H I J K L M N O P Q R S T U V W X Y Z A B C D E F G
```

a. Encode the message: LUKE SKYWALKER

b. Decode the message: CPJAVY MYHURLUZALPU

SOLUTION

a. Using the given encoding:

LUKE SKYWALKER
↓↓↓↓ ↓↓↓ ↓ ↓↓↓↓↓
SBRL ZRF D HSRLY

b. Using the inverse of the given correspondences:

CP J A V Y MYHURLUZALPU
↓↓↓↓↓↓ ↓↓↓↓↓↓↓↓↓↓↓↓
VICTOR FRANKENSTEIN

There is a simple encoding that uses modular arithmetic. (Before proceeding, you might want to review Section 1.3.) The encoding is called a **Caesar cypher** because Julius Caesar used such a cypher more than 2000 years ago. A Caesar cypher is a cyclical shifting of the letters of the alphabet a fixed number of positions. If we shift three positions, we obtain the encoding:

```
A B C D E F G H I J K L M N O P Q R S T U V W X Y Z
↓ ↓ ↓ ↓ ↓ ↓ ↓ ↓ ↓ ↓ ↓ ↓ ↓ ↓ ↓ ↓ ↓ ↓ ↓ ↓ ↓ ↓ ↓ ↓ ↓ ↓
D E F G H I J K L M N O P Q R S T U V W X Y Z A B C
```

A Caesar cypher can be described mathematically as follows: Associate with the letters A, B, C, . . . , Y, Z the integers 1, 2, 3, . . . , 25, 0. We use 0 in the correspondence rather than 26 because $26 \equiv 0 \bmod 26$. See Figure 5.29.

```
A  B  C  D  E  F  G  H  I  J  K  L  M  N  O  P  Q  R  S  T  U  V  W  X  Y  Z
|  |  |  |  |  |  |  |  |  |  |  |  |  |  |  |  |  |  |  |  |  |  |  |  |  |
1  2  3  4  5  6  7  8  9  10 11 12 13 14 15 16 17 18 19 20 21 22 23 24 25 0
```

Figure 5.29

If the alphabet is to be shifted b positions, the letter y corresponding to x is given by

$$f(x) = y \equiv (x + b) \bmod 26$$

where we interpret the right side of the congruence as the least residue modulo 26. The Caesar cypher just noted is, therefore, described by

$$f(x) = y \equiv (x + 3) \bmod 26$$

To find the image of the letter G, note that G is the seventh letter of the alphabet.

$$f(7) = y \equiv (7 + 3) \bmod 26 = 10$$

The tenth letter of the alphabet is J, so G is encoded as J. The Caesar cypher of Example 5.44 can be described by $f(x) = y \equiv (x + 7) \bmod 26$.

If we suspect a code is a Caesar cypher, how can we decode it? The method of the next example will usually work.

EXAMPLE 5.45

We suspect the following message is encoded using a Caesar cypher. Decode it.

ZK NRJ SVRLKP KYRK BZCCVU KYV SVRJK

SOLUTION

Take one word of the message and under each letter continue the alphabet from that word on until you come to a row containing a word you recognize. Usually, a long word is better than a short one:

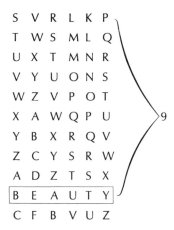

Hence, to decode, shift the alphabet nine letters. To encode, shift the alphabet $26 - 9 = 17$ letters. We leave the decoding of the rest of the message as exercise 3 in Exercises: Case Study 5C.

The Caesar cypher can be generalized by using the congruence $f(x) = y \equiv (ax + b) \bmod 26$. For this to be an encoding, however, f^{-1} must exist. To find f^{-1}, we must solve the congruence $y \equiv (ax + b) \bmod 26$ for x in terms of y (recall Note 5.15). First,

$$(ax + b) \equiv y \bmod 26 \qquad \text{[Theorem 1.4(ii)]}$$

Then,

$$ax \equiv y - b \bmod 26 \qquad \text{[Theorem 1.5(ii)]}$$

To solve this congruence for x we must be able to divide both sides by a. To guarantee this, we require $\gcd(a, 26) = 1$. See Theorem 1.8 and Corollary 1.1. We now define a **modular encoding** by

$$f(x) = y \equiv (ax + b) \bmod 26$$

where $\gcd(a, 26) = 1$.

EXAMPLE 5.46

Using the modular encoding $f(x) = y \equiv (3x + 2) \bmod 26$,

a. encode DARTH VADER   b. decode GKEDLQJ UZEDE

SOLUTION

a.
$$\begin{array}{l} \text{DARTH VADER} \\ \downarrow\downarrow\downarrow\downarrow\downarrow\ \downarrow\downarrow\downarrow\downarrow\downarrow \\ \text{NEDJZ PENQD} \end{array}$$

b. Since $y \equiv (3x + 2) \bmod 26$, $3x \equiv (y - 2) \bmod 26$. Now, multiply both sides of this congruence by 9, obtaining

$$9 \cdot 3x \equiv 9(y - 2) \bmod 26$$

or

$$27x \equiv 9(y - 2) \bmod 26$$

But,

$$27x \equiv x \bmod 26$$

so

$$x \equiv 9(y - 2) \bmod 26 \qquad \text{[Theorem 1.4(iii)]} \quad (1)$$

Why multiply both sides by 9? We multiply both sides by 9 because 9 is the solution to the congruence

$$3z \equiv 1 \bmod 26$$

or

$$3z \equiv 27 \bmod 26$$

or

$$z \equiv 9 \bmod 26$$

Modulo 26, 9 acts like the multiplicative inverse of 3.

The decoding equation (1) can now be used to decode our message. G is the seventh letter of the alphabet, thus $x \equiv 9(7 - 2) \bmod 26 \equiv 45 \bmod 26 \equiv 19 \bmod 26$. The letter corresponding to G is, then, S, the nineteenth letter of the alphabet. Continuing, we obtain

$$
\begin{array}{l}
\text{GKEDLQJ UZEDE} \\
\downarrow\downarrow\downarrow\downarrow\downarrow\downarrow\downarrow \; \downarrow\downarrow\downarrow\downarrow\downarrow \\
\text{SCARLET OHARA}
\end{array}
$$

EXAMPLE 5.47 _____

What is the decoding equation for the modular encoding $f(x) = y \equiv (7x + 5) \bmod 26$?

SOLUTION

$$y \equiv (7x + 5) \bmod 26$$

so

$$7x + 5 \equiv y \bmod 26$$

or

$$7x \equiv (y - 5) \bmod 26 \qquad\qquad\qquad (2)$$

We must now multiply both sides of this congruence by a number. To decide which number, we solve $7z \equiv 1 \bmod 26$. So,

$$7z \equiv 1 \bmod 26 \equiv 27 \bmod 26 \equiv 53 \bmod 26$$
$$\equiv 79 \bmod 26 \equiv 105 \bmod 26$$

Dividing by 7, $z \equiv 15 \bmod 26$. Therefore, we multiply both sides of (2) by 15, obtaining

$$15 \cdot 7x \equiv 15(y - 5) \bmod 26$$

or

$$x \equiv 15(y - 5) \bmod 26$$

If a coded message is suspected of being constructed using a modular encoding, how can it be decoded when a and b are not known? The decoding method used for Caesar cyphers in Example 5.45 cannot be used for modular encodings. If, however, we can decode two letters of the modular encoding, it is sometimes possible to decode the message.

In standard English, certain letters occur with greater frequency than others. The letter E is most frequent, followed by T, then N, and so on. The least frequently used letter is Z. Also, certain pairs of letters, like AN and HE, occur more frequently than other pairs. By making a frequency count of the letters of the given message and of some pairs of letters, we can make a tentative decoding of two letters. This will yield a tentative decoding function. We then decode the entire message using this decoding function. If the message makes sense, we are done. If not, we must make a further analysis of the text. We illustrate with a simple example.

EXAMPLE 5.48

The following message was constructed using a modular encoding. Decode it.

CFFYB RFFYB OZCCOF CRFFCB QZSA

SOLUTION

The most frequently occurring letter in the message is F, and the next most frequently occurring letter is C. If $f(x) = y \equiv (ax + b) \bmod 26$ is the encoding function for the message, then we guess that $f(E) = F$ and $f(T) = C$.

Since E and F are the fifth and sixth letters of the alphabet, and T and C are the twentieth and third, we are guessing that $f(5) = 6$ and $f(20) = 3$. We write the congruence defining f as $ax + b \equiv y \bmod 26$ and substitute the pairs of values of x and y. This yields the simultaneous congruences

$$5a + b \equiv 6 \bmod 26$$
$$20a + b \equiv 3 \bmod 26$$

Simultaneous congruences may be solved by methods similar to those used for solving simultaneous equations. To eliminate b, subtract the first congruence from the second obtaining

$$15a \equiv -3 \text{ mod } 26$$

We want to add some multiple of the modulus 26 to the right side to obtain a multiple of 15. Adding $3 \cdot 26 = 78 \equiv 0 \text{ mod } 26$ to the right side, we obtain

$$15a \equiv 75 \text{ mod } 26$$

Dividing by 15,

$$a \equiv 5 \text{ mod } 26$$

Since we are looking for least residues, $a = 5$. To find b, substitute $a = 5$ in the first of the simultaneous congruences:

$$5 \cdot 5 + b \equiv 6 \text{ mod } 26$$

or

$$b \equiv -19 \text{ mod } 26 \equiv 7 \text{ mod } 26$$

Therefore, we choose $b = 7$. The encoding function is

$$f(x) = y \equiv (5x + 7) \text{ mod } 26$$

The corresponding decoding function is

$$x \equiv 21(y - 7) \text{ mod } 26$$

(See exercise 4 in Exercises: Case Study 5C.) The decoded message is: TEENY WEENY LITTLE TWEETY BIRD.

EXERCISES: CASE STUDY 5C

1. Consider the following message:

AT MORN AT NOON AT TWILIGHT DIM
MARIA THOU HAST HEARD MY HYMN

   a. Use the Caesar cypher $f(x) = y \equiv (x + 10) \text{ mod } 26$ to encode the message.
   b. Use the modular encoding $f(x) = y \equiv (9x + 8) \text{ mod } 26$ to encode the message.
2. The following message was encoded using a Caesar cypher. Decode it.

ITAEQ IAAPE FTQEQ MDQ U FTUZW U WZAI

3. Complete decoding the message of Example 5.45.
4. Show that the decoding function of $f(x) = y \equiv (5x + 7) \text{ mod } 26$ is $x \equiv 21(y - 7) \text{ mod } 26$.
5. a. What is the decoding function for $f(x) = y \equiv (9x + 2) \text{ mod } 26$?
   b. Use the decoding function derived in part a to decode the message:

AVKZ FEMVZ ZVHGIMV SGXLUH AEXLGA THUKWQ

6. Decode the following message, which was encoded using a modular encoding:

JMVE IVD VKKDHKE VKKDHKE IVD JMVE

Problems and Projects

7. **a.** Write a program that accepts a message in English and encodes it using the encoding function of exercise 4.

   b. Write a program that accepts an encoded message and decodes it using the decoding function of exercise 4. Note that you must decide what to do with blanks.

REFERENCES

Dale, N., and Orshalick, D. *Introduction to Pascal and Structured Design*. Lexington, Mass.: D. C. Heath, 1983.

Fisher, J. L. *Application Oriented Algebra*. New York: T. Y. Crowell, 1977.

Foerster, P. A. *Algebra and Trigonometry: Functions and Applications*. Reading, Mass.: Addison-Wesley, 1980.

Goodman, S., and Hedetniemi, S. T. *Introduction to the Design and Analysis of Algorithms*. New York: McGraw-Hill, 1977.

Horowitz, E., and Sahni, S. *Fundamentals of Computer Algorithms*. Potomac, Md.: Computer Science Press, 1978.

Leithold, L. *The Calculus with Analytic Geometry*. 4th ed. New York: Harper and Row, 1980.

Stanat, D. F., and McAllister, D. F. *Discrete Mathematics in Computer Science*. Englewood Cliffs, N.J.: Prentice-Hall, 1977.

Starkey, J. D., and Ross, R. J. *Fundamental Programming*. St. Paul, Minn.: West Publishing, 1984.

CHAPTER 6

Many mathematics and computer science problems involve collections of variables that all measure the same quantity but for different individuals or items. For example, suppose we want to store on a computer the final exam grade for each of 35 students. One way to handle this problem is to create 35 variables and store one final exam grade in each variable's storage position. Another way is to create a single variable having the capacity to store 35 quantities simultaneously. In mathematics, this type of variable is called a **vector** variable. There are also variables that can store a table of numbers. For example, suppose we want to store the final exam grade for each of the 35 students in each of their five classes. Here we are storing a table having five rows (one row for each class) and 35 entries in each row (one entry for each student). A variable that stores a table of numbers is called a **matrix** variable. A vector is actually a matrix having a single row or a single column. In this chapter, we will study matrices, operations on them, and functions involving matrices. In the case studies at the end of the chapter, we will see how vectors and matrices are used in computer applications.

6.1   MATRICES AND ARRAYS

There are many situations where we need to store a table of numbers. One such situation is for inventory problems. Suppose a shoe store carries 80 shoe styles and each shoe comes in nine sizes (for example, 5 through 13). The store's inventory could be listed in a very orderly way in a table having 80 rows and 9 columns. If there is a 3 in row 28, column 4, this means the store has 3 pairs of style 28 shoes in size 8 (column 4 corresponds to size 8). See Figure 6.1.

A **matrix** is a rectangular table of numbers. The numbers are called **entries** (or **elements**) of the matrix. The **dimension** of a matrix is the number of rows and columns it contains. The dimension is written in the form:

(Number of rows) × (Number of columns)

where the × is read "by." In our shoe store example, the inventory matrix of Figure 6.1 (page 293) has 80 rows and 9 columns. Thus, it is an 80 × 9 matrix (read "80 by 9" matrix).

VECTORS AND MATRICES

EXAMPLE 6.1

Phil owns four gas stations, each of which sells three types of gas: regular, unleaded, and premium. The number of gallons of each type of gas sold by his stations last week are 3200, 6800, and 1100 for station 1; 4450, 7230, and 1245 for station 2; 3800, 6950, and 900 for station 3; and 5700, 9280, and 1360 for station 4. If rows correspond to stations,

a. What is the dimension of his sales matrix?

b. Write the sales matrix.

	Size								
Style	5	6	7	8	9	10	11	12	13
1	4	1	3	0	2	3	1	0	0
2	1	2	1	1	0	3	2	1	0
3	1	1	3	4	4	2	3	0	2
.	.	.	.	.	.	.	.	.	.
.	.	.	.	.	.	.	.	.	.
.	.	.	.	.	.	.	.	.	.
.	.	.	.	.	.	.	.	.	.
28	5	2	1	③	1	0	2	1	1
.	.	.	.	.	.	.	.	.	.
.	.	.	.	.	.	.	.	.	.
.	.	.	.	.	.	.	.	.	.
80	3	1	4	1	2	2	1	3	1

Figure 6.1

SOLUTION

a. His sales matrix is a 4×3 matrix.

b. The matrix is
$$\begin{bmatrix} 3200 & 6800 & 1100 \\ 4450 & 7230 & 1245 \\ 3800 & 6950 & 900 \\ 5700 & 9280 & 1360 \end{bmatrix}.$$

Throughout the next few sections, we explore properties of matrices and operations involving matrices (*matrices* is the plural form of *matrix*).

A matrix is named by using a boldface capital letter. Entries within the matrix are indicated by attaching two *subscripts* (the row number followed by the column number) to the lowercase letter corresponding to the matrix name.

EXAMPLE 6.2

For the matrix

$$\mathbf{A} = \begin{bmatrix} 6 & -1 & 3 \\ 4 & 2 & 5 \\ 7 & 1 & 8 \end{bmatrix}$$

list a_{23}, a_{31}, and a_{12}.

SOLUTION

a_{23} is the entry in the second row and third column of $\mathbf{A}$. Thus $a_{23} = 5$. Similarly, $a_{31} = 7$, and $a_{12} = -1$.

Occasionally, it is more convenient to express the entries of a matrix using a different format. This is particularly true when we are stating properties or proving theorems for matrices. Thus, we sometimes use $[\mathbf{A}]_{ij}$ to denote a_{ij}. In Example 6.2, for instance, we could have written $[\mathbf{A}]_{23} = 5$.

Suppose Phil, of Example 6.1, uses last week's sales figures as an estimate of average weekly sales. How many gallons of each type of gas can he expect each of his stations to sell during a 6-week period? To determine this, simply multiply each entry of his sales matrix

$$\begin{bmatrix} 3200 & 6800 & 1100 \\ 4450 & 7230 & 1245 \\ 3800 & 6950 & 900 \\ 5700 & 9280 & 1360 \end{bmatrix}$$

by 6 to get

$$\begin{bmatrix} 19200 & 40800 & 6600 \\ 26700 & 43380 & 7470 \\ 22800 & 41700 & 5400 \\ 34200 & 55680 & 8160 \end{bmatrix}$$

This is an example of an operation called **scalar multiplication,** which involves multiplying a number (called a **scalar**) times a matrix. If Phil waits six weeks and constructs the sale matrix for each week, he can find the actual total sales for the six week period by adding the six sales matrices elementwise. This is an example of **matrix addition.** The algebraic operations of addition, subtraction, and scalar multiplication for matrices are described in Note 6.1.

Note 6.1   Algebraic Operations on Matrices

1. **Equality:**   Two matrices **A** and **B** are equal if they have the same dimension and corresponding entries are equal, that is, $a_{ij} = b_{ij}$ for all pairs (i, j).
2. **Addition:**   Two matrices may be added only if they have the same dimension. To add two matrices, add corresponding entries. Symbolically, the (i, j) entry of the sum is $[\mathbf{A} + \mathbf{B}]_{ij} = a_{ij} + b_{ij}$.
3. **Subtraction:**   Two matrices may be subtracted only if they have the same dimension. To subtract two matrices, subtract corresponding entries. Thus $[\mathbf{A} - \mathbf{B}]_{ij} = a_{ij} - b_{ij}$.
4. **Scalar Multiplication:**   To multiply a matrix by a scalar, multiply each entry of the matrix by the scalar. Thus $[k\mathbf{A}]_{ij} = k \cdot a_{ij}$.

We now consider some examples that illustrate these operations.

EXAMPLE 6.3 _____

Write the 2×3 matrix **A** with entries $a_{11} = 5$, $a_{22} = -7$, and $a_{13} = 4$, and whose other entries are all 1s.

SOLUTION

$$\mathbf{A} = \begin{bmatrix} 5 & 1 & 4 \\ 1 & -7 & 1 \end{bmatrix}$$

EXAMPLE 6.4 _____

Write the 3×3 matrix **B** such that $b_{ij} = i + j$.

SOLUTION

$$B = \begin{bmatrix} 2 & 3 & 4 \\ 3 & 4 & 5 \\ 4 & 5 & 6 \end{bmatrix}$$

EXAMPLE 6.5

Write the 4×2 matrix C where $c_{ij} = i^2 + 3ij - 2j$.

SOLUTION

We calculate the entries: $c_{11} = 1^2 + 3(1)(1) - 2(1) = 2$, $c_{12} = 1^2 + 3(1)(2) - 2(2) = 3$, and $c_{21} = 2^2 + 3(2)(1) - 2(1) = 8$. Similarly, we find $c_{22} = 12$, $c_{31} = 16$, $c_{32} = 23$, $c_{41} = 26$, and $c_{42} = 36$. Thus,

$$C = \begin{bmatrix} 2 & 3 \\ 8 & 12 \\ 16 & 23 \\ 26 & 36 \end{bmatrix}$$

EXAMPLE 6.6

Let $A = \begin{bmatrix} 2 & -1 & 4 \\ -3 & 7 & 10 \end{bmatrix}$, $B = \begin{bmatrix} 3 & 5 & -1 \\ 12 & 2 & 6 \end{bmatrix}$, and $C = \begin{bmatrix} 1 & 1 & -4 \\ 2 & 8 & 17 \end{bmatrix}$.
Calculate each of the following.

a. $A + B$ b. $A - C$ c. $3B$ d. $A + 2B - C$ e. $B + 5C$

SOLUTION

a. $A + B = \begin{bmatrix} 2 & -1 & 4 \\ -3 & 7 & 10 \end{bmatrix} + \begin{bmatrix} 3 & 5 & -1 \\ 12 & 2 & 6 \end{bmatrix}$

$= \begin{bmatrix} 2+3 & -1+5 & 4+(-1) \\ -3+12 & 7+2 & 10+6 \end{bmatrix} = \begin{bmatrix} 5 & 4 & 3 \\ 9 & 9 & 16 \end{bmatrix}$

b. $A - C = \begin{bmatrix} 2 & -1 & 4 \\ -3 & 7 & 10 \end{bmatrix} - \begin{bmatrix} 1 & 1 & -4 \\ 2 & 8 & 17 \end{bmatrix}$

$= \begin{bmatrix} 2-1 & -1-1 & 4-(-4) \\ -3-2 & 7-8 & 10-17 \end{bmatrix} = \begin{bmatrix} 1 & -2 & 8 \\ -5 & -1 & -7 \end{bmatrix}$

c. $3B = 3\begin{bmatrix} 3 & 5 & -1 \\ 12 & 2 & 6 \end{bmatrix} = \begin{bmatrix} 3 \cdot 3 & 3 \cdot 5 & 3(-1) \\ 3 \cdot 12 & 3 \cdot 2 & 3 \cdot 6 \end{bmatrix}$

$= \begin{bmatrix} 9 & 15 & -3 \\ 36 & 6 & 18 \end{bmatrix}$

d. $A + 2B - C = \begin{bmatrix} 2 & -1 & 4 \\ -3 & 7 & 10 \end{bmatrix} + 2\begin{bmatrix} 3 & 5 & -1 \\ 12 & 2 & 6 \end{bmatrix} - \begin{bmatrix} 1 & 1 & -4 \\ 2 & 8 & 17 \end{bmatrix}$

$= \begin{bmatrix} 2+6-1 & -1+10-1 & 4-2+4 \\ -3+24-2 & 7+4-8 & 10+12-17 \end{bmatrix} = \begin{bmatrix} 7 & 8 & 6 \\ 19 & 3 & 5 \end{bmatrix}$

e. $B + 5C = \begin{bmatrix} 3 & 5 & -1 \\ 12 & 2 & 6 \end{bmatrix} + 5\begin{bmatrix} 1 & 1 & -4 \\ 2 & 8 & 17 \end{bmatrix}$

$= \begin{bmatrix} 3+5 & 5+5 & -1-20 \\ 12+10 & 2+40 & 6+85 \end{bmatrix} = \begin{bmatrix} 8 & 10 & -21 \\ 22 & 42 & 91 \end{bmatrix}$

EXAMPLE 6.7

Find D so that

$$2D + \begin{bmatrix} -5 & 8 \\ 1 & -3 \end{bmatrix} = \begin{bmatrix} 4 & 7 \\ -1 & 5 \end{bmatrix}$$

SOLUTION

D must be a 2×2 matrix, so let $D = \begin{bmatrix} x & y \\ z & w \end{bmatrix}$. Then $2\begin{bmatrix} x & y \\ z & w \end{bmatrix} + \begin{bmatrix} -5 & 8 \\ 1 & -3 \end{bmatrix} = \begin{bmatrix} 4 & 7 \\ -1 & 5 \end{bmatrix}$ yields $\begin{bmatrix} 2x-5 & 2y+8 \\ 2z+1 & 2w-3 \end{bmatrix} = \begin{bmatrix} 4 & 7 \\ -1 & 5 \end{bmatrix}$. Thus, using the definition of equality of matrices, $2x - 5 = 4$, $2y + 8 = 7$, $2z + 1 = -1$, and $2w - 3 = 5$. We solve each equation and get $x = 9/2$, $y = -1/2$, $z = -1$, and $w = 4$. Therefore,

$$D = \begin{bmatrix} 9/2 & -1/2 \\ -1 & 4 \end{bmatrix}$$

ALTERNATIVE SOLUTION

Begin by subtracting $\begin{bmatrix} -5 & 8 \\ 1 & -3 \end{bmatrix}$ from both sides of $2D + \begin{bmatrix} -5 & 8 \\ 1 & -3 \end{bmatrix} = \begin{bmatrix} 4 & 7 \\ -1 & 5 \end{bmatrix}$. This gives $2D = \begin{bmatrix} 4 & 7 \\ -1 & 5 \end{bmatrix} - \begin{bmatrix} -5 & 8 \\ 1 & -3 \end{bmatrix}$. We perform the subtraction and get $2D = \begin{bmatrix} 9 & -1 \\ -2 & 8 \end{bmatrix}$. Finally, multiply both sides by $\frac{1}{2}$ and get $D = \begin{bmatrix} 9/2 & -1/2 \\ -1 & 4 \end{bmatrix}$. Note that the usual methods for solving linear equations—adding something (in this case, a matrix) to both sides of an equation, multiplying both sides by a nonzero constant, and so on—are applicable to matrices.

Many of the familiar properties that hold for real numbers also hold for matrices provided the matrices have the same dimensions (see Note 6.2).

Note 6.2   Basic Properties of Matrix Addition and Scalar Multiplication

If **A**, **B**, and **C** have the same dimensions, the following laws hold:

1. Associative property of addition: $(\mathbf{A} + \mathbf{B}) + \mathbf{C} = \mathbf{A} + (\mathbf{B} + \mathbf{C})$.
2. Commutative property of addition: $\mathbf{A} + \mathbf{B} = \mathbf{B} + \mathbf{A}$.
3. Distributive properties of scalar multiplication:
   a. $k(\mathbf{A} + \mathbf{B}) = k\mathbf{A} + k\mathbf{B}$
   b. $(k + h)\mathbf{A} = k\mathbf{A} + h\mathbf{A}$
4. Properties of the zero matrix (the matrix having all entries equal to zero):
   a. $\mathbf{A} + \mathbf{0} = \mathbf{A}$ b. $k \cdot \mathbf{0} = \mathbf{0}$ c. $0 \cdot \mathbf{A} = \mathbf{0}$

The properties in Note 6.2 are not difficult to prove. For example, a proof of the commutative property for addition of matrices is as follows: Suppose **A** and **B** are matrices with the same dimension. To prove $\mathbf{A} + \mathbf{B} = \mathbf{B} + \mathbf{A}$, we must show that corresponding entries are equal, that is, $[\mathbf{A} + \mathbf{B}]_{ij} = [\mathbf{B} + \mathbf{A}]_{ij}$ for all pairs (i, j). By Note 6.1, part 2, $[\mathbf{A} + \mathbf{B}]_{ij} = a_{ij} + b_{ij} = b_{ij} + a_{ij} = [\mathbf{B} + \mathbf{A}]_{ij}$. At the crucial step, to get $a_{ij} + b_{ij} = b_{ij} + a_{ij}$, we use the fact that a_{ij} and b_{ij} are real numbers. Since addition of real numbers is commutative (see Section 1.4), $a_{ij} + b_{ij} = b_{ij} + a_{ij}$ for all pairs (i, j). Thus, $[\mathbf{A} + \mathbf{B}]_{ij} = [\mathbf{B} + \mathbf{A}]_{ij}$ for all pairs (i, j), and since corresponding entries are equal, $\mathbf{A} + \mathbf{B} = \mathbf{B} + \mathbf{A}$.

Vectors

A **vector** is a list of numbers. The list is written either horizontally in a row giving a **row vector,** or vertically in a column giving a **column vector.** Thus, a vector is a special type of matrix. A row vector with n entries is a $1 \times n$ matrix, and a column vector with m entries is an $m \times 1$ matrix. The **order** of a vector is the number of entries it contains. A vector is named using a boldfaced lowercase variable and individual entries are referred to by attaching a single subscript on the variable to indicate the position of the entry in the list.

EXAMPLE 6.8 ——————————————————————————————

Suppose $\mathbf{v} = [16 \quad 1.3 \quad -4 \quad 2 \quad 1 \quad 12]$. What is

a. v_3 b. v_4 c. i if $v_i = 1.3$ d. the order of **v**?

SOLUTION

a. v_3 is the third element of **v**, so $v_3 = -4$.

b. $v_4 = 2$.

c. v_2 is the only element equal to 1.3, so the subscript $i = 2$.

d. **v** has order 6.

EXAMPLE 6.9

Write the row vectors **b** and **c** of size 4 where $b_i = 3i - 1$ and $c_j = j^2 - 4j + 2$.

SOLUTION

Since $b_i = 3i - 1$, we have $b_1 = 3(1) - 1 = 2$, $b_2 = 3(2) - 1 = 5$, $b_3 = 3(3) - 1 = 8$, and $b_4 = 3(4) - 1 = 11$. Thus, $\mathbf{b} = \begin{bmatrix} 2 & 5 & 8 & 11 \end{bmatrix}$.

Since $c_j = j^2 - 4j + 2$, we have $c_1 = 1^2 - 4(1) + 2 = -1$.

$c_2 = 2^2 - 4(2) + 2 = -2$, $c_3 = 3^2 - 4(3) + 2 = -1$, and $c_4 = 4^2 - 4(4) + 2 = 2$. Thus, $\mathbf{c} = \begin{bmatrix} -1 & -2 & -1 & 2 \end{bmatrix}$.

Variables can also occur in individual entries, and we may need to find the values of these variables.

EXAMPLE 6.10

Find x and y in each of the following:

a. $2 \begin{bmatrix} 3+x \\ 4-2y \end{bmatrix} = \begin{bmatrix} 5 \\ -13 \end{bmatrix}$

b. $\begin{bmatrix} 4x+7 \\ 8-3x \end{bmatrix} + \begin{bmatrix} 1-y \\ 6x \end{bmatrix} = \begin{bmatrix} 10 \\ 23 \end{bmatrix}$

SOLUTION

a. $2 \begin{bmatrix} 3+x \\ 4-2y \end{bmatrix} = \begin{bmatrix} 5 \\ -13 \end{bmatrix}$ gives $\begin{bmatrix} 6+2x \\ 8-4y \end{bmatrix} = \begin{bmatrix} 5 \\ -13 \end{bmatrix}$. Thus, $6 + 2x = 5$ and $8 - 4y = -13$ by the definition of equality of matrices. So $2x = -1$, which gives $x = -1/2$, and $-4y = -21$, which gives $y = 21/4$. Thus, $x = -1/2$ and $y = 21/4$.

b. $\begin{bmatrix} 4x+7 \\ 8-3x \end{bmatrix} + \begin{bmatrix} 1-y \\ 6x \end{bmatrix} = \begin{bmatrix} 10 \\ 23 \end{bmatrix}$. Adding the vectors on the left side. we get

$\begin{bmatrix} 4x+8-y \\ 8+3x \end{bmatrix} = \begin{bmatrix} 10 \\ 23 \end{bmatrix}$. Thus, $4x + 8 - y = 10$ and $8 + 3x = 23$. We solve these simultaneously. The second equation gives $3x = 15$, so $x = 5$. Substitute this into the first equation: $4x + 8 - y = 10$, giving

$$4(5) + 8 - y = 10$$
$$28 - y = 10$$
$$-y = -18$$

so $y = 18$. Thus, $x = 5$ and $y = 18$.

Special Types of Matrices

Finally, we define four special classes of matrices. A matrix is a **square matrix** if the number of rows equals the number of columns. Thus, matrix **B** of Example 6.4 is a square matrix. The **diagonal** of a matrix **A** consists of all the entries a_{ii}, that is, the entries whose row and column numbers are equal. The diagonal of matrix **B** of Example 6.4 consists of the entries 2, 4, 6. A **diagonal matrix** is a square matrix whose only nonzero entries are diagonal entries. That is, $a_{ij} = 0$ if $i \neq j$.

Another important matrix occurring in many applications is the identity matrix. It is a special type of diagonal matrix. The **identity matrix** of order n is the $n \times n$ matrix having 1s on the diagonal and 0s elsewhere. It is denoted by $\mathbf{I_n}$.

EXAMPLE 6.11

Construct the identity matrices of orders 1, 2, and 3.

SOLUTION

$$\mathbf{I_1} = [1] \qquad \mathbf{I_2} = \begin{bmatrix} 1 & 0 \\ 0 & 1 \end{bmatrix} \qquad \mathbf{I_3} = \begin{bmatrix} 1 & 0 & 0 \\ 0 & 1 & 0 \\ 0 & 0 & 1 \end{bmatrix}$$

The fourth type of matrix we discuss is the $m \times n$ **zero matrix** $\mathbf{0_{mn}}$, which has all entries equal to zero. The zero matrix need not be square. Thus, the 2×3 zero matrix is

$$\mathbf{0_{23}} = \begin{bmatrix} 0 & 0 & 0 \\ 0 & 0 & 0 \end{bmatrix}$$

EXERCISES 6.1

1. Which of the following should be represented as vectors and which are more appropriate as scalars:
   a. a person's age
   b. the average temperature for each day of the year in a given city
   c. a point plotted in the (x, y)-coordinate plane.
2. Give an example of a vector that is simultaneously a row vector and a column vector.
3. Write the vector **v** of size 4 having the following properties: $v_1 = 10$, $v_3 = -17$, $v_2 = 5v_1 + v_3$, and $v_4 = v_1 - v_2 + 2v_3$.
4. Evaluate:

   a. $\begin{bmatrix} 3 \\ -1 \\ 10 \end{bmatrix} + \begin{bmatrix} 12 \\ 3 \\ -2 \end{bmatrix}$ **b.** $\begin{bmatrix} 7 \\ 4 \\ -5 \end{bmatrix} - \begin{bmatrix} 8 \\ -2 \\ 15 \end{bmatrix}$

   c. $3\begin{bmatrix} 1 \\ 2 \\ 5 \end{bmatrix} - 2\begin{bmatrix} 2 \\ 4 \\ -1 \end{bmatrix}$ **d.** $4\begin{bmatrix} 8 & -3 & 17 \end{bmatrix} + 5\begin{bmatrix} -6 & 1 & 11 \end{bmatrix}$

e. $\begin{bmatrix} 5 & -2 & 8 & 4 \end{bmatrix} + \begin{bmatrix} 2 & 7 & -5 \end{bmatrix}$

5. If $\mathbf{u} = \begin{bmatrix} -3 \\ 8 \\ 6 \end{bmatrix}$, $\mathbf{v} = \begin{bmatrix} 2 \\ 0 \\ 5 \end{bmatrix}$, and $\mathbf{w} = \begin{bmatrix} 4 \\ 4 \\ -9 \end{bmatrix}$, evaluate:

a. $\mathbf{u} + \mathbf{v} - \mathbf{w}$ **b.** $6\mathbf{u} - 5\mathbf{v}$ **c.** $2\mathbf{w} - 7\mathbf{u}$

d. $5\mathbf{v} - 8\mathbf{w}$ **e.** $(1/2)\mathbf{v} + (3/2)\mathbf{w}$ **f.** $0.1\mathbf{u} + 0.2\mathbf{v} + 0.3\mathbf{w}$

6. Construct each of the following vectors:

  a. row vector $\mathbf{u}$ of order 3 where $u_i = i^3 + 3i$.

  b. column vector $\mathbf{v}$ of order 4 where $v_j = 2j - j^2 + 5$.

  c. row vector $\mathbf{w}$ of order 3 where $w_i = \max\{2i + 3,\ 5i - 2\}$.

  d. row vector $\mathbf{x}$ of order 4 where $x_i = 1/(i + 2) + 3/i^2$.

7. Solve for the variables in each of the following:

a. $\begin{bmatrix} 2x+5 \\ y-3 \end{bmatrix} = \begin{bmatrix} 11 \\ 7 \end{bmatrix}$ **b.** $\begin{bmatrix} x+3 \\ 2y+1 \end{bmatrix} + \begin{bmatrix} 5x+1 \\ 4y-2 \end{bmatrix} = \begin{bmatrix} 10 \\ -4 \end{bmatrix}$

c. $\begin{bmatrix} 3x+4 \\ 5y+2 \end{bmatrix} = \begin{bmatrix} 4x-6 \\ 3y+16 \end{bmatrix}$ **d.** $\begin{bmatrix} 4x-3y \\ x+2y \end{bmatrix} = \begin{bmatrix} -8 \\ 9 \end{bmatrix}$

e. $\begin{bmatrix} x+2y-5 \\ 3x-y+2 \end{bmatrix} = \begin{bmatrix} -12 \\ 16 \end{bmatrix}$ **f.** $\begin{bmatrix} x+2y \\ y-4w \\ 2w \end{bmatrix} = \begin{bmatrix} 9 \\ 25 \\ -6 \end{bmatrix}$

8. State the dimension of each of the following matrices:

a. $\begin{bmatrix} 3 \\ -8 \end{bmatrix}$ **b.** $\begin{bmatrix} -3 & 1 & 4 & 7 & 8 \\ 1 & 2 & 3 & -1 & 11 \end{bmatrix}$

c. $\begin{bmatrix} 12 & -1 \\ 3 & 6 \\ -2 & 2 \end{bmatrix}$ **d.** $\begin{bmatrix} 5 & 5 & -1 & 0 \end{bmatrix}$

9. A company has three factories. Their production output is listed in the following matrix with the columns representing the four weeks in February:

$$\begin{bmatrix} 320 & 430 & 190 & 318 \\ 212 & 189 & 300 & 260 \\ 290 & 450 & 385 & 273 \end{bmatrix}$$

  a. How many items did factory 3 produce in the second week of February?

  b. How many items did factory 1 produce in the last week of February?

  c. What was factory 2's total output for the month?

  d. How many items did the company produce during the second week of February?

10. For the matrix

$$A = \begin{bmatrix} 8 & -1 & -3 & 7 \\ -3 & 9 & 4 & 2 \\ 0 & 1 & -2 & 5 \end{bmatrix}$$

list

a. a_{13} **b.** a_{22} **c.** a_{32} **d.** j if $a_{2j} = -3$

11. Evaluate:

a. $\begin{bmatrix} 2 & -1 & 3 \\ 0 & 5 & 7 \end{bmatrix} + \begin{bmatrix} 8 & 11 & -15 \\ -2 & -3 & 9 \end{bmatrix}$ **b.** $\begin{bmatrix} 5 & -5 \\ -2 & 3 \\ 0 & 1 \end{bmatrix} - 2\begin{bmatrix} 3 & 1 \\ 7 & -2 \\ 8 & -5 \end{bmatrix}$

c. $\begin{bmatrix} 4 & 2 & 7 \\ 13 & -5 & 8 \end{bmatrix} - \begin{bmatrix} 2 & 1 \\ 10 & 0 \end{bmatrix}$

12. Construct the identity matrices $\mathbf{I_4}$ and $\mathbf{I_5}$.

13. Construct each of the following matrices:
  a. 2×4 matrix $\mathbf{A}$ where $a_{ij} = i - 2j$
  b. 3×3 matrix $\mathbf{B}$ where $b_{ij} = 3i - 4ij + j^2$
  c. 3×2 matrix $\mathbf{C}$ where $c_{ij} = \max\{i, j\}$
  d. 2×3 matrix $\mathbf{D}$ where $d_{ij} = 2i + 7j$

14. For $\mathbf{A} = \begin{bmatrix} -5 & 2 \\ -1 & 0 \\ 2 & -6 \end{bmatrix}$, $\mathbf{B} = \begin{bmatrix} 3 & -6 \\ 0 & 1 \\ 5 & 8 \end{bmatrix}$, and $\mathbf{C} = \begin{bmatrix} 13 & -2 \\ 11 & 9 \\ 7 & -1 \end{bmatrix}$ evaluate:

  a. $\mathbf{B} - 2\mathbf{A}$ **b.** $\mathbf{C} + 3\mathbf{A} - \mathbf{B}$ **c.** $\mathbf{A} + \mathbf{C}$
  d. $-4\mathbf{A}$ **e.** $(1/2)\mathbf{A} + (3/2)\mathbf{B}$

15. Using the matrices of exercise 14, $k = 2$, and $h = 3$, illustrate:
  a. the associative property of addition
  b. the commutative property of addition
  c. the distributive properties of scalar multiplication

16. Find a matrix $\mathbf{E}$ so that

$$5\mathbf{E} - \begin{bmatrix} 1 & -3 & 4 \\ -3 & 1 & 7 \end{bmatrix} = \begin{bmatrix} -21 & 5 & -9 \\ 8 & 11 & 43 \end{bmatrix}$$

6.2 MATRIX MULTIPLICATION

In the last section, we studied the elementary matrix operations: addition, subtraction, and scalar multiplication. In this section, we examine matrix multiplication (this is different from scalar multiplication) and finding the transpose of a matrix.

Suppose at the beginning of the school year, you stock up on school supplies by purchasing six pens, eight notebooks, five pencils, and a ruler. The prices are 45¢ for a pen, 90¢ for a notebook, 10¢ for a pencil, and 40¢ for a ruler. We calculate the total amount spent (assume no sales tax) as follows:

$$
\begin{aligned}
6 \cdot 45¢ &= \$\ 2.70 \\
8 \cdot 90¢ &= \$\ 7.20 \\
5 \cdot 10¢ &= \$\ 0.50 \\
1 \cdot 40¢ &= \$\ 0.40 \\
\hline
&\ \ \$10.80
\end{aligned}
$$

We can write the computation as a matrix multiplication as follows:

$$[6 \quad 8 \quad 5 \quad 1] \begin{bmatrix} 0.45 \\ 0.90 \\ 0.10 \\ 0.40 \end{bmatrix} = [10.80]$$

Note that this is a row vector times a column vector. In matrix multiplication, we multiply a "row into a column pairwise and add." By *pairwise* we mean the first entry of the row vector times the first entry of the column vector, the second entry of the row vector times the second entry of the column vector, and so on. The following paired entries have identical underscores:

$$[\underline{6} \quad \underline{\underline{8}} \quad \underline{\underline{\underline{5}}} \quad \underline{\underline{\underline{\underline{1}}}}] \begin{bmatrix} \underline{0.45} \\ \underline{\underline{0.90}} \\ \underline{\underline{\underline{0.10}}} \\ \underline{\underline{\underline{\underline{0.40}}}} \end{bmatrix} = [(6)(0.45) + (8)(0.90) + (5)(0.10) + (1)(0.40)]$$

$$= [2.70 + 7.20 + 0.50 + 0.40] = [10.80]$$

Note two items:

1. In order to perform the pairwise multiplications, the number of columns in the first matrix (vector) must equal the number of rows in the second matrix (vector).
2. If we can multiply two matrices, the answer is a matrix, not a number.

EXAMPLE 6.12 _____

Perform the following multiplications:

a. $[2 \quad -5] \begin{bmatrix} 9 \\ 4 \end{bmatrix}$ b. $[3 \quad 7 \quad 2] \begin{bmatrix} 8 \\ -1 \\ 6 \end{bmatrix}$

SOLUTION

a. We multiply pairwise and add $2 \cdot 9 + (-5) \cdot 4 = 18 - 20 = -2$. Thus

$$[2 \quad -5] \begin{bmatrix} 9 \\ 4 \end{bmatrix} = [-2]$$

b. $(3 \cdot 8) + (7 \cdot (-1)) + (2 \cdot 6) = 24 - 7 + 12 = 29$. Thus

$$[3 \quad 7 \quad 2] \begin{bmatrix} 8 \\ -1 \\ 6 \end{bmatrix} = [29]$$

_____

We are now ready to extend this process from vectors to matrices. The rules for multiplying matrices are given in Note 6.3.

Note 6.3   Matrix Multiplication

1. In order to multiply **AB**, the number of columns of **A** must equal the number of rows of **B**. The answer matrix will have the same number of rows as **A** and the same number of columns as **B**.

2. To find the (i, j) entry of the answer matrix, multiply row i of **A** into column j of **B** pairwise and add.

Rule 2 of Note 6.3 reduces matrix multiplication to multiplying vectors, which we know how to do from Example 6.12. In Figure 6.2, we indicate how to find the $(2, 3)$ entry of the product **AB**. Multiply row 2 of **A** into column 3 of **B**. The multiplication can be performed because the number of columns in **A** is 2, which equals the number of rows in **B**. By rule 1 of Note 6.3, the answer matrix **AB** is a 3×4 matrix (why?). The $(2, 3)$ entry is found by multiplying row 2 of **A** into column 3 of **B**. Thus, the $(2, 3)$ entry is $(5 \cdot 7) + ((-2) \cdot 3) = 35 - 6 = 29$.

$$A = \begin{bmatrix} 1 & 3 \\ \hline 5 & -2 \\ \hline 6 & 4 \end{bmatrix} \quad B = \begin{bmatrix} 2 & 8 & 7 & 8 \\ 3 & 1 & 3 & 5 \end{bmatrix} \quad AB = \begin{array}{c} \\ 1 \\ 2 \\ 3 \end{array} \begin{array}{cccc} 1 & 2 & 3 & 4 \\ \begin{bmatrix} \cdot & \cdot & \cdot & \cdot \\ \cdot & \cdot & 29 & \cdot \\ \cdot & \cdot & \cdot & \cdot \end{bmatrix} \end{array}$$

$$(5 \cdot 7) + ((-2) \cdot 3) = 35 - 6 = 29$$

Figure 6.2

EXAMPLE 6.13

If **A** and **B** have the following dimensions, which products **AB** can be computed? For those products that can be computed, what is the dimension of the answer matrix?

a. **A** is 4×3 and **B** is 3×2 b. **A** is 3×2 and **B** is 3×2

c. **A** is 2×4 and **B** is 4×4 d. **A** is 3×1 and **B** is 1×2

SOLUTION

To answer this type of problem, list the dimensions next to one another. For the product **AB** to be defined, the middle two numbers must be equal. If the middle numbers are equal, the dimension of **AB** is given by the two outside numbers.

$$m \times n \quad k \times h$$

must be
equal

answer matrix

Thus, if $n = k$, the answer matrix is an $m \times h$ matrix.

a. $4 \times 3 \quad 3 \times 2$ The answer matrix is a 4×2 matrix.

b. $3 \times 2 \quad 3 \times 2$ The multiplication cannot be performed.

not equal

c. $2 \times 4 \quad 4 \times 4$ The answer matrix is a 2×4 matrix.

d. $3 \times 1 \quad 1 \times 2$ The answer matrix is a 3×2 matrix.

EXAMPLE 6.14

Calculate $\begin{bmatrix} 11 & 8 \\ 5 & -2 \\ 16 & 4 \end{bmatrix} \begin{bmatrix} 3 & 1 \\ -2 & 6 \end{bmatrix}.$

SOLUTION

We first check that the multiplication can be performed: $3 \times 2 \quad 2 \times 2$; the answer matrix is a 3×2 matrix. To get the $(1, 1)$ entry, multiply row 1 of the first matrix into column 1 of the second: $(11 \cdot 3) + (8 \cdot (-2)) = 33 - 16 = 17$. To get the $(1, 2)$ entry, multiply row 1 of the first matrix into column 2 of the second: $(11 \cdot 1) + (8 \cdot 6) = 11 + 48 = 59$. To get the $(2, 1)$ entry, multiply row 2 of the first matrix into column 1 of the second: $(5 \cdot 3) + ((-2)(-2)) = 15 + 4 = 19$. In general, to get the (i, j) entry, multiply row i of the first matrix into column j of the second. Thus

$$\begin{bmatrix} 11 & 8 \\ 5 & -2 \\ 16 & 4 \end{bmatrix} \begin{bmatrix} 3 & 1 \\ -2 & 6 \end{bmatrix} = \begin{bmatrix} 11 \cdot 3 + 8(-2) & 11 \cdot 1 + 8 \cdot 6 \\ 5 \cdot 3 + (-2)(-2) & 5 \cdot 1 + (-2) \cdot 6 \\ 16 \cdot 3 + 4 \cdot (-2) & 16 \cdot 1 + 4 \cdot 6 \end{bmatrix} = \begin{bmatrix} 17 & 59 \\ 19 & -7 \\ 40 & 40 \end{bmatrix}$$

EXAMPLE 6.15

Evaluate each of the following:

a. $\begin{bmatrix} 5 & 3 & 7 \\ -1 & 0 & 6 \end{bmatrix} \begin{bmatrix} 5 \\ -2 \\ 10 \end{bmatrix}$ b. $\begin{bmatrix} 1 & -4 & 0 \\ 3 & 0 & 8 \end{bmatrix} \begin{bmatrix} 5 & -2 \\ 1 & 6 \\ 11 & 3 \end{bmatrix}$

SOLUTION

a. $2 \times 3 \quad 3 \times 1$; the dimension of the answer matrix is 2×1.

$$\begin{bmatrix} 5 & 3 & 7 \\ -1 & 0 & 6 \end{bmatrix} \begin{bmatrix} 5 \\ -2 \\ 10 \end{bmatrix} = \begin{bmatrix} (5 \cdot 5) + (3(-2)) + (7 \cdot 10) \\ (-1 \cdot 5) + (0(-2)) + (6 \cdot 10) \end{bmatrix}$$

$$= \begin{bmatrix} 25 - 6 + 70 \\ -5 + 0 + 60 \end{bmatrix} = \begin{bmatrix} 89 \\ 55 \end{bmatrix}$$

b. The answer matrix has dimension 2×2.

$$\begin{bmatrix} 1 & -4 & 0 \\ 3 & 0 & 8 \end{bmatrix} \begin{bmatrix} 5 & -2 \\ 1 & 6 \\ 11 & 3 \end{bmatrix}$$

$$= \begin{bmatrix} (1 \cdot 5) + ((-4) \cdot 1) + (0 \cdot 11) & (1 \cdot (-2)) + ((-4) \cdot 6) + (0 \cdot 3) \\ (3 \cdot 5) + (0 \cdot 1) + (8 \cdot 11) & (3 \cdot (-2)) + (0 \cdot 6) + (8 \cdot 3) \end{bmatrix}$$

$$= \begin{bmatrix} 5 - 4 + 0 & -2 - 24 + 0 \\ 15 + 0 + 88 & -6 + 0 + 24 \end{bmatrix} = \begin{bmatrix} 1 & -26 \\ 103 & 18 \end{bmatrix}$$

Matrix multiplication can be used to express systems of equations (or inequalities) in compact form. Many computer systems have a packaged program available that will solve a system of linear equations provided the information appears in a certain format. The format usually involves rewriting the system of equations as a matrix equation:

(Matrix of coefficients) · (Column vector of variables)

= (Column vector of constants)

Symbolically, we have $\mathbf{Ax} = \mathbf{b}$. We illustrate this with several examples.

EXAMPLE 6.16

Write the following system as a matrix equation:

$$\begin{aligned} 3x + y - 2z &= 7 \\ 4x \quad\quad + 5z &= -2 \\ 8x - 3y + 4z &= 17 \end{aligned}$$

SOLUTION

We list:

(Coefficient matrix) · (Column vector of variables)

= (Column vector of constants)

$$\begin{bmatrix} 3 & 1 & -2 \\ 4 & 0 & 5 \\ 8 & -3 & 4 \end{bmatrix} \begin{bmatrix} x \\ y \\ z \end{bmatrix} = \begin{bmatrix} 7 \\ -2 \\ 17 \end{bmatrix}$$

To verify this is correct, multiply the left two matrices to get:

$$\begin{bmatrix} 3x + y - 2z \\ 4x + 0y + 5z \\ 8x - 3y + 4z \end{bmatrix} = \begin{bmatrix} 7 \\ -2 \\ 17 \end{bmatrix}$$

which by definition of matrix equality is equivalent to

$$3x + y - 2z = 7$$
$$4x + 5z = -2$$
$$8x - 3y + 3z = 17$$

EXAMPLE 6.17

Write the system of equations corresponding to the matrix equation:

$$\begin{bmatrix} 8 & -3 & 0 & 5 \\ 2 & 1 & 6 & -1 \\ 0 & 13 & -2 & 0 \end{bmatrix} \begin{bmatrix} x_1 \\ x_2 \\ x_3 \\ x_4 \end{bmatrix} = \begin{bmatrix} 12 \\ -3 \\ 14 \end{bmatrix}$$

SOLUTION

We multiply row 1 into column 1 to get the $(1, 1)$ entry, row 2 into column 1 to get the $(2, 1)$ entry, and so on. Thus, we get

$$\begin{bmatrix} 8x_1 - 3x_2 + 5x_4 \\ 2x_1 + x_2 + 6x_3 - x_4 \\ 13x_2 - 2x_3 \end{bmatrix} = \begin{bmatrix} 12 \\ -3 \\ 14 \end{bmatrix}$$

which by the definition of matrix equality is equivalent to the system:

$$8x_1 - 3x_2 + 5x_4 = 12$$
$$2x_1 + x_2 + 6x_3 - x_4 = -3$$
$$13x_2 - 2x_3 = 14$$

Just as we can write a system of equations in the form of a matrix equation $\mathbf{Ax} = \mathbf{b}$, we can write a system of inequalities using the matrix inequality $\mathbf{Ax} \le \mathbf{b}$ or $\mathbf{Ax} \ge \mathbf{b}$. (Matrix inequality is defined by $\mathbf{M} \le \mathbf{N}$ if and only if $m_{ij} \le n_{ij}$ for all pairs i, j. The inequality $\mathbf{M} \ge \mathbf{N}$ is defined similarly.) Note there is only one inequality sign for the matrix inequality, and the matrix multiplication takes place on the left side of the inequality sign (as it does for matrix equations). Therefore, to translate a system of inequalities into matrix form, we may first have to manipulate some inequalities. We use the standard rules for manipulating inequalities and alter the inequalities until: (1) all variables are on the left, and (2) all inequality signs in the system are the same.

EXAMPLE 6.18

Write the matrix inequality corresponding to the system:

$$3x + y \le 4 \qquad\qquad (1)$$
$$y \le 2x - 5 - z \qquad\qquad (2)$$
$$x + 4z \ge 7 + 11y \qquad\qquad (3)$$

SOLUTION

We first rewrite the system with all variables on the left by adding or subtracting appropriate terms to both sides: For (2) subtract $2x$ from both sides and add z to both sides; for (3) subtract $11y$ from both sides. The system becomes:

$$3x + \quad y \qquad \le \quad 4 \tag{1a}$$
$$-2x + \quad y + \quad z \le -5 \tag{2a}$$
$$x - 11y + 4z \ge \quad 7 \tag{3a}$$

We next fix the direction of the sign in (3a), which we alter by multiplying both sides of the inequality by -1. (Recall that when multiplying or dividing both sides of an equality by a negative number, the direction of the inequality changes.) Our system now becomes:

$$3x + \quad y \qquad \le \quad 4 \tag{1b}$$
$$-2x + \quad y + \quad z \le -5 \tag{2b}$$
$$-x + 11y - 4z \le -7 \tag{3b}$$

We can, therefore, represent the original system with the matrix inequality:

$$\begin{bmatrix} 3 & 1 & 0 \\ -2 & 1 & 1 \\ -1 & 11 & -4 \end{bmatrix} \begin{bmatrix} x \\ y \\ z \end{bmatrix} \le \begin{bmatrix} 4 \\ -5 \\ -7 \end{bmatrix}$$

The Transpose

The transpose is a very simple yet useful concept. If A is an $m \times n$ matrix, then the **transpose** of A is the $n \times m$ matrix whose first row is the first column of A, whose second row is the second column of A, and so on. Thus, the transpose of A is obtained by interchanging the rows and columns of A. The transpose of A is denoted A^t. Using subscripts, the (i, j) entry of A^t is the (j, i) entry of A: $[A^t]_{ij} = a_{ji}$.

EXAMPLE 6.19 _____

Find the transpose of

$$A = \begin{bmatrix} 3 & 1 & -5 \\ 12 & 7 & 6 \end{bmatrix}$$

SOLUTION

Since A is a 2×3 matrix, the dimension of A^t is 3×2. The first row of A^t is the first column of A, that is, 3   12. The second row of A^t is the second column of A, that is, 1   7. The third row of A^t is the third column of A. Thus, we have

$$A^t = \begin{bmatrix} 3 & 12 \\ 1 & 7 \\ -5 & 6 \end{bmatrix}$$

EXAMPLE 6.20

Determine each of the following:

a. $\begin{bmatrix} 5 & -1 & 9 \\ -3 & 0 & 2 \\ 3 & 1 & 6 \\ 15 & -2 & 8 \end{bmatrix}^{t}$ b. $\begin{bmatrix} 2 & 1 & 7 \\ 8 & -3 & 4 \end{bmatrix}\begin{bmatrix} 1 & 2 & 6 \end{bmatrix}^{t}$

SOLUTION

a. The transpose has dimension 3×4, and the rows of the transpose correspond to the columns of the given matrix. Thus, we have

$$\begin{bmatrix} 5 & -3 & 3 & 15 \\ -1 & 0 & 1 & -2 \\ 9 & 2 & 6 & 8 \end{bmatrix}$$

b. We first apply the transpose operation. The t applies only to the second matrix (the row vector that becomes a column vector). We then multiply:

$$\begin{bmatrix} 2 & 1 & 7 \\ 8 & -3 & 4 \end{bmatrix}\begin{bmatrix} 1 & 2 & 6 \end{bmatrix}^{t} = \begin{bmatrix} 2 & 1 & 7 \\ 8 & -3 & 4 \end{bmatrix}\begin{bmatrix} 1 \\ 2 \\ 6 \end{bmatrix}$$

$$= \begin{bmatrix} (2 \cdot 1)+(1 \cdot 2)+(7 \cdot 6) \\ (8 \cdot 1)+((-3) \cdot 2) + (4 \cdot 6) \end{bmatrix} = \begin{bmatrix} 46 \\ 26 \end{bmatrix}$$

An important use of the transpose is in linear programming problems. Since linear programming is beyond the scope of this book, we discuss this use of the transpose only in the Problems and Projects section of the exercises (see exercise 15). Note 6.4 states several properties of the transpose.

Note 6.4   Properties of the Transpose

The transpose satisfies the following properties:

1. $(A^{t})^{t} = A$ 2. $(A + B)^{t} = A^{t} + B^{t}$
3. $(AB)^{t} = B^{t}A^{t}$ 4. $(kA)^{t} = kA^{t}$

EXERCISES 6.2

1. Calculate each of the following:

a. $\begin{bmatrix} 3 & -5 & 6 \\ 2 & 0 & -4 \end{bmatrix}\begin{bmatrix} 5 & -7 \\ 8 & 11 \\ 0 & 2 \end{bmatrix}$ b. $\begin{bmatrix} 11 & 12 \\ -3 & 0 \\ 5 & -6 \end{bmatrix}\begin{bmatrix} 1 & 2 \\ -3 & 12 \end{bmatrix}$

c. $\begin{bmatrix} 5 & 8 & -1 & 10 \end{bmatrix} \begin{bmatrix} 2 \\ 4 \\ -3 \\ 17 \end{bmatrix}$

2. Fill in the dimension of the answer matrix **AB**, or state as undefined.

A	**B**	**AB**
5×7	7×3	
4×2	2×1	
3×2	2×3	
2×3	2×2	
7×7	7×5	

3. If **A** has dimension 4×9 and **BA** has dimension 6×9, what is the dimension of **B**?

4. If **B** has dimension 5×2 and **AB** is square, what is the dimension of:
   a. AB   **b.** $\mathbf{B}^t$ **c. BA**   **d. A**?

5. Simplify:

   a. $\begin{bmatrix} 4 & -3 & -3 \\ -1 & 0 & 4 \end{bmatrix} \begin{bmatrix} 5 & 4 \\ 3 & -2 \\ -1 & 0 \end{bmatrix} + \begin{bmatrix} 8 & 17 \\ -10 & 31 \end{bmatrix}$

   b. $\begin{bmatrix} -7 & 12 \end{bmatrix} \begin{bmatrix} 1 & -3 & 5 & -7 \\ -2 & 4 & -6 & 8 \end{bmatrix} - \begin{bmatrix} 9 & -10 & 11 & -12 \end{bmatrix}$

6. Matrix multiplication is not commutative because **AB** may be defined while **BA** is undefined. For example, if **A** is a 3×2 matrix and **B** is a 2×5 matrix, then **AB** is a 3×5 matrix, whereas **BA** is undefined.
   a. Prove: If **AB** and **BA** both exist, then **AB** and **BA** are square matrices.
   b. Show by example that **AB** and **BA** may both exist while **AB** has a different dimension than **BA**.

7. Let $\mathbf{A} = \begin{bmatrix} 1 & 2 \\ -4 & 1 \end{bmatrix}$ and $\mathbf{B} = \begin{bmatrix} 2 & 3 \\ -2 & 5 \end{bmatrix}$. Show $\mathbf{AB} \neq \mathbf{BA}$.

8. Find a 2×2 matrix **A** such that $\mathbf{A} + \mathbf{A} = \mathbf{AA}$.

9. Write the system below as a matrix equation:

$$\begin{aligned} 3x + y - 2z + 7w &= 12 \\ x - y \quad\;\; + 3w &= 18 \\ -2x \quad\;\; + 3z - 8w &= -4 \end{aligned}$$

10. Determine each of the following:

   a. $\begin{bmatrix} 4 & -3 & 6 \\ 1 & -1 & 0 \end{bmatrix}^t$ **b.** $\begin{bmatrix} 3 & 12 & 2 & -2 \\ -6 & 5 & -6 & 1 \end{bmatrix}^t$

11. a. Prove (1), (2), and (4) from Note 6.4. That is, show (1) $(\mathbf{A}^t)^t = \mathbf{A}$, (2) $(\mathbf{A} + \mathbf{B})^t = \mathbf{A}^t + \mathbf{B}^t$, and (4) $(k\mathbf{A})^t = k\mathbf{A}^t$. [*Hint:* Prove the (i, j) entry on the left side equals the (i, j) entry on the right side.]

b. Let

$$\mathbf{A} = \begin{bmatrix} 2 & 3 \\ -5 & 1 \end{bmatrix} \quad \text{and} \quad \mathbf{B} = \begin{bmatrix} -5 & 0 \\ 8 & 6 \end{bmatrix}$$

i. Show $(\mathbf{AB})^t \neq \mathbf{A}^t\mathbf{B}^t$.

ii. Show $(\mathbf{AB})^t = \mathbf{B}^t\mathbf{A}^t$, thereby verifying Note 6.4, part 3.

12. Translate each of the following systems of inequalities into a matrix inequality of the form $\mathbf{Ax} \leq \mathbf{b}$.

a. $5x - 2y \quad\quad \leq 3 + 7z$
   $x + 4y - \ z \geq -3$
   $\quad\quad y + 2z \leq 4x - 2$

b. $-x - 3y + 11z \geq 10$
   $8x \quad\quad\quad\quad \leq 4 - 3z$
   $3x \quad\quad - \ 5z \geq -2 + 9y$

Problems and Projects

13. Matrix $\mathbf{A}$ is **symmetric** if $\mathbf{A} = \mathbf{A}^t$. This implies $\mathbf{A}$ is square and $a_{ij} = a_{ji}$.
   a. If $\mathbf{A}$ and $\mathbf{B}$ are $n \times n$ symmetric matrices, show $\mathbf{A} + \mathbf{B}$ is symmetric.
   b. If $\mathbf{C}$ is any $n \times n$ matrix, show $\mathbf{C} + \mathbf{C}^t$ is symmetric.

14. a. Consider any 2×2 matrix $\mathbf{A} = \begin{bmatrix} a & b \\ c & d \end{bmatrix}$. Show $\mathbf{AI}_2 = \mathbf{A}$ and $\mathbf{I}_2\mathbf{A} = \mathbf{A}$, where $\mathbf{I}_2$ is the identity matrix of order 2.

   b. For any 3×3 matrix $\mathbf{B}$, show $\mathbf{I}_3\mathbf{B} = \mathbf{B}$ and $\mathbf{BI}_3 = \mathbf{B}$.

15. (Linear Programming). An important use of matrices is in linear programming. In this area of mathematics, the object is to find a solution to the following type of problem: Maximize (or minimize) a given linear function subject to given linear inequalities involving the variables. These problems are written using vectors and matrices. Let $\mathbf{A}$ be an $m \times n$ matrix, $\mathbf{u}$ be a $1 \times n$ now vector of real numbers, $\mathbf{v}$ be an $m \times 1$ column vector of real numbers, $\mathbf{x}$ be an $n \times 1$ column vector of variables, and $\mathbf{y}$ be an $m \times 1$ column vector of variables. The maximizing linear program has the form:

$$\text{Maximize } \mathbf{p} = \mathbf{ux}$$
$$(*) \quad \text{Subject to} \quad \mathbf{x} \geq \mathbf{0}$$
$$\mathbf{Ax} \leq \mathbf{v}$$

The minimizing problem has the form:

$$\text{Minimize } \mathbf{c} = \mathbf{v}^t\mathbf{y}$$
$$(**) \quad \text{Subject to} \quad \mathbf{y} \geq O$$
$$\mathbf{A}^t\mathbf{y} \geq \mathbf{u}^t$$

The minimizing problem (**) is called the *dual* of the maximizing problem (*) since (*) has a solution if and only if (**) has a solution (consult the *duality* *theorem* in Chapter 6 of Kolman, 1980). Consider the linear program:

$$\text{Maximize } p = 2x_1 + 3x_2$$
$$\text{Subject to} \quad x_1 \geq 0$$
$$x_2 \geq 0$$
$$5x_1 + 2x_2 \leq 20$$
$$7x_1 + 6x_2 \leq 30$$
$$x_2 \leq 4$$

a. Write the linear program using matrices.
b. Write the minimizing problem that is the dual of the problem in part a using matrices.
c. Rewrite the minimizing problem of part b without using matrices.

6.3   THE INVERSE OF A MATRIX AND DETERMINANTS

In this section, we examine two additional operations, which are defined only for square matrices. Thus, assume that matrices mentioned throughout this section are square matrices unless stated otherwise.

We encountered the identity matrix in Section 6.1. The role I_n plays for $n \times n$ matrices (see Note 6.5) is similar to the role 1 plays for real numbers (see Section 1.4).

Note 6.5   Multiplication Properties of I_n

The identity matrix of order n, I_n, has the following properties:

1. If **B** is an $n \times m$ matrix, $I_n B = B$.
2. If **C** is an $m \times n$ matrix, $CI_n = C$.
3. If **A** is an $n \times n$ matrix, $AI_n = I_n A = A$.

Any nonzero real number x has a multiplicative inverse x^{-1} such that $x \cdot x^{-1} = x^{-1} \cdot x = 1$. The number x^{-1} is the reciprocal of x: for example, $3^{-1} = 1/3$, $(5/6)^{-1} = 6/5$, and so on. We may ask: Do all nonzero $n \times n$ matrices **A** have an inverse A^{-1} such that $AA^{-1} = A^{-1}A = I_n$? The answer is no; however, there are infinitely many matrices that do have an inverse. We now develop a method for finding the inverse of a square matrix **A**. Two features of this method are: (1) it involves only simple arithmetic, and (2) if **A** does not have an inverse, the method discovers this fact.

Gauss–Jordan Reduction

The Gauss–Jordan reduction method is used in many contexts. Before we show how to find the inverse, we shall look at a more familiar situation involving the reduction process: solving a system of equations. (We will consider more general systems of equations in the exercises.) We first use the familiar *adding equations method*.

EXAMPLE 6.21 _____

Solve the system of equations

$$E_1:\quad 8x + 5y = 1$$
$$E_2:\quad 4x - 3y = 17$$

SOLUTION

We label each step in the process. We want to add multiples of the equations E_1 and E_2 so as to eliminate a variable.

Step 0
$$E_1:\qquad 8x + 5y = \quad 1$$
$$E_2:\qquad 4x - 3y = \quad 17$$

Step 1   Multiply E_2 by -2:
$$E_1:\qquad 8x + 5y = \quad 1$$
$$-2 \cdot E_2:\quad -8x + 6y = -34$$

Step 2   Add:
$$E_1 + (-2 \cdot E_2):\qquad 11y = -33$$

Step 3   Divide both sides by 11:
$$(E_1 - 2E_2)/11:\qquad y = \quad -3$$

Step 4   Substitute $y = -3$ into E_1, $8x + 5(-3) = 1$, which gives $8x - 15 = 1$, or $8x = 16$. Thus, $x = 2$. The solution is $x = 2$, $y = -3$.

_____

The basic operations used in Example 6.21 were: (1) adding a multiple of one equation to a multiple of another, and (2) multiplying both sides of an equation by a nonzero constant. Now consider a matrix formulation for the problem in Example 6.21. The system can be rewritten in matrix form as

$$\begin{bmatrix} 8 & 5 \\ 4 & -3 \end{bmatrix} \begin{bmatrix} x \\ y \end{bmatrix} = \begin{bmatrix} 1 \\ 17 \end{bmatrix}$$

The Gauss–Jordan reduction process uses an augmented matrix. Suppose a matrix equation has the form $\mathbf{Az = b}$, where $\mathbf{z}$ is a column vector of variables and $\mathbf{b}$ is a column vector of real numbers. The **augmented matrix** is the matrix formed by attaching the extra column $\mathbf{b}$ to matrix $\mathbf{A}$. For our example, the augmented matrix is

$$\begin{bmatrix} 8 & 5 & \vdots & 1 \\ 4 & -3 & \vdots & 17 \end{bmatrix}$$

It is customary to separate **A** from **b** by a vertical dashed line as we have done. In the reduction process using matrices, there are three row operations (see Note 6.6). Two of these row operations correspond to operations (1) and (2) listed for Example 6.21.

Note 6.6   Operations Used in the Reduction Process

The following operations may be performed on an augmented matrix:

1. Add element-wise a multiple of one row to a multiple of another row.

2. Multiply each entry in a row by a nonzero constant.

3. Change the order of the rows.

Note that these operations are performed like operations on vectors, that is, element-wise.

The goal of Gauss–Jordan reduction is to get the augmented matrix in the form $[\mathbf{I} \mid \mathbf{c}]$, where $\mathbf{I}$ is an identity matrix. When in this form, the entries of $\mathbf{c}$ are the values of the corresponding variables, which solve the given system of equations.

EXAMPLE 6.22 _____

Use Gauss–Jordan reduction to solve the system of equations: $8x + 5y = 1$
$4x - 3y = 17$

SOLUTION

Written as a matrix equation, the system is $\begin{bmatrix} 8 & 5 \\ 4 & -3 \end{bmatrix} \begin{bmatrix} x \\ y \end{bmatrix} = \begin{bmatrix} 1 \\ 17 \end{bmatrix}$ from which we obtain the augmented matrix. We label the rows to the left of the matrix.

$$
\begin{array}{c}
R_1 \\
R_2
\end{array}
\begin{bmatrix}
8 & 5 & \vdots & 1 \\
4 & -3 & \vdots & 17
\end{bmatrix}
$$

$$
\begin{array}{c}
R_1 - 2R_2 = R_3 \\
R_2
\end{array}
\begin{bmatrix}
0 & 11 & \vdots & -33 \\
4 & -3 & \vdots & 17
\end{bmatrix}
\quad
\begin{array}{l}
\text{That is, multiply row } R_2 \text{ by 2} \\
\text{and subtract it from row } R_1.
\end{array}
$$

$$
\begin{array}{c}
(1/11)R_3 = R_4 \\
R_2
\end{array}
\begin{bmatrix}
0 & 1 & \vdots & -3 \\
4 & -3 & \vdots & 17
\end{bmatrix}
\quad
\begin{array}{l}
\text{Multiply each entry} \\
\text{of row } R_3 \text{ by 1/11.}
\end{array}
$$

$$
\begin{array}{c}
R_4 \\
3R_4 + R_2 = R_5
\end{array}
\begin{bmatrix}
0 & 1 & \vdots & -3 \\
4 & 0 & \vdots & 8
\end{bmatrix}
$$

$$
\begin{array}{c}
R_4 \\
(1/4)R_5 = R_6
\end{array}
\begin{bmatrix}
0 & 1 & \vdots & -3 \\
1 & 0 & \vdots & 2
\end{bmatrix}
$$

Now interchange rows R_4 and R_6.

$$\begin{array}{c} R_6 \\ R_4 \end{array} \left[\begin{array}{cc|c} 1 & 0 & 2 \\ 0 & 1 & -3 \end{array} \right]$$

We stop because we have the identity matrix $\mathbf{I}_2$ to the left of the dotted line. This is the augmented matrix of the system $\begin{aligned} x &= 2 \\ y &= -3 \end{aligned}$

Hence the solution!

___

When dealing with problems having more than two variables or equations, you must have a systematic way of converting the matrix to the left of the dotted line into the identity matrix. We will work column by column from left to right, fixing one column at a time. Our goal, which is not always achievable (see exercise 13), is to get a 1 in the diagonal position of the column in which we are working and 0s in all the other positions of that column. The column entry to be converted to 1 at a given step is called the **pivot entry** and will be circled. The row (column) containing the pivot entry is called the **pivot row (pivot column).** To transform a given column, proceed according to Note 6.7.

Note 6.7

Transforming a column by Gauss–Jordan reduction:

1. To convert the pivot entry to 1, do one of the following:
   a. Multiply the pivot row by a nonzero number.
   b. Add a multiple of another row, located farther down than the pivot row, to the pivot row.
   c. Perform both (a) and (b).
2. To convert the rest of the pivot column to 0s, add to each row an appropriate multiple of the new pivot row obtained in step 1.

The *appropriate multiple* mentioned in step 2 of Note 6.7 is the additive inverse of the number in the pivot column being converted to 0.

EXAMPLE 6.23 _____

Use Gauss–Jordan reduction to find the solution to the system

$$\begin{aligned} x - 3y + 4z &= 9 \\ 2x + 5y &= 3 \\ -x + y + 2z &= -4 \end{aligned}$$

SOLUTION

The augmented matrix is:

$$\begin{array}{c} R_1 \\ R_2 \\ R_3 \end{array} \left[\begin{array}{ccc|c} \textcircled{1} & -3 & 4 & 9 \\ 2 & 5 & 0 & 3 \\ -1 & 1 & 2 & -4 \end{array} \right]$$

Our first pivot is the $(1, 1)$ entry in the matrix and is circled. Since the pivot entry is already 1, we can omit step 1 of Note 6.7. We now apply step 2 twice to obtain 0s in the two other positions in column 1:

$$\begin{array}{c} R_1 \\ -2R_1 + R_2 = R_4 \\ R_3 \end{array} \left[\begin{array}{ccc|c} \textcircled{1} & -3 & 4 & 9 \\ 0 & 11 & -8 & -15 \\ -1 & 1 & 2 & -4 \end{array} \right]$$

$$\begin{array}{c} R_1 \\ R_4 \\ R_1 + R_3 = R_5 \end{array} \left[\begin{array}{ccc|c} 1 & -3 & 4 & 9 \\ 0 & \textcircled{11} & -8 & -15 \\ 0 & -2 & 6 & 5 \end{array} \right]$$

The second pivot is the $(2, 2)$ entry in the matrix, as circled above. We now apply step 1(b) of Note 6.7:

$$\begin{array}{c} R_1 \\ 5R_5 + R_4 = R_6 \\ R_5 \end{array} \left[\begin{array}{ccc|c} 1 & -3 & 4 & 9 \\ 0 & \textcircled{1} & 22 & 10 \\ 0 & -2 & 6 & 5 \end{array} \right]$$

Now apply step 2 of Note 6.7 to each of the nonpivot entries of column 2. We combine the two applications of step 2:

$$\begin{array}{c} 3R_6 + R_1 = R_7 \\ R_6 \\ 2R_6 + R_5 = R_8 \end{array} \left[\begin{array}{ccc|c} 1 & 0 & 70 & 39 \\ 0 & 1 & 22 & 10 \\ 0 & 0 & \textcircled{50} & 25 \end{array} \right]$$

To get 1 in the last pivot position, which is circled above, multiply the pivot row by $1/50$, that is, use step 1(a) of Note 6.7:

$$\begin{array}{c} R_7 \\ R_6 \\ (1/50)R_8 = R_9 \end{array} \left[\begin{array}{ccc|c} 1 & 0 & 70 & 39 \\ 0 & 1 & 22 & 10 \\ 0 & 0 & \textcircled{1} & 1/2 \end{array} \right]$$

Now apply step 2 of Note 6.7 to the other entries in the pivot column. Again, we combine the two applications of step 2.

$$\begin{array}{c} -70R_9 + R_7 = R_{10} \\ -22R_9 + R_6 = R_{11} \\ R_9 \end{array} \left[\begin{array}{ccc|c} 1 & 0 & 0 & 4 \\ 0 & 1 & 0 & -1 \\ 0 & 0 & 1 & 1/2 \end{array} \right]$$

The solution is, therefore, $x = 4$, $y = -1$, $z = 1/2$.

The Inverse

We will now present a method for finding the inverse of a square matrix if it exists. The details on why the method works can be found in Kolman (1980). We shall be satisfied to illustrate the method through several examples.

Note 6.8   Finding the Inverse of a Matrix

To find the inverse of an $n \times n$ matrix $\mathbf{A}$, form the $n \times 2n$ matrix $\mathbf{B}$ having $\mathbf{I_n}$ to the right of $\mathbf{A}$. Perform Gauss–Jordan reduction on $\mathbf{B}$ with the goal of converting $\mathbf{A}$ to $\mathbf{I_n}$. If you succeed, the rightmost n columns will contain $\mathbf{A}^{-1}$. You can check your work by multiplying $\mathbf{AA}^{-1}$, which must give $\mathbf{I_n}$. If at any time during the procedure, the first n positions in a row are all 0s, stop; $\mathbf{A}$ does not have an inverse.

We indicate the procedure in Figure 6.3.

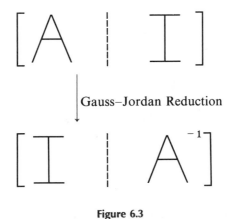

Gauss–Jordan Reduction

Figure 6.3

EXAMPLE 6.24

Find the inverse of

$$\mathbf{A} = \begin{bmatrix} -5 & 3 \\ 4 & 2 \end{bmatrix}$$

SOLUTION

We place $\mathbf{I_2}$ to the right of $\mathbf{A}$ and use the reduction process. Note that the reduction process instructions apply to the whole matrix, not just the $\mathbf{A}$ part.

$$R_1 \begin{bmatrix} \boxed{-5} & 3 & | & 1 & 0 \\ R_2 & 4 & 2 & | & 0 & 1 \end{bmatrix}$$

$$\begin{array}{c} -R_1 - R_2 = R_3 \\ R_2 \end{array} \begin{bmatrix} \boxed{1} & -5 & | & -1 & -1 \\ 4 & 2 & | & 0 & 1 \end{bmatrix}$$

$$\begin{array}{c} R_3 \\ -4R_3 + R_2 = R_4 \end{array} \begin{bmatrix} 1 & -5 & | & -1 & -1 \\ 0 & \boxed{22} & | & 4 & 5 \end{bmatrix}$$

$$\begin{array}{c} R_3 \\ (1/22)R_4 = R_5 \end{array} \begin{bmatrix} 1 & -5 & | & -1 & -1 \\ 0 & \boxed{1} & | & 4/22 & 5/22 \end{bmatrix}$$

$$\begin{array}{c} 5R_5 + R_3 = R_6 \\ R_5 \end{array} \begin{bmatrix} 1 & 0 & | & -2/22 & 3/22 \\ 0 & 1 & | & 4/22 & 5/22 \end{bmatrix}$$

Therefore,

$$\mathbf{A}^{-1} = \begin{bmatrix} -2/22 & 3/22 \\ 4/22 & 5/22 \end{bmatrix}$$

Check: We check to verify that $\mathbf{AA}^{-1} = \mathbf{I_2}$ and $\mathbf{A}^{-1}\mathbf{A} = \mathbf{I_2}$.

$$\mathbf{AA}^{-1} = \begin{bmatrix} -5 & 3 \\ 4 & 2 \end{bmatrix} \begin{bmatrix} -2/22 & 3/22 \\ 4/22 & 5/22 \end{bmatrix}$$

$$= \begin{bmatrix} (10/22)+(12/22) & (-15/22)+(15/22) \\ (-8/22)+(8/22) & (12/22)+(10/22) \end{bmatrix} = \begin{bmatrix} 1 & 0 \\ 0 & 1 \end{bmatrix}$$

$$\mathbf{A}^{-1}\mathbf{A} = \begin{bmatrix} -2/22 & 3/22 \\ 4/22 & 5/22 \end{bmatrix} \begin{bmatrix} -5 & 3 \\ 4 & 2 \end{bmatrix}$$

$$= \begin{bmatrix} (10/22)+(12/22) & (-6/22)+(6/22) \\ (-20/22)+(20/22) & (12/22)+(10/22) \end{bmatrix} = \begin{bmatrix} 1 & 0 \\ 0 & 1 \end{bmatrix}$$

For $n > 2$, the check is tedious. Fortunately, it can be cut in half because of a theorem which states that if $\mathbf{AB} = \mathbf{I}$, then $\mathbf{BA}$ also equals $\mathbf{I}$. That is, $\mathbf{B} = \mathbf{A}^{-1}$. Thus, we need only check that $\mathbf{AA}^{-1} = \mathbf{I}$.

EXAMPLE 6.25 _____

Find the inverse of

$$\mathbf{A} = \begin{bmatrix} -2 & -1 & 1 \\ 7 & 3 & 1 \\ 1 & 0 & 7 \end{bmatrix}$$

and check.

SOLUTION

We place $\mathbf{I_3}$ to the right of $\mathbf{A}$ and begin the reduction:

$$\begin{array}{c} R_1 \\ R_2 \\ R_3 \end{array} \left[\begin{array}{ccc|ccc} \boxed{-2} & -1 & 1 & 1 & 0 & 0 \\ 7 & 3 & 1 & 0 & 1 & 0 \\ 1 & 0 & 7 & 0 & 0 & 1 \end{array}\right]$$

$$\begin{array}{c} 3R_1 + R_2 = R_4 \\ R_2 \\ R_3 \end{array} \left[\begin{array}{ccc|ccc} \boxed{1} & 0 & 4 & 3 & 1 & 0 \\ 7 & 3 & 1 & 0 & 1 & 0 \\ 1 & 0 & 7 & 0 & 0 & 1 \end{array}\right]$$

Now apply step 2 of Note 6.7 to the other two entries in the pivot column. We combine the two applications of step 2:

$$\begin{array}{c} R_4 \\ -7R_4 + R_2 = R_5 \\ -R_4 + R_3 = R_6 \end{array} \left[\begin{array}{ccc|ccc} 1 & 0 & 4 & 3 & 1 & 0 \\ 0 & \boxed{3} & -27 & -21 & -6 & 0 \\ 0 & 0 & 3 & -3 & -1 & 1 \end{array}\right]$$

Step 1(a) of Note 6.7 will convert the pivot entry to 1; since the other entries in column 2 are already 0, step 2 need not be applied in this case.

$$\begin{array}{c} R_4 \\ (1/3)R_5 = R_7 \\ R_6 \end{array} \left[\begin{array}{ccc|ccc} 1 & 0 & 4 & 3 & 1 & 0 \\ 0 & 1 & -9 & -7 & -2 & 0 \\ 0 & 0 & \boxed{3} & -3 & -1 & 1 \end{array}\right]$$

$$\begin{array}{c} R_4 \\ R_7 \\ (1/3)R_6 = R_8 \end{array} \left[\begin{array}{ccc|ccc} 1 & 0 & 4 & 3 & 1 & 0 \\ 0 & 1 & -9 & -7 & -2 & 0 \\ 0 & 0 & \boxed{1} & -1 & -1/3 & 1/3 \end{array}\right]$$

Finally, we convert the nonpivot entries from column 3 to 0s using step 2 of Note 6.7:

$$\begin{array}{c} -4R_8 + R_4 = R_9 \\ 9R_8 + R_7 = R_{10} \\ R_8 \end{array} \left[\begin{array}{ccc|ccc} 1 & 0 & 0 & 7 & 7/3 & -4/3 \\ 0 & 1 & 0 & -16 & -5 & 3 \\ 0 & 0 & 1 & -1 & -1/3 & 1/3 \end{array}\right]$$

Therefore,

$$\mathbf{A}^{-1} = \left[\begin{array}{ccc} 7 & 7/3 & -4/3 \\ -16 & -5 & 3 \\ -1 & -1/3 & 1/3 \end{array}\right]$$

Check: We need check only that $\mathbf{AA}^{-1} = \mathbf{I}_3$.

$$\mathbf{AA}^{-1} = \left[\begin{array}{cc} -2 & -1 \\ 7 & 3 \\ 1 & 0 \end{array}\right] \left[\begin{array}{cccc} 1 & 7 & 7/3 & -4/3 \\ 1 & -16 & -5 & 3 \\ 7 & -1 & -1/3 & 1/3 \end{array}\right]$$

$$= \left[\begin{array}{ccc} -14+16-1 & (-14/3)+5-(1/3) & (8/3)-3+(1/3) \\ 49-48-1 & (49/3)-15-(1/3) & (-28/3)+9+(1/3) \\ 7+0-7 & (7/3)+0-(7/3) & (-4/3)+0+(7/3) \end{array}\right] = \left[\begin{array}{ccc} 1 & 0 & 0 \\ 0 & 1 & 0 \\ 0 & 0 & 1 \end{array}\right]$$

The inverse has several important properties.

Theorem 6.1

a. The inverse of a matrix is unique.

b. $(\mathbf{A}^{-1})^{-1} = \mathbf{A}$.

c. If $\mathbf{A}$ and $\mathbf{B}$ have the same dimension and each has an inverse, then the product $\mathbf{AB}$ has an inverse and $(\mathbf{AB})^{-1} = \mathbf{B}^{-1}\mathbf{A}^{-1}$.

Proof: (We prove part a; parts b and c are left as exercises). Suppose $\mathbf{C}$ and $\mathbf{D}$ are both inverses of $\mathbf{A}$. We will show that $\mathbf{C} = \mathbf{D}$; that is, there is only one inverse. If $\mathbf{C}$ is an inverse of $\mathbf{A}$, then $\mathbf{CA} = \mathbf{AC} = \mathbf{I}$. If $\mathbf{D}$ is an inverse of $\mathbf{A}$, then $\mathbf{DA} = \mathbf{AD} = \mathbf{I}$. Recall from Note 6.5 that $\mathbf{IX} = \mathbf{X}$ for any matrix $\mathbf{X}$ provided $\mathbf{IX}$ is defined. Thus $\mathbf{C} = \mathbf{CI} = \mathbf{C}(\mathbf{AD}) = (\mathbf{CA})\mathbf{D} = \mathbf{ID} = \mathbf{D}$. Thus, $\mathbf{C} = \mathbf{D}$.

We now consider an example where we apply Gauss–Jordan reduction to a matrix that does not have an inverse. In this case, we eventually get all 0s in some row on the side we are trying to convert to $\mathbf{I}$.

EXAMPLE 6.26 _____

Find the inverse of

$$\mathbf{A} = \begin{bmatrix} 5 & 4 & 6 \\ 0 & 2 & 3 \\ 4 & 4 & 6 \end{bmatrix}$$

SOLUTION

$$\begin{array}{c} R_1 \\ R_2 \\ R_3 \end{array}\left[\begin{array}{ccc|ccc} ⑤ & 4 & 6 & 1 & 0 & 0 \\ 0 & 2 & 3 & 0 & 1 & 0 \\ 4 & 4 & 6 & 0 & 0 & 1 \end{array}\right]$$

$$\begin{array}{c} R_1 - R_3 = R_4 \\ R_2 \\ R_3 \end{array}\left[\begin{array}{ccc|ccc} ① & 0 & 0 & 1 & 0 & -1 \\ 0 & 2 & 3 & 0 & 1 & 0 \\ 4 & 4 & 6 & 0 & 0 & 1 \end{array}\right]$$

$$\begin{array}{c} R_4 \\ R_2 \\ -4R_4 + R_3 = R_5 \end{array}\left[\begin{array}{ccc|ccc} 1 & 0 & 0 & 1 & 0 & -1 \\ 0 & ② & 3 & 0 & 1 & 0 \\ 0 & 4 & 6 & -4 & 0 & 5 \end{array}\right]$$

Since the (3, 2) entry is a multiple of the pivot entry, there is no way to convert the 2 to 1 using a nice multiple of R_5. Therefore, we use the reciprocal of the pivot entry to convert it to 1 (that is, apply part 1(a) of Note 6.7).

$$\begin{array}{c} R_4 \\ (1/2)R_2 = R_6 \\ R_5 \end{array}\left[\begin{array}{ccc|ccc} 1 & 0 & 0 & 1 & 0 & -1 \\ 0 & ① & 3/2 & 0 & 1/2 & 0 \\ 0 & 4 & 6 & -4 & 0 & 5 \end{array}\right]$$

We use part 2 of Note 6.7 to convert the $(3, 2)$ entry to 0.

$$\begin{array}{c} R_4 \\ R_6 \\ -4R_6 + R_5 = R_7 \end{array} \left[\begin{array}{ccc|ccc} 1 & 0 & 0 & 1 & 0 & -1 \\ 0 & 1 & 3/2 & 0 & 1/2 & 0 \\ 0 & 0 & 0 & -4 & -2 & 5 \end{array}\right]$$

At this point, a row of 0s has appeared on the left side of the dashed line. By Note 6.8, we conclude **A** does not have an inverse. There is no way to convert the left side to an identity matrix using row operations.

There are certain times when knowing the inverse of a matrix is quite useful. For example, if we have many systems of equations involving the same matrix of coefficients, we would be able to find the solutions by simple matrix multiplications. This follows from the property stated in Note 6.9.

Note 6.9 Solving a Matrix Equation

If $\mathbf{Ax} = \mathbf{b}$, and **A** has an inverse, then the solution to the matrix equation is $\mathbf{x} = \mathbf{A}^{-1}\mathbf{b}$.

To see why Note 6.9 is true, multiply each side of the equation $\mathbf{Ax} = \mathbf{b}$ (from the front—remember multiplication is not commutative) by $\mathbf{A}^{-1}$. We get $\mathbf{A}^{-1}\mathbf{Ax} = \mathbf{A}^{-1}\mathbf{b}$, which gives $\mathbf{Ix} = \mathbf{A}^{-1}\mathbf{b}$, so $\mathbf{x} = \mathbf{A}^{-1}\mathbf{b}$.

EXAMPLE 6.27 _____

In Example 6.25 we found the inverse of

$$\mathbf{A} = \begin{bmatrix} -2 & 1 & 1 \\ 7 & 3 & 1 \\ 1 & 0 & 7 \end{bmatrix} \text{ to be } \mathbf{A}^{-1} = \begin{bmatrix} 7 & 7/3 & -4/3 \\ -16 & -5 & 3 \\ -1 & -1/3 & 1/3 \end{bmatrix}. \text{ Find the solution to}$$

the system of equations

$$\begin{array}{rcrcrcr} -2x & + & y & + & z & = & 8 \\ 7x & + & 3y & + & z & = & -5 \\ x & & & + & 7z & = & 3 \end{array}$$

SOLUTION

The system is equivalent to

$$\begin{bmatrix} -2 & 1 & 1 \\ 7 & 3 & 1 \\ 1 & 0 & 7 \end{bmatrix} \begin{bmatrix} x \\ y \\ z \end{bmatrix} = \begin{bmatrix} 8 \\ -5 \\ 3 \end{bmatrix}.$$

Thus, the solution is

$$\begin{bmatrix} x \\ y \\ z \end{bmatrix} = \mathbf{A}^{-1}\mathbf{b} = \begin{bmatrix} 7 & 7/3 & -4/3 \\ -16 & -5 & 3 \\ -1 & -1/3 & 1/3 \end{bmatrix} \begin{bmatrix} 8 \\ -5 \\ 3 \end{bmatrix}$$

$$= \begin{bmatrix} 56-(35/3)-4 \\ -128+25+9 \\ -8+(5/3)+1 \end{bmatrix} = \begin{bmatrix} 121/3 \\ -94 \\ -16/3 \end{bmatrix}$$

The Determinant Function

The determinant is a function defined for square matrices. The domain of this function is the set of all square matrices, and its range is the set of real numbers. The determinant function is denoted by **det** or by the absolute value sign. Thus $\det(\mathbf{A})$ means the same as $|\mathbf{A}|$. To define $\det(\mathbf{A})$ we need two concepts, the first of which is that of an elementary product. For an $n \times n$ matrix $\mathbf{A}$, an **elementary product** is a product of n entries, with each entry from a different row and column of $\mathbf{A}$. For a 3×3 matrix, $a_{11}a_{23}a_{32}$ is an elementary product. Note that the row numbers $(1, 2, 3)$ of the three entries in the product are all different and the column numbers $(1, 3, 2)$ are all different.

To determine all the elementary products for a matrix $\mathbf{A}$, it is useful to know how many such products there are. If we list the factors of an elementary product in order according to the row they come from, the row numbers are $1, 2, 3, \ldots, n$. The column numbers of successive factors would then correspond to a permutation of the numbers $1, 2, 3, \ldots, n$. For example, the elementary product $a_{11}a_{23}a_{32}$ corresponds to the permutation $1, 3, 2$. In this way, there is a 1-1 correspondence between the elementary products of an $n \times n$ matrix and the permutations of $1, 2, 3, \ldots, n$. Therefore Note 6.10 is true.

Note 6.10

An $n \times n$ matrix has $n!$ elementary products.

EXAMPLE 6.28

List the elementary products of a 3×3 matrix $\mathbf{A}$.

SOLUTION

By Note 6.10, the matrix $\mathbf{A}$ has $3! = 6$ elementary products. Since each product corresponds to a permutation of the digits 1, 2, 3, we first list the permutations: $(1, 2, 3)$, $(1, 3, 2)$, $(2, 1, 3)$, $(2, 3, 1)$, $(3, 1, 2)$, $(3, 2, 1)$. Each permutation lists the successive column numbers of an elementary product. Thus, the elementary products are $a_{11}a_{22}a_{33}$, $a_{11}a_{23}a_{32}$, $a_{12}a_{21}a_{33}$, $a_{12}a_{23}a_{31}$, $a_{13}a_{21}a_{32}$, and $a_{13}a_{22}a_{31}$.

The second concept we need to define the determinant function is the sign of an elementary product. The column numbers in an elementary product correspond to a permutation of $1, 2, \ldots, n$. The sign of an elementary product is determined by how many pairs of numbers are not in increasing order in this permutation. For the elementary product $a_{13}a_{21}a_{32}$, the column numbers of the factors in the product determine the permutation $(3, 1, 2)$. The pairs of numbers from this permutation (taken in the same order that they appear in the permutation) are $(3, 1)$, $(3, 2)$, and $(1, 2)$, which are found by pairing the first with the second number, the first with the third number, and the second with the third number. The **sign of an elementary product** is $+$ if an even number of pairs of numbers in the corresponding permutation is not in increasing order; otherwise the sign is $-$. For the permutation $(3, 1, 2)$, two of the three pairs are not in increasing order, namely $(3, 1)$ and $(3, 2)$. Thus, the sign of $a_{13}a_{21}a_{32}$ is $+$.

EXAMPLE 6.29

Determine the sign for each of the following elementary products: $a_{12}a_{21}a_{33}$, $b_{14}b_{23}b_{32}b_{41}$, $c_{11}c_{23}c_{34}c_{42}$.

SOLUTION

For $a_{12}a_{21}a_{33}$ the permutation is $(2, 1, 3)$. The pairs of numbers from this permutation are $(2, 1)$, $(2, 3)$, and $(1, 3)$. Of these, one is not in increasing order, so the sign of $a_{12}a_{21}a_{33}$ is $-$.

The permutation for $b_{14}b_{23}b_{32}b_{41}$ is $(4, 3, 2, 1)$. Since there are four numbers in the permutation, we will have $\binom{4}{2} = 6$ pairs. The pairs are $(4, 3)$, $(4, 2)$, $(4, 1)$, $(3, 2)$, $(3, 1)$, and $(2, 1)$. All six of these pairs are not in increasing order, so the sign of $b_{14}b_{23}b_{32}b_{41}$ is $+$.

The permutation for the product $c_{11}c_{23}c_{34}c_{42}$ is $(1, 3, 4, 2)$. From this we get pairs $(1, 3)$, $(1, 4)$, $(1, 2)$, $(3, 4)$, $(3, 2)$, and $(4, 2)$. Two of these pairs $(3, 2)$ and $(4, 2)$ are not in increasing order. Since there is an even number of pairs not in increasing order, the sign of $c_{11}c_{23}c_{34}c_{42}$ is $+$.

We are now in a position to define the determinant of a matrix.

Note 6.11

The **determinant** det $(\mathbf{A})$ of a square matrix $\mathbf{A}$ is the sum of all the signed elementary products of $\mathbf{A}$.

For a 2×2 matrix $\mathbf{A}$, the elementary products are $a_{11}a_{22}$ and $a_{12}a_{21}$. The sign of the first of these is $+$, and the sign of the second is $-$. Thus $\det(\mathbf{A}) = a_{11}a_{22} - a_{12}a_{21}$.

EXAMPLE 6.30

Find the determinant for each of the following:

a. $\mathbf{A} = \begin{bmatrix} 3 & -5 \\ 2 & 7 \end{bmatrix}$ b. $\mathbf{B} = \begin{bmatrix} 1 & 15 \\ 3 & 19 \end{bmatrix}$ c. $\mathbf{C} = \begin{bmatrix} 60 & -2 \\ -6 & 1/5 \end{bmatrix}$

SOLUTION

a. $a_{11}a_{22} - a_{12}a_{21} = (3 \cdot 7) - ((-5) \cdot 2) = 21 + 10 = 31$. Thus $\det(\mathbf{A}) = 31$.
b. $b_{11}b_{22} - b_{12}b_{21} = (1 \cdot 19) - (15 \cdot 3) = 19 - 45 = -26$. Thus $\det(\mathbf{B}) = -26$.
c. $c_{11}c_{22} - c_{12}c_{21} = (60 \cdot (1/5)) - ((-2)(-6)) = 12 - 12 = 0$. Thus $\det(\mathbf{C}) = 0$.

One method of remembering the signs of the permutations for 2×2 and 3×3 matrices uses *downward ovals*. Each downward oval corresponds to a signed elementary product. To find the product, multiply the matrix entries in the oval. A downward oval pointing toward the right gives a positive elementary product, and one pointing toward the left gives a negative elementary product. The downward ovals for a 2×2 matrix and a 3×3 matrix are displayed in Figure 6.4. For the 3×3 matrix, we recopy the first two columns to the right of the matrix to get all six elementary products using downward ovals.

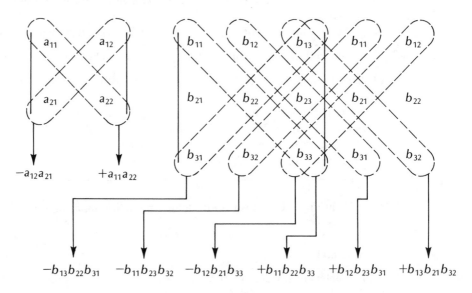

Figure 6.4

We can check the pairs of numbers in the corresponding permutations for the terms in Figure 6.4 to see that the signs are correct.

EXAMPLE 6.31

Evaluate the following:

a. $\begin{vmatrix} 4 & -1 & 2 \\ 3 & 8 & 6 \\ -3 & 2 & 7 \end{vmatrix}$ b. $\begin{vmatrix} 1/2 & 3 & -5 \\ 9 & 4 & -6 \\ 8 & -10 & 1/3 \end{vmatrix}$

SOLUTION

Recall that the absolute value sign indicates the determinant. We display the downward ovals for these in Figure 6.5. Thus, we have (writing the + products first):

a. $(4 \cdot 8 \cdot 7) + ((-1) \cdot 6 \cdot (-3)) + (2 \cdot 3 \cdot 2) - (2 \cdot 8 \cdot (-3)) - (4 \cdot 6 \cdot 2) - ((-1) \cdot 3 \cdot 7)$
$= 224 + 18 + 12 + 48 - 48 + 21 = 275$
The determinant is 275.

b. $((1/2) \cdot 4 \cdot (1/3)) + (3 \cdot (-6) \cdot 8) + ((-5) \cdot 9 \cdot (-10)) - ((-5) \cdot 4 \cdot 8)$
$- ((1/2)(-6)(-10)) - (3 \cdot 9 \cdot (1/3)) = 2/3 - 144 + 450 + 160 - 30 - 9 = 427\frac{2}{3}$
Thus the determinant is $427\frac{2}{3}$.

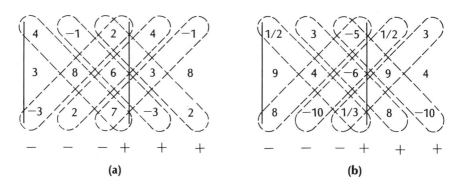

Figure 6.5

The method of downward ovals does not extend to matrices larger than 3 × 3. For example, with a 4 × 4 matrix, downward ovals would include a total of eight elementary products, but we know there are 4! = 24 elementary products for a 4 × 4 matrix. For matrices larger than 3 × 3, the standard method used is *cofactor expansion* (see exercise 11), which enables us to write the determinant of an $n \times n$ matrix as a sum of n determinants of $(n - 1) \times (n - 1)$ matrices.

> **Note 6.12   Properties of Determinants**
>
> 1. $\det(\mathbf{A}^t) = \det(\mathbf{A})$.
> 2. If $\mathbf{A}$ and $\mathbf{B}$ are $n \times n$ matrices, then $\det(\mathbf{AB}) = \det(\mathbf{A}) \det(\mathbf{B})$.
> 3. If $\mathbf{A}$ is a square matrix, then $\mathbf{A}$ has an inverse if and only if $\det(\mathbf{A}) \neq 0$.
> 4. If $\mathbf{A}$ has an inverse, then $\det(\mathbf{A}^{-1}) = 1/\det(\mathbf{A})$.

Determinants are also related to solving systems of equations. A system $\mathbf{Ax} = \mathbf{b}$ of n equations having n variables has a unique solution if and only if $\det(\mathbf{A}) \neq 0$. In fact, the solution can be calculated using determinants by a procedure called Cramer's rule (see exercise 12). We stress that Cramer's rule applies only to a system of n equations in n unknowns where $\det(\mathbf{A}) \neq 0$. It is generally less efficient than Gauss–Jordan reduction and is much less efficient than Gauss–Jordan reduction when $n > 5$. Gauss–Jordan reduction requires about n^3 multiplications, whereas calculating the determinant of an $n \times n$ matrix using the definition of the determinant requires about $n!$ multiplications. For a 20×20 system, Gauss–Jordan reduction requires about 8000 multiplications. Today's computers can do these easily in 1 sec. If we use Cramer's rule to solve the same system, we would have to calculate the determinant of its coefficient matrix. To calculate the determinant of a 20×20 matrix using the definition of the determinant would require about $20! \approx 2.4 \times 10^{18}$ multiplications. This would take a computer over 190 years. [Today's fastest computers can do 400 million operations per second (*Newsweek*, p. 63, July 4, 1983).]

In view of the complexity inherent in calculating the determinant of a matrix, how can we use a computer to calculate the determinant efficiently? Surely not by using the definition! The usual method is to use Gauss–Jordan reduction to reduce the matrix to diagonal form and then evaluate the determinant of the resulting matrix. Calculating the determinant in this way requires about n^4 multiplications.

EXERCISES 6.3

1. Write the augmented matrix and solve using Gauss–Jordan reduction:

   a. $2x - 3y = 26$
   $5x + 2y = 27$

   b. $2x + 3y = -2$
   $-5x + 6y = 18$

   c. $7x - 2y - 3z = 15$
   $-4x + 2y + z = 1$
   $3x + y - z = 7$

   d. $x + 8y + 3z = 8$
   $5x + 4y = 11$
   $-2x + 9z = -17$

2. Find the inverse for each of the following or state that it doesn't exist if such is the case:

   a. $\begin{bmatrix} -9 & 2 \\ 4 & -1 \end{bmatrix}$ **b.** $\begin{bmatrix} 4 & 7 \\ -2 & -3 \end{bmatrix}$ **c.** $\begin{bmatrix} 8 & 4 \\ -6 & -3 \end{bmatrix}$

d. $\begin{bmatrix} 2 & 1 & -1 \\ 4 & 1 & 0 \\ -1 & -1 & 1 \end{bmatrix}$ **e.** $\begin{bmatrix} 3 & -1 & 2 \\ 2 & 3 & 5 \\ 3 & 1 & -4 \end{bmatrix}$ **f.** $\begin{bmatrix} 4 & -2 & -2 \\ 14 & 3 & 1 \\ 11 & 2 & 1 \end{bmatrix}$

3. Without using Gauss–Jordan reduction

a. Verify that $\begin{bmatrix} 1 & 1 & 2 \\ 2 & 1 & -1 \\ 0 & 1 & -2 \end{bmatrix}^{-1} = \begin{bmatrix} -1/7 & 4/7 & -3/7 \\ 4/7 & -2/7 & 5/7 \\ 2/7 & -1/7 & -1/7 \end{bmatrix}$.

b. Verify that $\begin{bmatrix} 3/2 & -1/2 & -1/2 \\ -1/2 & 1/2 & 1/2 \\ -1/2 & 3/2 & 1/2 \end{bmatrix}$ is not the inverse of $\begin{bmatrix} 1 & 1 & 0 \\ 0 & 1 & -1 \\ 1 & 2 & 1 \end{bmatrix}$.

4. Show that any matrix of the form $\begin{bmatrix} 0 & k \\ 1/k & 0 \end{bmatrix}$ is its own inverse.

5. Prove Theorem 6.1(b): $(\mathbf{A}^{-1})^{-1} = \mathbf{A}$. (*Hint:* Verify that $\mathbf{A}$ satisfies the condition that the inverse of $\mathbf{A}^{-1}$ must satisfy.)

6. Prove Theorem 6.1(c): If $\mathbf{A}$ and $\mathbf{B}$ have the same dimension and each has an inverse, then the product $\mathbf{AB}$ has an inverse and $(\mathbf{AB})^{-1} = \mathbf{B}^{-1}\mathbf{A}^{-1}$. (*Hint:* Verify that $\mathbf{B}^{-1}\mathbf{A}^{-1}$ satisfies the condition that the inverse of $\mathbf{AB}$ must satisfy.)

7. Given that $\begin{bmatrix} 5 & 2 & 2 \\ 3 & -3 & 1 \\ 2 & 7 & 1 \end{bmatrix}^{-1} = \begin{bmatrix} -5 & 6 & 4 \\ -1/2 & 1/2 & 1/2 \\ 27/2 & -31/2 & -21/2 \end{bmatrix}$ find the solutions

to the following systems using the appropriate matrix multiplications:

a. $5x + 2y + 2z = 1$ **b.** $15x + 6y + 6z = 9$
  $3x - 3y + z = -2$ $3x - 3y + z = 1$
  $2x + 7y + z = 0$ $2x + 7y + z = -3$
                         (*Hint:* First adjust equation 1.)

c. $-10x + 12y + 8z = 4$
  $-x + y + z = 0$
  $27x - 31y - 21z = 5$
  (*Hint:* Adjust all equations by a factor of 1/2.)

8. Suppose $\mathbf{A}$ is a 4×4 matrix.
  a. How many elementary products does $\mathbf{A}$ have?
  b. List all elementary products for $\mathbf{A}$.
  c. Determine a formula for $\det(\mathbf{A})$ by finding the sign of each elementary product from part b.

9. Evaluate each of the following:

a. $\begin{vmatrix} 4 & 17 \\ 3 & 14 \end{vmatrix}$ **b.** $\begin{vmatrix} 5/2 & 1/3 \\ 7/4 & 1/6 \end{vmatrix}$ **c.** $\begin{vmatrix} 12 & -11 \\ -3 & -4 \end{vmatrix}$

d. $\begin{vmatrix} 7 & 1 & 2 \\ -1 & 0 & 3 \\ 2 & -1 & 1 \end{vmatrix}$ **e.** $\begin{vmatrix} 5 & 8 & -3 \\ 1 & 2 & 0 \\ -1 & 2 & -1 \end{vmatrix}$ **f.** $\begin{vmatrix} 1/2 & 4/3 & -9 \\ -6 & 4 & 3 \\ 2/3 & 2 & -1/2 \end{vmatrix}$

10. Let $\mathbf{A} = \begin{bmatrix} 1 & 2 & -1 \\ 3 & 1 & 2 \\ 4 & -1 & 3 \end{bmatrix}$ and $\mathbf{B} = \begin{bmatrix} -1 & 5 & 2 \\ 1 & 0 & -1 \\ -2 & 1 & 3 \end{bmatrix}$.

Verify:
a. $\det(\mathbf{A}^t) = \det(\mathbf{A})$
b. $\det(\mathbf{AB}) = \det(\mathbf{A})\det(\mathbf{B})$

Problems and Projects

11. (Cofactor Expansion) The determinant of an $n \times n$ matrix can be written as the sum of n determinants of $(n-1) \times (n-1)$ matrices. The **minor** $\mathbf{M}_{ij}$ of a matrix $\mathbf{A}$ is the matrix obtained by deleting row i and column j from $\mathbf{A}$. The **cofactor** of an entry a_{ij} is $(-1)^{i+j}\det(\mathbf{M}_{ij})$. The determinant of $\mathbf{A}$ can be found by *expanding along* a given row or column. It can be shown that the following formulas hold:

i. (Expanding along row i): $\det(\mathbf{A}) = \sum_{j=1}^{n} (-1)^{i+j} \cdot a_{ij} \cdot \det(\mathbf{M}_{ij})$

ii. (Expanding along column j): $\det(\mathbf{A}) = \sum_{i=1}^{n} (-1)^{i+j} \cdot a_{ij} \cdot \det(\mathbf{M}_{ij})$

Let

$$\mathbf{A} = \begin{bmatrix} 3 & 1 & 0 & -1 \\ 1 & 2 & -1 & 0 \\ 2 & -1 & 0 & 2 \\ 0 & 0 & 1 & -1 \end{bmatrix}$$

a. Find the minor $\mathbf{M}_{23}$.
b. Find the cofactor of a_{42}.
c. Evaluate $\det(\mathbf{A})$ by expanding along row 1.
d. Evaluate $\det(\mathbf{A})$ by expanding along column 3.
e. Just by looking at matrix $\mathbf{A}$, determine which is the easiest row and which is the easiest column to expand along? Why?

12. (Cramer's Rule) A system $\mathbf{Ax} = \mathbf{b}$ of n equations in n unknowns can be solved using determinants if $\det(\mathbf{A}) \neq 0$. Let $\mathbf{A}_i$ be the matrix obtained by replacing column i of $\mathbf{A}$ with the column vector $\mathbf{b}$. Cramer's rule says that the value of the ith coordinate of the solution vector x is $x_i = \det(\mathbf{A}_i)/\det(\mathbf{A})$.

Use Cramer's rule to find the solution to the following:

a.
$x_1 + x_2 + 2x_3 = 1$
$2x_1 + x_2 - x_3 = -2$
$x_2 - 2x_3 = 5$

b.
$2x_1 + x_2 - x_3 = -3$
$4x_1 + x_2 = 8$
$-x_1 - x_2 + x_3 = 0$

13. Gauss–Jordan reduction can be used to solve a system of equations that has infinitely many solutions, as well as to identify one that has no solutions. When there are infinitely many solutions, the final form of the augmented matrix has an identity matrix in the upper-left corner. The rows below the identity

$$\begin{bmatrix} 1 & 0 & 2 & \vdots & 5 \\ 0 & 1 & -3 & \vdots & 8 \\ 0 & 0 & 0 & \vdots & 0 \end{bmatrix} \qquad \begin{bmatrix} 1 & 0 & 3 & \vdots & 2 \\ 0 & 1 & 1 & \vdots & -3 \\ 0 & 0 & 0 & \vdots & 4 \end{bmatrix}$$

(a)   (b)

Figure 6.6

matrix (if any) contain all 0s. An example is given in Figure 6.6(a). For that example, row 1 is read $x_1 + 2x_3 = 5$, and row 2 is read $x_2 - 3x_3 = 8$. We solve each of these in terms of x_3 and get the solution $x_1 = 5 - 2x_3$, $x_2 = 8 + 2x_3$, and $x_3 = x_3$. Thus, each variable is written as a function of x_3.

a. Solve the following system using Gauss–Jordan reduction:

$$\begin{aligned} -3x_1 + x_2 - 5x_3 &= -10 \\ x_1 \qquad\quad + x_3 &= 3 \\ 2x_1 + 3x_2 - 4x_3 &= 3 \end{aligned}$$

A system that has no solution is called **inconsistent**. When we apply Gauss–Jordan reduction to an inconsistent system, we eventually obtain a row whose initial positions all equal 0 and whose final entry is not 0. The system has no solution because that row is read

$$0x_1 + 0x_2 + \cdots + 0x_n = k \qquad (k \neq 0)$$

which is impossible since the left side equals 0 while the right side does not. See Figure 6.6(b).

b. Show that the following system is inconsistent:

$$\begin{aligned} x_1 + 2x_2 + 4x_3 &= -4 \\ x_2 + x_3 &= -3 \\ 2x_1 - x_2 + 3x_3 &= 9 \end{aligned}$$

14. Prove: If a row (column) of A consists entirely of 0s, then $\det(A) = 0$.

15. If $\det(AB) = 0$ and A and B are square matrices, prove either $\det(A) = 0$ or $\det(B) = 0$.

16. If $A = A^{-1}$, what is $\det(A)$?

17. If $A^{-1} = A^t$, what is $\det(A)$?

18. A diagonal matrix D is a square matrix with off-diagonal entries all 0:

$$D = \begin{bmatrix} d_{11} & 0 & & & 0 \\ 0 & d_{22} & & & \\ & & d_{33} & & \\ & & & & 0 \\ 0 & & & 0 & d_{nn} \end{bmatrix}$$

a. Show $\det(D) = d_{11} \cdot d_{22} \cdot d_{33} \cdots d_{nn}$.

b. What is $\det(I_n)$?

19. In exercise 6 of Exercises 6.2, we noted that for square matrices **A** and **B** it was not necessary that **AB** = **BA**. Is it always true that det (**AB**) = det (**BA**)? Justify your answer.

CASE STUDY 6A   STACKS, QUEUES, AND DEQUES

Arrays are one of the most elementary types of data structures. A one-dimensional array or vector is a special type of linear list. A **linear list** is a sequence of elements. For a vector, we can access any element in the list at a given time. For example, if **v** is a vector with ten entries, we can print v_1, v_5, and v_{10} in succession without having to access the entries between v_1 and v_5 and between v_5 and v_{10}. In this case study, we examine other types of linear lists—stacks, queues, and deques—where access is restricted to the ends of the lists.

Stacks

A **stack** is a linear list of which we have access to only one end, called the **top** of the stack. Stacks are also called **push-down lists.** This name comes from the spring-controlled device many cafeterias use to store trays. When a tray is placed on top, the rest of the stack is pushed down. A customer has access only to the top tray. The one short-coming of this model of a stack is that it gives the impression that the elements of the stack move around within the computer. They do not. Instead, a **pointer** (variable) is used to tell the location of the top of the stack at any time (see Figure 6.7).

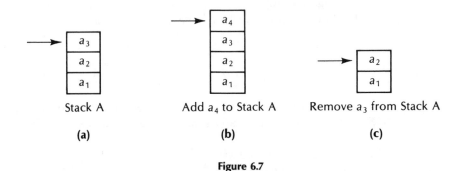

	Stack A	Add a_4 to Stack A	Remove a_3 from Stack A
	(a)	**(b)**	**(c)**

Figure 6.7

If an element a_4 is added to the stack [see Figure 6.7(b)] or if a_3 is removed from the stack [see Figure 6.7(c)], the value of the pointer is changed accordingly. In its simplest form, the value of the pointer is the subscript of the element on the top of the stack. In this case, when an element x is added to the stack (adding an element to a stack is called a **push**), the value of the pointer P becomes $P + 1$. If an element y is deleted from the stack (deleting an item from the top of the stack is called a **pop**), the value of the pointer P becomes $P - 1$.

Remark   When using stacks to program, we must guard against overflow and underflow. **Overflow** occurs if we try to add an element to a stack whose length has reached maximum capacity for that stack. **Underflow** occurs when we try to remove an item from an empty stack. Thus, we must test the length of the stack before we attempt to add to it or delete from it.

EXAMPLE 6.32

Write procedures for pop and push that test for underflow and overflow.

SOLUTION

The following program displays the procedures that allow us to add or delete elements from the stack:

```
program popnpush (input,output);
var  p,under,x,over,pcode,i: integer;
     s: array[1..10] of integer;
  procedure pop;
  begin
     if p>=1 then
        p:=p-1
     else
        begin
           under:=1;
           pcode:=999
        end
  end;

  procedure push;
  begin
     writeln('input an integer');
     read(x);
     if p<10 then
        begin
           p:=p+1;
           s[p]:=x
        end
     else
        begin
           over:= 1;
           pcode:=999
        end
  end;
begin
   writeln('type 222 to push, 444 to pop, 999 to end');
   read(pcode);
   while pcode<>999 do
      begin
         if pcode=222 then
            push
         else
            if pcode=444 then
               pop;
         if pcode<>999 then
            begin
               writeln('type 222 to push, 444 to pop, 999 to end');
               read(pcode)
            end
      end;
   if over=1 then
      writeln('stack overflow')
   else
      if under=1 then
         writeln('stack underflow')
      else
         begin
            writeln('current stack is as follows');
            for i:= 1 to p do
               writeln(s[i])
         end
end.
```

Stacks are used in various computer applications: (1) in translations of code by the computer from a high-level language to machine language, (2) in programming certain sorting procedures, (3) in many situations where recursion is used, and (4) in implementing subroutine calls. (See, for example, Tenenbaum and Augenstein, 1981.)

We illustrate the use of stacks with an example of sorting a sequence of numbers using stacks. The procedure uses two stacks A and B to rearrange a given sequence of numbers. The numbers are placed on the stacks one at a time so that in each stack the numbers at each step are nondecreasing as we go down the stack. The procedure ends with all elements in stack B in nondecreasing order.

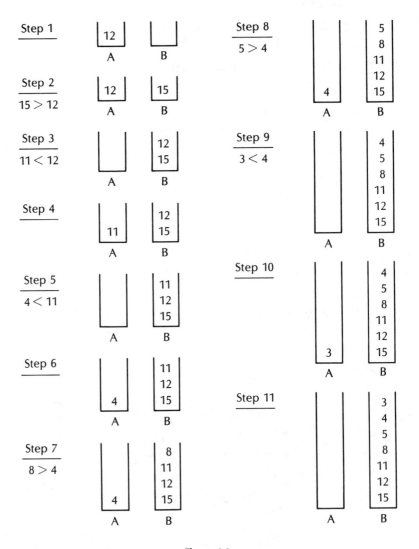

Figure 6.8

To be more specific, as each element X is read from the given sequence, we compare it to the element Y on the top of stack A. If $X \geq Y$, we place X in stack B. If $X < Y$, remove Y from stack A, place it in stack B, and compare X to the new top of stack A. Whenever A is empty, place X in A. When there are no more input items, unstack A item by item, placing each item on top of stack B.

We illustrate the procedure in Figure 6.8 with the sequence 12, 15, 11, 4, 8, 5, 3.

Can we always rearrange a sequence of numbers into nondecreasing order using two stacks? If we require that the numbers in each stack be in nondecreasing order at every step, the answer is no (see exercise 1). However, Note 6.13 holds.

Note 6.13

Using three stacks, it is always possible to rearrange a sequence of numbers into nondecreasing order within the stacks so that at every step the numbers in each stack are in nondecreasing order.

The 3-stack problem is related to the **Tower of Hanoi** puzzle. In this puzzle, there are three pegs: A, B, and C. On peg A there are n hoops having different diameters arranged in increasing diameter from top to bottom **(size order)**. The goal is to move the hoops to peg C so that they are still arranged in size order. There are two rules:

1. Only the top hoop from one of the pegs may be moved.

2. After each move, the hoops on each peg must be arranged in size order.

After several moves in the puzzle with seven hoops, the situation might appear as in Figure 6.9.

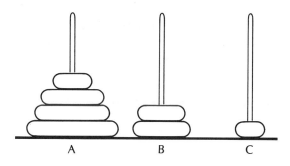

Figure 6.9

We note that at various stages in the puzzle, it is necessary to move hoops back onto peg A. We illustrate with an example.

EXAMPLE 6.33

Solve the Tower of Hanoi puzzle with four hoops.

SOLUTION

We number the hoops and list them as they appear from top to bottom on each peg.

Move	Peg A	Peg B	Peg C	Move	Peg A	Peg B	Peg C
0	1234			8		123	4
1	234	1		9		23	14
2	34	1	2	10	2	3	14
3	34		12	11	12	3	4
4	4	3	12	12	12		34
5	14	3	2	13	2	1	34
6	14	23		14		1	234
7	4	123		15			1234

Note 6.14

To solve the Tower of Hanoi puzzle having n hoops, $2^n - 1$ moves are required.

Queues

In a stack, the first item to enter is the last to leave. To remove a particular element from a stack, each element on top of it must be removed first. For this reason, a stack is frequently called a **LIFO** (last-in, first-out) mechanism. A **queue** is a linear list for which elements enter the list at one end (called the **rear**) and leave at the other (called the **front**). Thus, the first item to enter the queue is the first to leave. Queues, therefore, are referred to as **FIFO** (first-in, first-out) mechanisms. More generally, the earlier an item enters a queue, the earlier it leaves. Figure 6.10(b) depicts the queue in Figure 6.10(a)

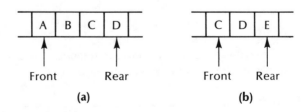

(a)   (b)

Figure 6.10

after two deletions and one addition. Note that two pointers are required: one to keep track of the front of the queue and one to keep track of the rear of the queue.

We encounter queues every day, the most common being the lines at the supermarket checkout counter and at the bank teller's window.

An important example of a queue is the phone queue. Many airlines and government agencies handle phone calls as follows:

1. When a call is received, it is put on a queue if the queue is not full (the length of the queue is determined by the number of calls the personnel can deal with in a "reasonable" amount of time).

2. An appropriate message is given to the caller.

3. As personnel become available, callers are taken off the queue and switched to the appropriate personnel.

4. If the queue were full, the caller would get a "busy" signal.

As with stacks, queues require us to test for underflow and overflow. Stacks have one end (the bottom) fixed. The other end (the top of the stack) moves up and down as additions and deletions are made. With a queue, however, neither end is fixed and its capacity may quickly be exceeded. To overcome this problem, a circular list can be used (see Figure 6.11). For example, suppose n positions $a_1, a_2, \ldots, a_n$ are allocated for the queue. In the **circular list,** a_1 is thought of as following a_n, and the front pointer, F, points *one space ahead* of the actual front of the queue. If the value of the rear pointer R equals the value of F, the queue is either empty or full. To decide which, a variable S is used as a signal. In our next example we show how the signal is used

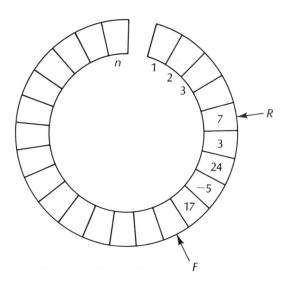

Figure 6.11   A Circular List

when adding an element to a queue. We leave the problem of deleting from a queue as an exercise (see exercise 7).

EXAMPLE 6.34

Write a procedure that adds an element to a queue after testing for overflow.

SOLUTION

The following program reads the elements and calls the procedure.

```
program queue (input,output);
var s,f,r,i,full: integer;
    x: real;
    q: array[1..10] of real;
  procedure addto;
  begin
        if s<>1 then
          begin
            if f=10 then
                f:=1
            else
                f:=f+1;
            q[f]:=x;
            if f=r then
                s:=1
            else
                s:=0
          end
        else
          begin
            full:=1;
            s:=1
          end
  end;

begin
    f:=0;
    s:=0;
    r:=10;
    while s<>1 do
        begin
            writeln;
            writeln('input a real number (9999 to end)');
            read(x);
            if x<>9999 then
                begin
                    addto;
                    for i:= 1 to f do
                        write(q[i])
                end
            else
                s:=1
        end;
    if full=1 then
        writeln('queue is full')
end.
```

Note that the value of the signal S is changed to 1 if, after adding to the queue, F and R are equal (that is, the front and rear pointers point to the same position in the list). When deleting from a queue, the value of S is changed to -1 when $F = R$.

Deques

A **deque** is a linear list in which additions and deletions can be made from either the front or the rear, thus making it more general than a queue or a stack. A deque, like a queue, requires two pointers F and R, and is generally stored as a circular list to save storage space.

To summarize, stacks allow additions and deletions, reference (access) only at the top, and require one pointer. Queues allow additions to the rear, deletions from the front, and accesses from either end. They require two pointers. Deques allow additions, deletions, and accesses from both the front and rear, and require two pointers. In each of these structures, there is no access available to internal elements.

To help understand the difference in the use of stacks, queues, and deques, we consider the problem of generating certain permutations. Suppose we input the letters A, B, C, D (in that order) one at a time. At the time of input, we have the choice of printing the letter or adding it to a stack and printing it later as it is deleted from the stack. Some permutations of the letters A, B, C, D can be printed in this way and some cannot. For example, the permutation CABD cannot be printed in this manner. If a queue were used instead of a stack, the permutation CABD could be printed. On the other hand, the permutation BDCA could be printed with a stack but not a queue. A deque is more effective than stacks or queues in generating permutations (see exercise 6).

In storing computer jobs awaiting execution, one deque can be used to do the work of two stacks. Jobs anticipated to be long or complicated are added to one end (say, end A) of the deque while simple jobs are added to the other end (end B). When large blocks of time and/or storage are available, a job is removed from end A of the deque and processed; otherwise, a job is removed from end B and processed.

An arraylike structure that is used to facilitate internal access in structures such as stacks, queues, and deques is the **linked list.** An element of a linked list has two fields: a *data field* and a *link field*. The **link field** is a pointer to the next element on the list. (See Figure 6.12). To access the third element, use the pointer FRONT to find the first element, the pointer of the first element to find the second, and the pointer of the second to find the third.

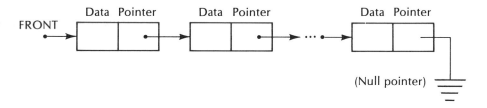

Figure 6.12

A still more general structure is a **doubly linked list,** in which each element has a data field, a forward pointer, and a backward pointer. These are used in generating linked stacks, linked queues, linked deques, and linked circular lists. For more information on these topics see Page and Wilson (1973) or Tremblay and Sorenson (1976).

EXERCISES: CASE STUDY 6A

1. Given the list 3, 8, 1, 5, 6, 2, 9.
   a. Show that the list cannot be rearranged into nondecreasing order with two stacks if we require that each stack be nondecreasing at each step.
   b. Rearrange the list as in part a using three stacks.
   c. Suppose we do not require that each stack be nondecreasing. Show all steps in rearranging the list using two stacks.
2. a. Explain why any list can be arranged into nondecreasing order using two stacks.
   b. Write a procedure to arrange any list into nondecreasing order using two stacks (be careful about empty stacks).
3. a. Solve the Tower of Hanoi puzzle having five hoops.
   b. How many moves are required? How many moves did you make? Is your solution optimal?
4. Give four examples of queues besides the three given in this section.
5. Suppose you start the Tower of Hanoi puzzle at the intermediate position:

Peg A	Peg B	Peg C
37	456	12

   a. Perform several more moves to get hoop 7 onto peg C.
   b. Without performing any moves after completing part a, use Note 6.14 to show that you would have needed 68 moves to complete the puzzle.
6. a. Suppose that letters A, B, C, D are input one at a time and are either printed immediately or placed on a linear list for later printing. Which of the 24 permutations of A, B, C, D *cannot* be printed if the linear list is:
      i. A stack   ii. A queue   iii. A deque
   b. Which of the 120 permutations of A, B, C, D, E *cannot* be generated using a deque?

Problems and Projects

7. Write a procedure for deleting an element from a queue represented as a circular list.
8. Write procedures for (a) adding and (b) deleting elements from a deque represented as a circular list.

CASE STUDY 6B   IMPLEMENTATION AND COMPARISON OF SORTING ALGORITHMS

In Chapter 2, we introduced the concepts of searching and sorting, and presented a simple sorting method, the method of successive minima. This method performs a sequential search on list $a_1, a_2, \ldots, a_n$ to find the smallest element, and switches it with a_1; searches the remaining list to find the next smallest element, and switches it with a_2; and so on. Having discussed vectors, we are now ready to examine several other sorting algorithms. In this case study, we look at insertion sort, bubble sort, and quicksort. We then compare the efficiencies of these algorithms.

Insertion Sort

With the insertion sort we begin by comparing a_2 with a_1. If they are out of order, they are switched. We then move to a_3. In general, we move along the list comparing an element a_i to the element on its left, a_{i-1}. If the two are in order, we continue along in the list. If a_i and a_{i-1} are out of order, a_i is moved to its left and inserted in the proper position in the list. Note that when working on a_i, the list $a_1, a_2, \ldots, a_{i-1}$ is already sorted into proper order.

EXAMPLE 6.35 _____

Write a program for insertion sort.

SOLUTION

```
program insertion (input,output);
var n,sw,i,j: integer;
    x: real;
    a: array[1..50] of real;
begin
    writeln('input a positive integer');
    read(n);
    if n>50 then
       writeln('array too large')
    else
       begin
          writeln('input',n,' numbers in any order');
          for i:=1 to n do
              read(a[i]);
          for i:=2 to n do
             begin
                x:=a[i];
                j:=i-1;
                sw:=0;
                while sw=0 do
                    begin
                       if x<a[j] then
                           begin
                              a[j+1]:=a[j];
                              j:=j-1;
                              if j=0 then
                                  sw:=1
                           end
                       else
                           sw:=1
                    end;
                a[j+1]:=x
             end;
          writeln('the sorted list is as follows');
          for i:=1 to n do
              write (a[i])
       end
end.
```

EXAMPLE 6.36 _____

Perform an insertion sort on the list 5, -1, 2, 3, 1, 7, 4.

SOLUTION

For each *i*, we display the list as it appears at the beginning of that step. We also count the number of comparisons made. Note that the first $i - 1$ elements are in order for each *i*.

I	List A	X	Number of Comparisons
2	5, −1, 2, 3, 1, 7, 4	−1	1
3	−1, 5, 2, 3, 1, 7, 4	2	2
4	−1, 2, 5, 3, 1, 7, 4	3	2
5	−1, 2, 3, 5, 1, 7, 4	1	4
6	−1, 1, 2, 3, 5, 7, 4	7	1
7	−1, 1, 2, 3, 4, 5, 7	4	3

Output: −1, 1, 2, 3, 4, 5, 7   Total: 13

Bubble Sort

The bubble sort makes several passes through the list comparing adjacent elements. If adjacent elements are out of order, they are switched. This process continues until one complete pass through the list is made without switches occurring. Note that on the *i*th pass, the sort needs to check only as far as the $(n - i + 1)$st element, because the largest $i - 1$ elements will already have been sorted to the end of the list.

EXAMPLE 6.37

Write a program for bubble sort.

SOLUTION

```
program bubble(input,output);
var n,i,pass,switch,k: integer;
    hold:real;
    a: array [1..50] of real;
begin
    writeln('input a positive integer');
    read(n);
    if n>50 then
        writeln('array is too large')
    else
        begin
            writeln('input',n,' real numbers in any order');
            for i:=1 to n do
                read(a[i]);
            pass:=1;
            switch:=0;
            while(pass<n)and(switch=0)do
```

```
            begin
               switch:=1;
               k:=n-pass;
               for i:=1 to k do
                  if a[i]>a[i+1]then
                     begin
                        hold:=a[i];
                        a[i]:=a[i+1];
                        a[i+1]:=hold;
                        switch:=0
                     end;
               pass:=pass+1
            end;
         writeln('the sorted list is as follows');
         for i:=1 to n do
            write(a[i])
      end
end.
```

EXAMPLE 6.38

Perform a bubble sort on the list 5, −1, 2, 3, 1, 7, 4.

SOLUTION

We display the list after each switch is made.

Pass	i	List A	Elements Switched
1	1	−1, 5, 2, 3, 1, 7, 4	5, −1
	2	−1, 2, 5, 3, 1, 7, 4	5, 2
	3	−1, 2, 3, 5, 1, 7, 4	5, 3
	4	−1, 2, 3, 1, 5, 7, 4	5, 1
	6	−1, 2, 3, 1, 5, 4, 7	7, 4
2	3	−1, 2, 1, 3, 5, 4, 7	3, 1
	5	−1, 2, 1, 3, 4, 5, 7	5, 4
3	2	−1, 1, 2, 3, 4, 5, 7	2, 1

Output after fourth pass: −1, 1, 2, 3, 4, 5, 7

The algorithm actually makes a fourth pass, at which time no switches are made and the algorithm terminates and prints the sorted list. The number of comparisons on the first, second, third, and fourth passes are 6, 5, 4, and 3, respectively. Thus, a total of 18 comparisons is made.

Quicksort

Since the quicksort involves recursive procedures, we will simply describe the algo-rithm without giving it in Pascal. For a pseudocode description, see Tenenbaum and Augenstein (1981).

Quicksort begins with the left-most element as a **pivot** and compares it to the right-most element. If they are out of order, they are switched. The process continues by comparing the pivot element with the most distant element to which it has not yet been compared, and switching these elements if they are out of order. This is done until all elements to the left of the pivot are less than the pivot, and all elements to the right of the pivot are greater than the pivot. The process continues with the subsets on either side of the pivot using new (left-most) pivot entries for each of those subsets, and comparing the pivot elements within their own subset.

Note that after pivoting has been completed with a particular element using quick-sort, that element is in its final position.

During the quicksort, various left and right subsets are produced. The standard procedure is to work on the left subset and stack the right subset (see Case Study 6A). Whenever there is no left subset, unstack a right subset and work on it.

EXAMPLE 6.39

Perform a quicksort on 5, −1, 2, 3, 1, 7, 4.

SOLUTION

We shall circle the pivot element, underline the element it has just switched with, and box all previous pivot entries.

							Compare:
⑤	−1	2	3	1	7	4	5 to 4
<u>4</u>	−1	2	3	1	7	⑤	5 to −1, 2, 3, 1, and then 7
④	−1	2	3	1	5	7	4 to 1 in subset {4, −1, 2, 3, 1}
<u>1</u>	−1	2	3	④	5	7	4 to −1, 2, and then 3
①	−1	2	3	4	5	7	1 to 3, 2, and then −1
−1	1	②	3	4	5	7	2 to 3
−1	1	2	3	4	5	7	

Output: −1, 1, 2, 3, 4, 5, 7

A total of 14 comparisons is made.

EXAMPLE 6.40

Perform a quicksort on 3, 4, 1, 2, 8, 0, 5.

SOLUTION

							Compare:
③	4	1	2	8	0	5	3 to 5, then 0
<u>0</u>	4	1	2	8	③	5	3 to 4
0	③	1	2	8	<u>4</u>	5	3 to 8, then 2
0	<u>2</u>	1	③	8	4	5	3 to 1
⓪	2	1	[3]	8	4	5	0 to 1 and then 2 in $\{0, 2, 1\}$
[0]	②	1	[3]	8	4	5	2 to 1
[0]	1	[2]	[3]	⑧	4	5	Unstack $\{8, 4, 5\}$; compare 8 to 5
[0]	1	[2]	[3]	<u>5</u>	4	⑧	8 to 4
[0]	1	[2]	[3]	⑤	4	[8]	5 to 4
[0]	1	[2]	[3]	4	[5]	8	

Output: 0, 1, 2, 3, 4, 5, 8

A total of 12 comparisons is made.

Efficiency Comparison

In Table 6.1, we list the order of efficiency for the average case (that is, a randomly generated array) and for the worst case of the various sorting algorithms.

For a derivation of the orders of efficiency in Table 6.1, see, for example, Scheid (1982). Since $\log_2 n < n$, we have $n \cdot \log_2 n < n^2$. Thus, quicksort is best in the average case. Does this mean we should always use quicksort? Not necessarily. Quicksort involves recursion. Languages such as BASIC and FORTRAN do not allow recursive procedures, while others such as PL/I, Pascal, and APL do. Writing a nonrecursive version of quicksort, although possible, would be considerably more difficult than writing a recursive one. It *could* be more efficient though!

	Order of Efficiency	
Sorting Algorithm	Average Case	Worst Case
Successive Minima	n^2	n^2
Insertion Sort	n^2	n^2
Bubble Sort	n^2	n^2
Quicksort	$n \cdot \log_2 n$	n^2

Table 6.1

A second consideration for choosing a sorting algorithm involves prior knowledge of the content of the array. For example, if an array happens to be completely sorted, bubble sort and insertion sort each require one pass through the list and $n - 1$ comparisons. Quicksort requires $n(n - 1)/2 \approx n^2$ comparisons. A perfectly sorted list is the worst case for quicksort! Suppose, for example, a list of clients has been filed alphabetically. We want to run a computer check and rearrange any misfiled items. Assuming the client list has been filed fairly accurately, which sorting procedure should we use? Although quicksort is best for the average case, insertion sort is much more efficient for this problem.

In Chapter 8, we discuss another sorting algorithm, heap sort, and compare its efficiency to the algorithms discussed in this case study.

EXERCISES: CASE STUDY 6B

1. Sort using the method of successive minima:
   a. $3, -1, 2, 8, 0, -3, 7$ **b.** $5, -2, -1, 5, 7, -2, 6, 1$
2. Sort the following lists using the insertion sort algorithm. Show the steps as in Example 6.36.
   a. $4, 7, -3, 1, 8, 2$ **b.** $1, 5, 6, -1, 7, -2$
   c. $7, 3, 2, 0, 1, 5, 4$ **d.** $1, 2, 3, 4, 5$
3. Sort the following lists using the bubble sort algorithm. Show the steps as in Example 6.38.
   a. $4, 7, -3, 1, 8, 2$ **b.** $1, 5, 6, -1, 7, -2$
   c. $7, 3, 2, 0, 1, 5, 4$ **d.** $1, 2, 3, 4, 5$
4. Sort the following lists using the quicksort algorithm. Show the steps as in Example 6.40.
   a. $4, 7, -3, 1, 8, 2$ **b.** $1, 5, 6, -1, 7, -2$
   c. $7, 3, 2, 0, 1, 5, 4$ **d.** $1, 2, 3, 4, 5$
5. Show that in a bubble sort of n elements using r passes, the number of comparisons is

$$\sum_{i=1}^{r} (n - i) = nr - \binom{r+1}{2}$$

6. Insert two counter variables COMPARE and ASSIGN into the insertion sort program and increment them when appropriate. COMPARE should count the number of comparisons and ASSIGN should count the number of assignment $(:=)$ statements performed (other than those involving COMPARE and ASSIGN) during program execution. RUN your program with the following data

$$17, 5, 2, 1.3, 31, 46, 21.4, 11, 0, -4.8, 32, 19, 21, 7, 6, 4, 1, 9$$

(Note that 17 is the value of N.)

7. Insert counter variables COMPARE and ASSIGN into the bubble sort program as in exercise 6. Run your program with the same data as exercise 6.
8. The **median** of a set of numbers is the number in the middle of the list when the numbers are in increasing order. If there is an even number of numbers on the list, the median is the average of the middle two numbers when the numbers are in in-

creasing order. Thus, to find the median of a set of numbers, we sort them and then extract the number in the center of the sorted list.

a. Adjust the insertion sort program to enable it to find the median of a set of numbers.

b. Run your program with the data in exercise 6.

c. Run your program with the data 8, -3, 4, 10, 1, 7.3, 2.4, -2, 9, 4. (*Note:* 8 is the value of *N.*)

REFERENCES

Garey, M. R., and Johnson, D. S. *Computers and Intractability: A Guide to the Theory of NP-Completeness.* New York: W. H. Freeman, 1979.

Kolman, B. *Introductory Linear Algebra with Applications.* 2d ed. New York: Macmillan, 1980.

Page, E. S., and Wilson, L. B. *Information Representation and Manipulation in a Computer.* London: Cambridge University Press, 1973.

Scheid, F. *Computers and Programming.* New York: McGraw-Hill, 1982.

Tenenbaum, A. M., and Augenstein, M. J. *Data Structures Using Pascal.* Englewood Cliffs, N.J.: Prentice-Hall, 1981.

Tremblay, J. P., and Sorenson, P. G. *An Introduction to Data Structures with Applications.* New York: McGraw-Hill, 1976.

There are three doors to my dining room. I want to control the chandelier light from any of three switches, one switch by each door. Can you design an electric circuit that will allow the chandelier light to be independently controlled from any of the three wall switches?

In this chapter we discuss Boolean algebra, which is a generalization of set algebra and the algebra of propositions (see Chapter 2). We shall also learn how Boolean algebra is used to construct and simplify electric circuits—a topic of great importance in the construction of computers.

7.1 BOOLEAN ALGEBRAS AND THEIR PROPERTIES

In Chapter 2 we discussed set theory and logic. For each equality involving intersections, unions, and complements of sets (for example, $(A \cup B)' = A' \cap B'$) there is a corresponding equivalence involving conjunction, disjunction, and negation of propositions ($\sim(p \vee q) \equiv \sim p \wedge \sim q$). The more important correspondences are shown in Table 7.1 where we assume in the set column that all sets are subsets of the universal set U, and in the logic column T is a proposition that is always true (a tautology) and F is a proposition that is always false (a contradiction).

Boolean Algebras

Both the algebra of sets and the algebra of propositions obey the same laws because they are both special cases of what is called a Boolean algebra. A **Boolean algebra** is a set K together with two binary operations, the **product** (or **meet**) $\cdot$ and the **sum** (or **join**) $+$, and a unary operation, **complement** $'$ defined on K, which satisfy the axioms in Note 7.1. As we do for numeric products, we shall use juxta-

BOOLEAN ALGEBRA

position to denote the product in a Boolean algebra. Thus, we usually write ab instead of $a \cdot b$. We also use the same precedence rules for evaluating a Boolean expression as we used for evaluating the truth value of propositions: Evaluate expressions in parentheses first, then evaluate complements, then products, and finally evaluate sums. (See page 85).

Sets	Logic	Property
1. a. $A \cup (B \cup C) = (A \cup B) \cup C$ b. $A \cap (B \cap C) = (A \cap B) \cap C$	$p \vee (q \vee r) \equiv (p \vee q) \vee r$ $p \wedge (q \wedge r) \equiv (p \wedge q) \wedge r$	Associative Laws
2. a. $A \cup B = B \cup A$ b. $A \cap B = B \cap A$	$p \vee q \equiv q \vee p$ $p \wedge q \equiv q \wedge p$	Commutative Laws
3. a. $A \cap (B \cup C) = (A \cap B) \cup (A \cap C)$ b. $A \cup (B \cap C) = (A \cup B) \cap (A \cup C)$	$p \wedge (q \vee r) \equiv (p \wedge q) \vee (p \wedge r)$ $p \vee (q \wedge r) \equiv (p \vee q) \wedge (p \vee r)$	Distributive Laws
4. a. $A \cup \varnothing = A$ b. $A \cap U = A$	$p \vee F \equiv p$ $p \wedge T \equiv p$	Identity Laws
5. a. $A \cup A' = U$ b. $A \cap A' = \varnothing$	$p \vee \sim p \equiv T$ $p \wedge \sim p \equiv F$	Complement Laws
6. a. $(A \cup B)' = A' \cap B'$ b. $(A \cap B)' = A' \cup B'$	$\sim(p \vee q) \equiv \sim p \wedge \sim q$ $\sim(p \wedge q) \equiv \sim p \vee \sim q$	De Morgan's Laws

Table 7.1

Note 7.1   Axioms of a Boolean Algebra K

For all $a, b, c \in K$ the following laws hold:

1. **Associative Laws:**
   i. $a + (b + c) = (a + b) + c$ ii. $a(bc) = (ab)c$

2. **Commutative Laws:**
   i. $a + b = b + a$ ii. $ab = ba$

3. **Distributive Laws:**
   i. $a + (bc) = (a + b)(a + c)$ ii. $a(b + c) = ab + ac$

4. **Identity Laws:**
   There exist two distinct elements 0 and 1, called zero and one, satisfying
   i. $a + 0 = a$ ii. $a \cdot 1 = a$

5. **Complement Laws:**
   i. $a + a' = 1$ ii. $a \cdot a' = 0$

Some of the laws of Boolean algebra are like the rules of the algebra of real numbers. If we interpret a, b, and c as real numbers, $+$ and $\cdot$ as addition and multiplication of real numbers, and 0 and 1 as the real numbers zero and one, the associative, commutative, and identity laws, and distributive law (ii) for a Boolean algebra are identical in form to their counterparts in the algebra of real numbers. Other laws of Boolean algebra, however, such as distributive law (i), are not true if interpreted as statements about real numbers. The set of real numbers, therefore, with $+$ and $\cdot$ as the two binary operations is not a Boolean algebra.

EXAMPLE 7.1 _____

Show that the algebra of subsets of a set and the algebra of propositions are Boolean algebras.

SOLUTION

Let S be any set and let $K = \mathscr{P}(S)$ be the power set of S. Then K is a Boolean algebra, where the meet is the intersection of two sets; the join is the union of two sets; the Boolean complement is the complement of a set; the zero element is the empty set $\varnothing$; and the unit element is S. The Sets column of Table 7.1 shows that K is a Boolean algebra.

Let L be the set of all propositions. Then L is a Boolean algebra, where the meet is the conjunction of two propositions; the join is the disjunction of two propositions; the Boolean complement is the negation of a proposition; the zero element is a proposition F that is always false; and the unit element is a proposition T that is always true. The logic column of Table 7.1 shows that L is a Boolean algebra.

EXAMPLE 7.2 _____

Let $B = \{0, 1\}$ with operations $+$ and $\cdot$ defined in Figure 7.1 and the complement defined by $1' = 0$ and $0' = 1$. Show that B is a Boolean algebra.

+	0	1		·	0	1
0	0	1		0	0	0
1	1	1		1	0	1

Figure 7.1

SOLUTION

It is easy to verify for B all the axioms of a Boolean algebra by using a table similar to the truth tables of logic. See Figure 7.2. In fact, we shall call the tables we construct truth tables. We verify distributive law (i) and leave the rest as an exercise. We must show that $a + (bc) = (a + b)(a + c)$.

The columns for $a + bc$ and $(a + b)(a + c)$ are identical. Thus, every combination of 0 and 1 for a, b, and c yields the same value on both sides of the distributive law (i). The law, therefore, is established for B.

Note that the sum and product tables for the Boolean algebra B are similar to the truth tables for conjunction and disjunction (see Chapter 2). If we interpret 1 as T, 0 as F, $+$ as disjunction ($\vee$) and $\cdot$ as conjunction ($\wedge$), the tables of Figure 7.1 become

$\vee$	F	T		$\wedge$	F	T
F	F	T		F	F	F
T	T	T		T	F	T

a	b	c	bc	$a + b$	$a + c$	$a + bc$	$(a + b)(a + c)$
0	0	0	0	0	0	0	0
0	0	1	0	0	1	0	0
0	1	0	0	1	0	0	0
0	1	1	1	1	1	1	1
1	0	0	0	1	1	1	1
1	0	1	0	1	1	1	1
1	1	0	0	1	1	1	1
1	1	1	1	1	1	1	1

Figure 7.2

which are equivalent to the truth tables for disjunction and conjunction, respectively.

The Boolean algebra B introduced in Example 7.2 is representative of all Boolean algebras in the sense that a theorem true in B is true for all other Boolean algebras, as stated in Note 7.2. In the remainder of this chapter, therefore, we shall restrict our attention to the Boolean algebra B. Also B is of special importance in the theory of switching circuits (see Case Study 7A).

Note 7.2

All theorems and properties established for the Boolean algebra B hold for all other Boolean algebras.

Boolean Expressions

The right and left sides of each axiom for a Boolean algebra are examples of Boolean expressions. More generally, a **Boolean variable** is a variable that can take on values in some Boolean algebra. A **Boolean expression** is any Boolean variable or expression built up from Boolean variables using the Boolean operations $\cdot$, $+$, $'$, and parentheses. For example, assuming x, y, and z are Boolean variables, $(x + y)'z + (y + z')$ is a Boolean expression.

We evaluate Boolean expressions by using the following precedence rules: Evaluate expressions in parentheses first, then evaluate complements, then products and, finally, sums. For example, to evaluate $x + (y + z)z'$ when $x = 0$, $y = 0$, and $z = 1$:

$$(y + z) = 0 + 1 = 1$$
$$z' = 0$$
$$(y + z)z' = 1 \cdot 0 = 0$$
$$x + (y + z)z' = 0 + 0 = 0$$

Two Boolean expressions E_1 and E_2 are **equivalent,** and we write $E_1 = E_2$, if for every possible combination of values of the Boolean variables in E_1 and E_2, both expressions evaluate to the same Boolean value. In the Boolean algebra B, we can show two expressions are equivalent by constructing a truth table as we did in Example 7.2. This technique is sometimes called **proof by perfect induction** (Note 7.3).

Note 7.3

To show that two Boolean expressions are equivalent, construct a truth table and show that the two columns corresponding to the expressions are identical.

We can use this technique to prove each part of Theorem 7.1.

Theorem 7.1

Let a, b, and c be any elements in a Boolean algebra. Then the following are true:

1. **Idempotent Laws:**
   i. $a + a = a$ ii. $a \cdot a = a$
2. **Absorption Laws:**
   i. $a + (ab) = a$ ii. $a(a + b) = a$
3. **Boundedness Laws:**
   i. $a + 1 = 1$ ii. $a \cdot 0 = 0$
4. **De Morgan's Laws:**
   i. $(a + b)' = a'b'$ ii. $(ab)' = a' + b'$
5. **Involution Law:** $(a')' = a$
6. **Uniqueness of the Complement:** For a given element a, if there is an element x such that $a + x = 1$ *and* $a \cdot x = 0$, then $x = a'$.

EXAMPLE 7.3 _____

Use perfect induction to prove De Morgan's law (i): $(a + b)' = a'b'$.

SOLUTION

See Figure 7.3.

a	b	a'	b'	$a + b$	$(a + b)'$	$a'b'$
0	0	1	1	0	1	1
0	1	1	0	1	0	0
1	0	0	1	1	0	0
1	1	0	0	1	0	0

Figure 7.3

EXAMPLE 7.4

Prove the uniqueness of the complement using perfect induction.

SOLUTION

Since we do not know what x is, we make a blank column for x. We know $a + x = 1$ and $a \cdot x = 0$. Thus, we obtain Figure 7.4(a). Since $a + x = 1$, the first two entries in the x column must be 1. See Figure 7.4(b). Since $a \cdot x = 0$, the last two entries in the x column must be 0. See Figure 7.4(c). Thus, x is always the complement of a; that is, $x = a'$.

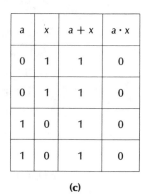

a	x	$a + x$	$a \cdot x$
0		1	0
0		1	0
1		1	0
1		1	0

(a)

a	x	$a + x$	$a \cdot x$
0	1	1	0
0	1	1	0
1		1	0
1		1	0

(b)

a	x	$a + x$	$a \cdot x$
0	1	1	0
0	1	1	0
1	0	1	0
1	0	1	0

(c)

Figure 7.4

Remark

The properties of Theorem 7.1 can also be derived algebraically from the axioms of a Boolean algebra. To prove idempotent law (ii) in this way we proceed as follows:

$$
\begin{aligned}
a \cdot a &= (a \cdot a) + 0 && \text{[Axiom 4(i)]} \\
&= (a \cdot a) + (a \cdot a') && \text{[Axiom 5(ii)]} \\
&= a(a + a') && \text{[Axiom 3(ii)]} \\
&= a \cdot 1 && \text{[Axiom 5(i)]} \\
&= a && \text{[Axiom 4(ii)]}
\end{aligned}
$$

Principle of Duality

If we take another look at the axioms of a Boolean algebra, we note a great deal of symmetry. Each axiom has two parts, each of which is similar to the other. To be more specific, the **dual** of any statement in a Boolean algebra is the statement obtained by interchanging · and +, and interchanging 0 and 1 in the original statement. For example, the dual of distributive law (ii) is distributive law (i). Each · is replaced by + and conversely. Likewise, the dual of complement law (i) is complement law (ii). The + is replaced by · and 1 by 0. Thus, each axiom pair in the definition of a Boolean algebra is a pair of dual statements. Because the dual of every axiom is true, the dual of every statement derivable from the axioms is also true. We, therefore, have Theorem 7.2.

Theorem 7.2   The Principle of Duality for Boolean Algebras

The dual of any theorem in a Boolean algebra is also a theorem.

In parts 1 through 4 of Theorem 7.1, each statement i is the dual of statement ii.

EXAMPLE 7.5

Prove De Morgan's law (ii), $(ab)' = a' + b'$.

SOLUTION

We proved De Morgan's law (i), $(a + b)' = a'b'$ in Example 7.3. By the principle of duality, the dual of this statement is also true; that is, $(ab)' = a' + b'$, which is De Morgan's law (ii). No further proof is necessary.

Note that De Morgan's laws can be used to "expand" Boolean expressions. For example,

$$x(y + z')' = x(y'(z')')   \text{[De Morgan's law (i)]}$$
$$= x(y'z)   \text{(Involution law)}$$

Also,

$$(x'y' + xy)' = (x'y')'(xy)'   \text{[De Morgan's law (i)]}$$
$$= ((x')' + (y')')(x' + y')   \text{[De Morgan's law (ii)]}$$
$$= (x + y)(x' + y')   \text{(Involution law)}$$

Remark   The principle of duality applies to the algebra of sets and the algebra of logic because they are Boolean algebras. Note that each pair of properties listed in Table 7.1 are duals.

EXERCISES 7.1

1. Let D be the set of all divisors of 30; that is, $D = \{1, 2, 3, 5, 6, 10, 15, 30\}$. Define the operations $+, \cdot$, and $'$ on D as follows:

$$a + b = \text{Least common multiple of } a \text{ and } b = \text{lcm}(a, b)$$
$$a \cdot b = \text{Greatest common divisor of } a \text{ and } b = \text{gcd}(a, b)$$
$$a' = 30/a$$

   a. Find $3 + 6$, $10 + 15$, and $5 + 1$.
   b. Find $5 \cdot 15$, $2 \cdot 6$, and $6 \cdot 15$.
   c. Find $3'$, $10'$, $1'$, and $30'$.
   d. Find $2 + (3 \cdot 5)$ and $(15' + 6)' \cdot 3$.
   e. Convince yourself that D, with the operations as defined, is a Boolean algebra. What are the zero and unit elements?

2. Let E be the set of all divisors of 28; that is, $E = \{1, 2, 4, 7, 14, 28\}$. Define the operations $+, \cdot$, and $'$ on E as follows:

$$a + b = \text{lcm}(a, b), \qquad a \cdot b = \text{gcd}(a, b), \qquad a' = 28/a$$

   Show that E, with these operations, is not a Boolean algebra. That is, find at least one axiom of a Boolean algebra that does not hold for E.

3. Write the duals of the following expressions:
   a. $a(bc')$ **b.** $(a' + b)' + ab'$

4. Use perfect induction and duality to prove Theorem 7.1, parts 1, 2, 3, and 5.

5. Use perfect induction to prove the following:
   a. $a(a' + b) = ab$ **b.** $ab' + c = bc + b'(a + c)$
   c. $(a + b)(a + b') = a$

6. State the dual of each theorem of exercise 5.

7. Find the complement for each of the following using De Morgan's laws:
   a. $x(y' + z')$ **b.** $y(xz + x'z')$ **c.** $x + x'y$

Problems and Projects

8. Let K be any Boolean algebra. A useful relation $\prec$ can be defined on the elements of K as follows: $x \prec y$ (read "x precedes y") if and only if $xy = x$.
   a. If K is the Boolean algebra of subsets of a set S, to what familiar relation on the subsets of S does $\prec$ correspond?
   b. If K is the Boolean algebra of propositions, to what familiar logical relation does $\prec$ correspond?
   c. Use the axioms and laws of Boolean algebra to prove the following properties of $\prec$ in an arbitrary Boolean algebra K:
      i. $x \prec x$ for all $x \in K$. (Reflexive Property)
      ii. If $x \prec y$ and $y \prec x$, then $x = y$. (Antisymmetric Property)
      iii. If $x \prec y$ and $y \prec z$, then $x \prec z$. (Transitive Property)

A relation satisfying properties (i), (ii), and (iii) is called a **partial order,** and a set (K in our instance) with a partial order defined on it is called a **partially ordered set** (or **poset**).

9. Let K be a Boolean algebra and $\prec$ the relation define in exercise 8. Given two elements x and y in K, we can define the **greatest lower bound** of x and y, **glb(x, y)** as that element g of K satisfying the properties:

   i. g precedes both x and y; that is, $g \prec x$ and $g \prec y$.

  ii. If an element precedes both x and y, it precedes g; that is, if $z \prec x$ and $z \prec y$, then $z \prec g$.

  a. Prove that each pair of elements x and y can have at most one glb.

  b. In the Boolean algebra of subsets of a set, how would you interpret the glb?

  c. In the Boolean algebra of propositions, how would you interpret the glb?

  d. In the Boolean algebra of exercise 1, how would you interpret the glb?

7.2 BOOLEAN FUNCTIONS

We can associate a function f, called a **Boolean function,** with a given Boolean expression E by using the expression as the formula for the function; that is, $f = E$. For example, we can associate with the Boolean expression $x + x'y$, the Boolean function $f(x, y) = x + x'y$. We obtain the value of the function for a given combination of values of the variables by substitution. Thus, $f(1, 1) = 1 + 1' \cdot 1 = 1 + 0 \cdot 1 = 1 + 0 = 1$. We can completely specify the Boolean function f by using a truth table, as shown in Figure 7.5.

Recall from Chapter 5 that a function is known if we know its value for each possible combination of input values. Thus, we can specify a Boolean function by giving its output for each combination of input values.

Figure 7.6 specifies a Boolean function of two variables.

We now ask the main question of this section: Given a Boolean function in tabular form, is there a Boolean expression for the function? The answer is yes.

x	y	x'	$x'y$	f
0	0	1	0	0
0	1	1	1	1
1	0	0	0	1
1	1	0	0	1

Figure 7.5

x	y	f
0	0	1
0	1	0
1	0	1
1	1	1

Figure 7.6

We describe in Note 7.4 how to derive such an expression, called the **complete sum-of-products** (or **disjunctive normal**) **form** of f.

Note 7.4   The Complete Sum-of-Products Form

1. Add a column to the function table. This column will consist of product terms.

2. Each entry in the new column will contain the product of all the variables, with a variable complemented if its input value in that row is 0, and not complemented if its input value in that row is 1. Each product in the new row is called a **minterm.**

3. The expression for f is now obtained by writing the sum of those minterms that lie in rows in which the value of f is 1.

For the function of Figure 7.6, we obtain the table in Figure 7.7.

x	y	f	Minterms
0	0	1	$x'y'$
0	1	0	$x'y$
1	0	1	xy'
1	1	1	xy

Figure 7.7

The function defined in Figure 7.6 can, therefore, be written as

$$f(x, y) = x'y' + xy' + xy$$

The expression obtained by the procedure of Note 7.4 is called the *complete* sum-of-products form because each term of the sum (each minterm) contains *every* variable, either complemented or not complemented. This is in contrast to an expression in **sum-of-products** form where each term of the sum is a product, not necessarily containing every variable. For example, the expression $x + x'y$ is in sum-of-products form but is not complete because the first term does not contain either of the variables y or y'.

Given a function f, the procedure of Note 7.4 yields an expression E_f. How do we know that $f = E_f$; that is, how do we know that for a given combination of variable values the function f and the expression E_f have the same value? To see that $f = E_f$, note that for a given combination of values for $x, y, \ldots$ either *all* minterms in the expression E_f are 0 (this occurs when $f = 0$) or *exactly* one minterm in E_f is 1 (this occurs when $f = 1$). For example, in the function f of Figure 7.7, consider the combination of values $x = 0$ and $y = 1$. Every minterm in $E_f = x'y' + xy' + xy$ is 0 for this combination of x and y values, and $f(0, 1) = 0$. However, consider the combination of values $x = 1$, $y = 0$. The minterm $xy' = 1$ for this combination of x and y values; the other two minterms in E_f equal 0, and $f(1, 0) = 1$.

As noted at the beginning of this section, any Boolean expression defines a function. We can find the complete sum-of-products form for the resulting function and, therefore, obtain a complete sum-of-products expression equivalent to the original expression.

EXAMPLE 7.6 _____

Find the complete sum-of-products form of $x(yz + y'z')$.

SOLUTION

Construct a table for the function (see Figure 7.8). We show in the product column only those minterms for which the expression evaluates to 1.

The complete sum-of-product form of $x(yz + y'z')$ is, therefore, $xy'z' + xyz$.

This expression could be obtained algebraically by applying the distributive law to the original expression.

_____

We summarize in Theorem 7.3.

Theorem 7.3

Any Boolean function (therefore, any Boolean expression) can be written uniquely as a sum of minterms.

x	y	z	y'	z'	yz	$y'z'$	$yz + y'z'$	$x(yz + y'z')$	Minterms
0	0	0	1	1	0	1	1	0	
0	0	1	1	0	0	0	0	0	
0	1	0	0	1	0	0	0	0	
0	1	1	0	0	1	0	1	0	
1	0	0	1	1	0	1	1	1	$xy'z'$
1	0	1	1	0	0	0	0	0	
1	1	0	0	1	0	0	0	0	
1	1	1	0	0	1	0	1	1	xyz

Figure 7.8

By "uniqueness" in Theorem 7.3 we mean unique up to a reordering of the factors in each minterm and a reordering of the minterms in the sum. Thus, xy is the same as yx, and $x + y$ is the same as $y + x$, and $xy'z' + xyz$ is the same as $xzy + z'xy'$.

Theorem 7.3 and the definition of equivalent expressions yield the following.

Corollary 7.1

Two Boolean expressions are equivalent if and only if they have the same complete sum-of-products form.

As noted in the previous section, a Boolean algebra possesses the property of duality. If, therefore, we can find the complete sum-of-products form for a Boolean expression, we should, by duality, be able to find its **complete product-of-sums** (or **conjunctive normal**) form. To do this, we dualize the procedure for finding the complete sum-of-products form.

Note 7.5   The Complete Product-of-Sums Form

1. Add a column to the table of the function (or expression). This column will consist of sum terms.

2. Each entry in the new column will contain the sum of all the variables, with a variable complemented if its input value is 1, and not complemented if its input value is 0. Each sum in the new column is called a **maxterm.**

3. The complete product-of-sums form of the function (or expression) is obtained by forming the product of those maxterms that lie in rows in which the function (or expression) value is 0.

We illustrate with two examples.

EXAMPLE 7.7 _____

Find the complete product-of-sums form of the function given in Figure 7.6.

SOLUTION

We repeat in Figure 7.9 the table with the new sum column.

x	y	f	Maxterms
0	0	1	$x + y$
0	1	0	$x + y'$
1	0	1	$x' + y$
1	1	1	$x' + y'$

Figure 7.9

The product-of-sums form of f is $f(x, y) = x + y'$ because there is only one 0 in the f column. Note that this form of f is much simpler than the complete sum-of-products form obtained earlier.

EXAMPLE 7.8 _____

Find the complete product-of-sums form for $x + xy + x'z$.

SOLUTION

The table is given in Figure 7.10.

x	y	z	x'	xy	$x'z$	$x + xy + x'z$	Maxterms
0	0	0	1	0	0	0	$x + y + z$
0	0	1	1	0	1	1	
0	1	0	1	0	0	0	$x + y' + z$
0	1	1	1	0	1	1	
1	0	0	0	0	0	1	
1	0	1	0	0	0	1	
1	1	0	0	1	0	1	
1	1	1	0	1	0	1	

Figure 7.10

We show in the table of Figure 7.10 only those maxterms corresponding to an expression value of 0. Thus,

$$x + xy + x'z = (x + y + z)(x + y' + z)$$

We now state the dual of Theorem 7.3 and its corollary.

Theorem 7.4   (Dual of Theorem 7.3)

Any Boolean function (or expression) can be written uniquely as a product of maxterms.

Corollary 7.2   (Dual of Corollary 7.1)

Two Boolean expressions are equivalent if and only if they have the same complete product-of-sums form.

Returning for a moment to Example 7.7, we see that a Boolean function can be represented by several expressions. The function f of that example can be written $f = x + y'$ or, as we saw earlier, $f = x'y' + xy' + xy$. Using the axioms of a Boolean algebra, we can write other equivalent expressions for f. For example, noting that $x + x' = 1$ [axiom 5(i)], we can write f as follows, using axiom 4(ii):

$$f = (x + y')(x + x')$$

or, using axioms 5(ii) and 4(i),

$$f = x'y' + xy' + xy + xx'$$

Generally, a Boolean function can be expressed in an infinite number of ways by equivalent Boolean expressions. This is analogous to the functions of standard algebra. The function $g(x) = 2x + 4$ can be written as $g(x) = 2(x + 2)$, or $2(x + 1) + 2$, and so on. In the next section we take the first step toward finding the simplest expression for a Boolean function.

EXERCISES 7.2

1. Obtain the complete sum-of-products and the complete product-of-sums forms for each function whose table is given in Figure 7.11.

x	y	f
0	0	0
0	1	1
1	0	0
1	1	1

(a)

x	y	z	g
0	0	0	1
0	0	1	0
0	1	0	1
0	1	1	1
1	0	0	0
1	0	1	0
1	1	0	1
1	1	1	0

(b)

x	y	z	h
0	0	0	1
0	0	1	1
0	1	0	0
0	1	1	0
1	0	0	1
1	0	1	1
1	1	0	0
1	1	1	0

(c)

Figure 7.11

2. Obtain the complete sum-of-products and the complete product-of-sums forms for each of the following expressions:
   a. $(x + y)'(x + z)$ **b.** $(xy')'(x' + xy'x)$ **c.** $x + y$
   d. $x(yz')'$ **e.** $x + x'y$
3. Use Corollary 7.1 or 7.2 to determine which of the following pairs of expressions are equivalent:
   a. $(xy + x'y')'$ and $x'y + xy'$
   b. $(x + y)y'$ and $(x + y)(xy)'$

Problems and Projects

4. In standard algebra, there are an infinite number of functions of one variable. For example, $x, x + 1, x + 2, x + 3$, and so on are all functions of one variable. There are also an infinite number of functions of two variables, of three variables, and so on. This is in contrast to Boolean functions in our Boolean algebra B: For a given number of variables, there are only a finite number of Boolean functions.
   a. Show there are only four Boolean functions of one variable. (*Hint:* We are asking, in effect, how many tables can be constructed for Boolean functions of one variable. How many rows does a table for a Boolean function of one variable have? There are two possible values for the function in each row. Now use the multiplication principle of Chapter 3.)
   b. Show there are 16 Boolean functions of two variables. (*Hint:* Count the number of possible tables as in part a.)
   c. How many Boolean functions are there of n variables?
5. Show that the following procedure yields the complete product-of-sums form for a Boolean function f.

   Step 1: Construct a table for the complement f' of the function f. The complement f' is the function with value 1 if and only if f has the value 0.
   Step 2: Find the complete sum-of-product form for f'.
   Step 3: Complement the complete sum-of-product form of f' using De Morgan's laws. The resulting expression is the complete product-of-sums form of f.

7.3   KARNAUGH MAPS AND PRIME IMPLICANTS

Karnaugh Maps

The complete sum-of-products form of a Boolean expression in two, three, or four variables can be represented using a pictorial device called a **Karnaugh map.** (The technique can be extended to expressions of five or six variables.) The true utility of Karnaugh maps lies in simplifying Boolean expressions. In this section we learn the basic facts about Karnaugh maps, how to construct them, and how to take the first step in simplifying Boolean expressions.

Figure 7.12 shows the Karnaugh map layouts for Boolean expressions of two, three, and four variables, which have 4, 8, and 16 cells, respectively. The 1s are not part of the maps. They will be used in Example 7.9. The number of cells in a Karnaugh map is given in Note 7.6.

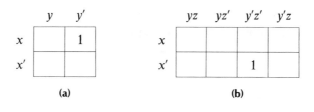

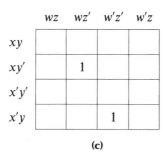

Figure 7.12

Note 7.6

The Karnaugh map for an expression of n variables has 2^n cells.

The 2-, 3-, and 4-variable Karnaugh maps have, therefore, 4, 8, and 16 cells, respectively. These cells correspond to the minterms of 2-, 3-, and 4-variable expressions. Note 7.7 describes how to find minterms.

Note 7.7

To find the minterm of a particular cell in a Karnaugh map, form the product of the variables listed at the beginning of the row and at the top of the column of the cell.

Note that we shall sometimes refer to a cell by the minterm it represents.

EXAMPLE 7.9 ────────────────────

Find the minterms of each cell containing a 1 in Figure 7.12.

SOLUTION

In Figure 7.12(a), the minterm of the cell containing 1 is xy'.
   In Figure 7.12(b), the minterm of the cell containing 1 is $x'y'z'$.

In Figure 7.12(c), the minterm of the cell in row 2 containing 1 is $xy'wz'$. The minterm of the cell containing 1 in row 4 is $x'yw'z'$.

The rows and columns of a Karnaugh map are not labeled haphazardly (see Note 7.8).

Note 7.8

Each row and column label in a Karnaugh map differs from its adjacent label in exactly one variable.

The column labels in Figure 7.12(b), for example, are yz, then yz' (the z variable changed from z to z'), then $y'z'$ (the y variable changed from y to y'), then $y'z$ (the z variable changed from z' to z).

You cannot label the rows of a 4-variable map in the order xy, $x'y'$, xy', $x'y$ because the first and second labels differ in two variables (as do the third and fourth).

To construct the Karnaugh map of a Boolean expression we proceed as described in Note 7.9.

Note 7.9

Given a Boolean expression E in complete sum-of-products form, we construct its Karnaugh map by placing a 1 in each cell that corresponds to a minterm of the expression.

EXAMPLE 7.10 _____

Construct the Karnaugh map of each of the following Boolean expressions:

a. $xy + x'y$ b. $x'yz' + x'yz + xyz' + xyz$
c. $xywz + xyw'z + x'yw'z + x'ywz$

SOLUTION

a. For the 2-variable expression $xy + x'y$ we have the map of Figure 7.13(a).

b. The 3-variable expression $x'yz' + x'yz + xyz' + xyz$ is the complete sum-of-products form of the function of Example 7.9. Its Karnaugh map is shown in Figure 7.13(b).

c. For the 4-variable expression $xywz + xyw'z + x'yw'z + x'ywz$ we have the Karnaugh map of Figure 7.13(c).

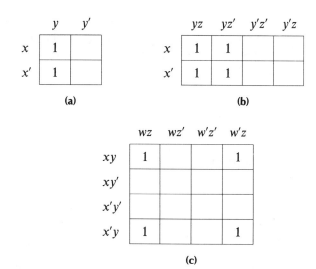

Figure 7.13

Two cells of a Karnaugh map are **adjacent** if their minterms differ in exactly one variable. For example, cells xyz' and $x'yz'$ are adjacent in Figure 7.12(b), but cells xyz' and $x'yz$ are not. A simple combinatorial argument (see exercise 6, of Exercises 7.3) gives Note 7.10.

Note 7.10

In an n-variable Karnaugh map, a cell has n adjacent cells.

The way in which the rows and columns of a Karnaugh map are labeled implies Note 7.11.

Note 7.11

The minterm of a cell in a Karnaugh map differs from the minterms of its adjacent cells by having exactly one variable complemented in one cell that is not complemented in the adjacent cell.

EXAMPLE 7.11 _____

List the cells adjacent to each cell containing a 1 in Figure 7.12.

SOLUTION

In Figure 7.12(a), the minterm of the cell containing 1 is xy'. Its adjacent cells are xy (differing from xy' in y) and $x'y'$ (differing from xy' in x).

In Figure 7.12(b), the minterm of the cell containing 1 is $x'y'z'$. Its adjacent cells are $x'yz'$ (differing in y), $xy'z'$ (differing in x), and $x'y'z$ (differing in z).

In Figure 7.12(c), the minterm of the cell containing 1 in row 2 is $xy'wz'$. Its adjacent cells are $xy'wz$ (differing in z), $xywz'$ (differing in y), $xy'w'z'$ (differing in w), and $x'y'wz'$ (differing in x).

We leave the listing of the cells adjacent to the cell containing 1 in row 4 as an exercise.

Geometrically, cells in a Karnaugh map are adjacent if they have a straight line boundary in common. Thus, adjacent cells are next to one another or directly above and below each other. This interpretation does, however, lead to a problem. If each cell in a 3-variable Karnaugh map has three adjacent cells, what cells are adjacent to the corner cell xyz of the map of Figure 7.12(b)? From the figure it is easy to find two adjacent cells, namely $x'yz$ and xyz'. From the definition of adjacency, the cell $xy'z$ should also be adjacent to xyz. This cell is at the other end of the first row. We must, therefore, agree on the statement in Note 7.12.

> **Note 7.12**
>
> In a 3-variable Karnaugh map, the cells at the ends of a given row are adjacent.

A similar analysis of the border cells in a 4-variable Karnaugh map gives Note 7.13.

> **Note 7.13**
>
> In a 4-variable Karnaugh map the cells at the ends of a given row are adjacent and the cells at the ends of a given column are adjacent.

EXAMPLE 7.12 ———————————————————————

In the 4-variable Karnaugh map of Figure 7.13(c), list the minterms of cells adjacent to the cells of $x'yw'z'$ and of $xyw'z$.

SOLUTION

The cells adjacent to $x'yw'z'$ are $x'ywz'$, $x'y'w'z'$, $x'yw'z$, and $xyw'z'$ (from the other end of column 3).

Note that $xyw'z$ is a corner cell. Its adjacent cells are $xyw'z'$, $xy'w'z$, $xywz$ (from the other end of row 1), and $x'yw'z$ (from the other end of column 4).

To see why adjacency of cells is important in Karnaugh maps, we shall investigate the result given in Note 7.14.

Note 7.14

Let E be any Boolean expression. Then

$$Ex + Ex' = E$$

The proof of Note 7.14 is simple:

$$Ex + Ex' = E(x + x') \qquad \text{[Distributive law (ii)]}$$
$$= E \cdot 1 \qquad \text{[Complement law (i)]}$$
$$= E \qquad \text{[Identity law (ii)}$$

Recall (Note 7.11) that adjacent cells differ by a complement in exactly one variable. To see how Note 7.14 can be used, suppose the Karnaugh map of an expression in three variables has 1s in the two adjacent cells xyz' and $xy'z'$. See Figure 7.14(a), where we have circled these cells. If we form the sum of these min-terms and apply Note 7.14 with $E = xz'$, we obtain

$$xyz' + xy'z' = (xz')y + (xz')y' = xz'$$

We can, therefore, replace $xyz' + xy'z'$ in any complete sum-of-products expression by xz' and obtain an equivalent expression.

We can extend our rule to other sets of adjacent cells. Consider the four cells $xyw'z'$, $xy'w'z'$, $xy'wz'$, and $xywz'$ in a 4-variable Karnaugh map. See Figure 7.14(b), where we have circled the four 1s. The sum of these minterms can be simplified as follows:

$xyw'z' + xy'w'z' + xy'wz' + xywz'$
$\quad = xw'z'(y + y') + xwz'(y + y')$ Note 7.14 twice. Once with
$\qquad\qquad\qquad\qquad\qquad\qquad\qquad E = xw'z'$ and once with
$\qquad\qquad\qquad\qquad\qquad\qquad\qquad E = xwz'$.

$\quad = xw'z' + xwz'$
$\quad = xz'(w' + w)$ Note 7.14 with $E = xz'$
$\quad = xz'$

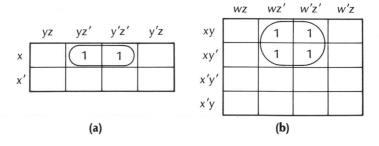

(a)   (b)

Figure 7.14

The sum of the four minterms corresponding to the four cells can be replaced in an expression by xz'.

These algebraic manipulations can be done graphically on a Karnaugh map by using basic rectangles. By a **basic rectangle** we mean one of the following:

1. A single cell containing 1
2. Two adjacent cells containing 1s
3. Four adjacent cells containing 1s which form either a 1×4 rectangle or a 2×2 rectangle
4. Eight adjacent cells containing 1s which form a 2×4 rectangle

It is important to remember that a basic rectangle must have one, two, four, or eight cells.

Basic rectangles of one, two, and four cells in 2- and 3-variable Karnaugh maps are shown in Figure 7.15. Note in Figure 7.15(f) that the two cells in row 1 form a basic rectangle.

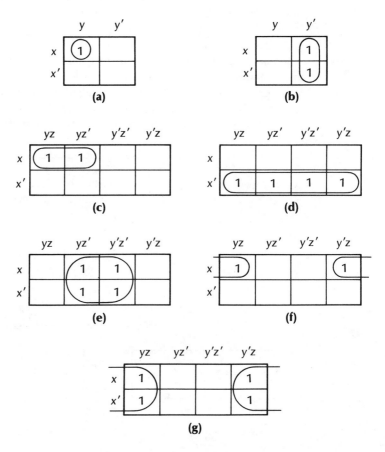

Figure 7.15

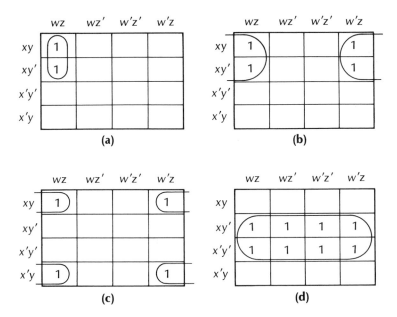

Figure 7.16

Basic rectangles of two, four, and eight cells in 4-variable Karnaugh maps are shown in Figure 7.16. Note that the basic rectangles in Figure 7.16(b) and (c) exist because cells at the opposite ends of rows and columns are considered adjacent.

The definition of a basic rectangle allows one basic rectangle to include another. The basic rectangle enclosed by the dashed oval in Figure 7.17(a) is contained in the basic rectangle enclosed in the solid oval.

A **maximal rectangle** in a Karnaugh map is a basic rectangle that is not included in any larger basic rectangle. The basic rectangle enclosed in the solid oval in Figure 7.17(a) is a maximal rectangle. There are two overlapping maximal rectangles in Figure 7.17(b). Recall that a basic rectangle must have one, two, four, or eight cells. Thus, the three cells containing 1s in Figure 7.17(b) do not qualify as a basic rectangle.

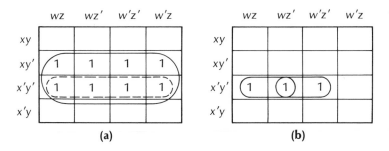

Figure 7.17

Prime Implicants of Boolean Expressions

Now, let E be a Boolean expression in complete sum-of-products form. A maximal rectangle in the Karnaugh map of E represents a sum of terms. As noted, the sum for this rectangle can be simplified and replaced by a single term. This term, which corresponds to a maximal rectangle in the Karnaugh map of E, is called a **prime implicant** of E.

Note 7.15 shows how to find the prime implicant corresponding to a maximal rectangle.

Note 7.15

The prime implicant corresponding to a maximal rectangle is found by forming the product of the variables that do not change from complemented to uncomplemented in the maximal rectangle.

Therefore, the prime implicant corresponding to the maximal rectangle in Figure 7.17(a) is y'. The variables x, z, and w do not appear in the prime implicant because they appear in both complemented and uncomplemented form in the rectangle.

Before considering a final example, we make an observation in Note 7.16 concerning the number of variables in a prime implicant.

Note 7.16

Suppose the number of variables in a Karnaugh map and the number of cells in a maximal rectangle are given. Then an entry in Table 7.2 tells the number of variables in the corresponding prime implicant.

	Number of Cells in Maximal Rectangle			
	1	2	4	8
2	2	1	—	—
3	3	2	1	—
4	4	3	2	1

Number of Variables in the Karnaugh Map

Table 7.2

For example, an expression whose maximal rectangle has four cells in a Karnaugh map with four variables has a prime implicant with two variables. To summarize, the sum of the minterms corresponding to the cells in a maximal rectangle can be "collapsed" to (replaced by) a single term—namely, the prime implicant corresponding to the maximal rectangle.

EXAMPLE 7.13

Find the prime implicants of each of the following expressions. Which terms of each sum can be replaced by the corresponding prime implicants?

a. $xyz' + x'yz' + x'yz + x'y'z' + x'y'z$

b. $xy'zw + x'y'zw + xy'z'w + x'y'z'w + xyzw' + xyz'w'$

SOLUTION

a. The Karnaugh map for this expression is shown in Figure 7.18(a), where we have enclosed the two maximal rectangles in ovals. By Note 7.16, the 2-cell maximal rectangle has a 2-variable prime implicant. By Note 7.15, this prime implicant is yz' because y and z' do not change in the 2-cell maximal rectangle, but x does.

  Similarly, the 4-cell maximal rectangle has a 1-variable prime implicant, namely x'.

  The sum $xyz' + x'yz'$ can be replaced by yz'.

  The sum $x'yz + x'yz' + x'y'z' + x'y'z$ can be replaced by x'.

b. The Karnaugh map for this expression is shown in Figure 7.18(b), where we have enclosed two maximal rectangles in ovals. By Note 7.16, the 4-cell maximal rectangle has a 2-variable prime implicant which, by Note 7.15, is $y'w$.

  Similarly, the 2-cell maximal rectangle has the 3-variable prime implicant xyw'.

  The sum $xy'zw + x'y'zw + xy'z'w + x'y'z'w$ can be replaced by $y'w$. The sum $xyzw' + xyz'w'$ can be replaced by xyw'.

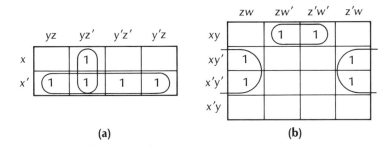

(a)   (b)

Figure 7.18

In the next section we use Karnaugh maps and maximal rectangles to obtain the simplest form of a Boolean expression.

EXERCISES 7.3

1. In the Karnaugh maps of Figure 7.19, list all the cells adjacent to each cell containing a 1 and write the minterms of those cells.

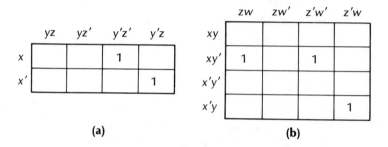

Figure 7.19

2. Complete Example 7.11 (that is, list all cells adjacent to the cell containing a 1 in row 4).
3. Use ovals to enclose the basic rectangles in the Karnaugh maps of Figure 7.20.

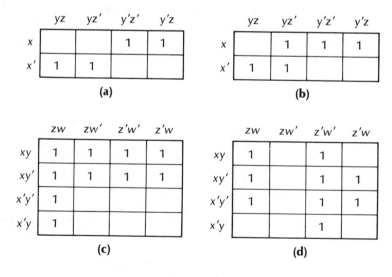

Figure 7.20

4. Construct Karnaugh maps for the following expressions and find the prime implicants of each. State which terms each prime implicant can replace.
   a. $x'yz + xyz' + x'yz'$
   b. $xyz + x'yz' + x'yz + xy'z + x'y'z$
   c. $x'y'zw + x'y'zw' + x'yzw + x'yzw' + xyz'w + x'yz'w$
   d. $xyzw + x'yzw + x'yz'w + xyz'w + x'y'zw + x'y'zw' + x'y'z'w' + x'y'z'w$

Problems and Projects

5. Prove Note 7.6.
6. Prove Note 7.10.

7.4 MINIMIZING BOOLEAN FUNCTIONS AND EXPRESSIONS

A given Boolean function or expression has many equivalent forms. Sometimes the complete sum-of-products form of an expression is simpler than the complete product-of-sums form, and sometimes the opposite is true. And sometimes neither form is simpler than the other, as the next example shows.

EXAMPLE 7.14 _____

Find the complete sum-of-products and the complete product-of-sums forms for the function f given in Figure 7.21.

x	y	z	f	Minterms	Maxterms
0	0	0	1	$x'y'z'$	
0	0	1	0		$x + y + z'$
0	1	0	1	$x'yz'$	
0	1	1	0		$x + y' + z'$
1	0	0	0		$x' + y + z$
1	0	1	0		$x' + y + z'$
1	1	0	1	xyz'	
1	1	1	1	xyz	

Figure 7.21

SOLUTION

From the minterm column, the sum-of-products form is

$$f = x'y'z' + x'yz' + xyz' + xyz$$

From the maxterm column the product-of-sums form is

$$f = (x + y + z')(x + y' + z')(x' + y + z)(x' + y + z')$$

We could use the algebraic rule of Note 7.14 to simplify the sum-of-products form of the function of Example 7.14 as follows:

$$f = x'y'z' + x'yz' + xyz' + xyz$$
$$= x'z' + xyz' + xyz \qquad \text{Note 7.14 on the first two terms with } E = x'z'.$$
$$= x'z' + xy \qquad \text{Note 7.14 on the last two terms with } E = xy.$$

This representation of f is considerably simpler than either of the forms derived in Example 7.14. A Boolean expression in complete sum-of-products form can always be simplified by repeated application of Note 7.14, which is what we just did. However, a simpler method is to use the Karnaugh map for the expression and find the prime implicants of the expression.

EXAMPLE 7.15 _____

Construct a Karnaugh map for the function of Example 7.14, find the prime implicants, and write a simple expression for f.

SOLUTION

The Karnaugh map for the function is given in Figure 7.22. Two maximal rectangles are enclosed in ovals. Their prime implicants are the terms xy and $x'z'$. Thus, $xyz + xyz'$ can be replaced by xy, and $x'yz' + x'y'z$ can be replaced by $x'z'$. Therefore, $f = xy + x'z'$.

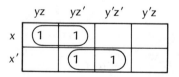

Figure 7.22

There is a third maximal rectangle in the Karnaugh map of Figure 7.22. This is shown, together with the two other maximal rectangles, in Figure 7.23.

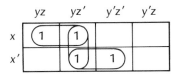

Figure 7.23

Therefore, we can write f as

$$f = xy + x'z' + yz'$$

That is, a function can be written in several ways as a sum of prime implicants. Which sum of prime implicants shall we choose? We choose the sum that is in some sense minimal. Before proceeding, we define precisely what we mean by the simplest form of a Boolean expression.

By a **literal** we mean either a variable or its complement. The literals of the expression $xy + xy'$ are x (twice), y, and y'. Note that in counting the number of literals in an expression we include multiple occurrences of a literal. Thus, $xy + xy'$ has *four* literals, x being counted twice.

Let E_1 and E_2 be equivalent sum-of-products (not necessarily complete) Boolean expressions. E_1 is **simpler** than E_2 if (1) E_1 contains fewer literals, or (2) E_1 contains fewer terms than E_2.

EXAMPLE 7.16

Which of the following equivalent expressions is simpler: $E_1 = xy'wz + y'wz' + xy'w'z'$ or $E_2 = xy'w + y'wz' + xy'z'$?

SOLUTION

They both have three terms, so to decide which is simpler we must count the number of literals in E_1 and E_2. E_1 contains 11 literals and E_2 contains 9. Hence, E_2 is simpler than E_1.

A Boolean expression E is in **minimal sum-of-products** form, or simply **minimal-sum** form, if it is in sum-of-products form and there is no simpler equivalent expression in sum-of-products form.

Minimal-sum form is what we shall mean by **simplest form** for a Boolean expression or function because such a form contains the fewest terms possible and the fewest literals possible for a sum-of-products form. Although there are other kinds of "simplest" forms for Boolean expressions depending on the allowed operations and the allowed form of the expression, we shall not consider these in this text.

We now show, in Note 7.17, how to find the minimal-sum form of a Boolean expression.

> **Note 7.17**
>
> To find the minimal-sum form of the Boolean expression E, proceed as follows:
>
> 1. Construct the Karnaugh map of E.
> 2. Find the smallest number of maximal rectangles that together include *all* the cells of E.
> 3. Form the sum of the prime implicants corresponding to the maximal rectangles found in step 2.

The tricky part of the procedure is step 2. We consider some examples.

EXAMPLE 7.17 _____

Find the minimal-sum form of

$$E = x'yz + x'yz' + xyz' + xy'z' + xy'z$$

SOLUTION

There are four maximal rectangles in this Karnaugh map, as shown by the four ovals in Figure 7.24(a). However, only three are necessary to include all the cells of E, as shown in Figure 7.24(b). The expression E can, therefore, be written as:

$$E = x'y + yz' + xy'$$

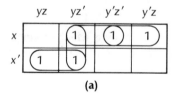

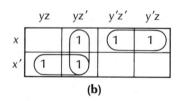

(a)   (b)

Figure 7.24

Remark   The minimal-sum form of an expression is not unique. Another minimal-sum form for the expression E of Example 7.17 is $E = x'y + xz' + xy'$. Can you explain how to get this expression for E?

EXAMPLE 7.18 _____

Find the minimal-sum form of:

$$E = xy'zw' + xy'z'w' + x'y'z'w' + x'yzw + x'yzw' + x'yz'w + x'y'zw' + x'y'z'w$$

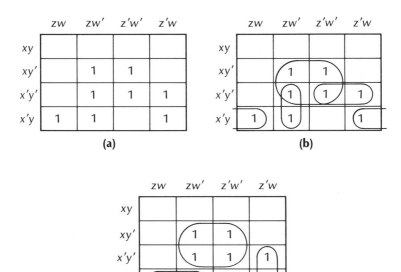

Figure 7.25

SOLUTION

The Karnaugh map of E is shown in Figure 7.25(a).

There are several ways to include all the cells of E in maximal rectangles. Figure 7.25(b) shows how to cover the cells of E using four maximal rectangles. However, the cells of E can be covered using only three maximal rectangles, as shown in Figure 7.25(c). Therefore, $E = x'yz + x'z'w + y'w'$.

We now state in Note 7.18 how to simplify a Boolean expression.

Note 7.18

To simplify a Boolean expression or function:

1. Convert the expression or function to complete sum-of-products form.
2. Transform the resulting expression to minimal-sum form as described in Note 7.17.

EXAMPLE 7.19 _____

Simplify $E = (x + y)(y' + z)(x' + z)$.

x	y	z	x'	y'	$x + y$	$y' + z$	$x' + z$	E	Minterms
0	0	0	1	1	0	1	1	0	
0	0	1	1	1	0	1	1	0	
0	1	0	1	0	1	0	1	0	
0	1	1	1	0	1	1	1	1	$x'yz$
1	0	0	0	1	1	1	0	0	
1	0	1	0	1	1	1	1	1	$xy'z$
1	1	0	0	0	1	0	0	0	
1	1	1	0	0	1	1	1	1	xyz

Figure 7.26

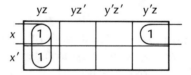

Figure 7.27

SOLUTION

First, construct a table for the expression, find its minterms, and write E in complete sum-of-products form. See Figure 7.26. According to the figure, $E = x'yz + xy'z + xyz$.

Now construct a Karnaugh map for the complete sum-of-products form of E. The map and maximal rectangles are shown in Figure 7.27. From the figure, the simplest form of E is

$$E = yz + xz$$

Can we obtain a minimal-product form of an expression E or function f? The answer is yes, and it can be derived using what we have already learned (see Note 7.19).

> **Note 7.19**
>
> To obtain the **minimal-product form** of an expression E or a function f:
>
> 1. Convert the expression or function to complete sum-of-products form.
> 2. Construct the Karnaugh map of the expression but put 0s in the cells that do not correspond to the minterms of the expression. We call this the complementary Karnaugh map.
> 3. Find the smallest number of maximal rectangles that together include all the 0 cells.
> 4. Form the sum of the prime implicants corresponding to the maximal rectangles found in step 3. This gives an expression for E'.
> 5. Use De Morgan's laws to complement E', giving E.

We illustrate with an example.

EXAMPLE 7.20

Find the minimal-product form of $E = x'y' + yz$.

SOLUTION

The table for E is given in Figure 7.28, where we have written the terms corresponding to values of 0 for the expression.

x	y	z	x'	y'	$x'y'$	yz	$x'y' + yz$	Nonminterms
0	0	0	1	1	1	0	1	
0	0	1	1	1	1	0	1	
0	1	0	1	0	0	0	0	$x'yz'$
0	1	1	1	0	0	1	1	
1	0	0	0	1	0	0	0	$xy'z'$
1	0	1	0	1	0	0	0	$xy'z$
1	1	0	0	0	0	0	0	xyz'
1	1	1	0	0	0	1	1	

Figure 7.28

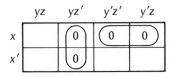

Figure 7.29

The complementary Karnaugh map is given in Figure 7.29.

The resulting expression for E' is $E' = yz' + xy'$. Using De Morgan's laws, we have the minimal-product form:

$$E = (E')' = (yz' + xy')' = (yz')' \cdot (xy')'$$
$$= (y' + z)(x' + y)$$

___

EXERCISES 7.4

1. Convert the following expressions to minimal-sum form:
   a. $(x' + z)(x' + y' + z')(x + y')$ **b.** $x'y(y'z + yz')$
   c. $(x' + z)(xy + xz + yz)$ **d.** $(x + y' + z)(x' + y + w)$
2. Convert the following expressions to minimal-product form:
   a. $x + xy + x'y$ **b.** $x'y(y'z + yz')$
   c. $(x' + y')xy'z$ **d.** $xy(x' + w)(y' + z)$
3. Find the minimal-sum form for the functions f_1, f_2, and f_3 whose outputs are given in Figure 7.30.
4. Find the minimal-product form for the functions of exercise 3.

x	y	z	f_1	f_2	f_3
0	0	0	1	1	0
0	0	1	0	1	0
0	1	0	1	0	0
0	1	1	0	0	0
1	0	0	0	1	0
1	0	1	1	0	1
1	1	0	0	0	0
1	1	1	0	1	1

Figure 7.30

Problems and Projects

5. a. State a definition for the minimal-product form of an expression similar to the one given in the text for the minimal-sum form.

   b. Justify the procedure of Note 7.19.

CASE STUDY 7A   SWITCHING THEORY

Switching theory is the application of Boolean algebra to the control and operation of devices. Specifically, we shall deal with digital devices. A **digital device** has a discrete control and response, such as a push-button telephone. Pressing a specific sequence of buttons rings a specific number. Digital devices may be contrasted with **analog devices,** which have continuous control and response. A light dimmer switch is an analog device. The brightness of the light varies continuously as you turn the knob on the switch.

A digital device has a finite number of inputs and outputs, each of which can assume a finite number of values. A digital device for which each input and output can assume only one of two values is called a **binary device.** Binary devices are of particular importance in electronics and the design of computers because of the binary nature of switches (they are either on or off) and of computers (see Case Study 1A).

Digital devices may be further characterized as sequential or combinational. A **combinational device** is a digital device in which the output is a function of input values only. A push-button telephone is such a device. A **sequential device** is a digital device in which the output is a function of the input values, the time, and past history. Thus, sequential devices combine combinational devices with clocks and memory. A traffic light controller is an example of a sequential device. The proportion of time the controller sets a light to green is a function of (1) how many cars (a digital input) have passed the intersection in the previous five minutes (past history) and (2) the time of day.

In this case study we shall restrict our attention to binary combinational devices.

Gates

A **gate** is a binary electronic circuit operating on one or more input signals to produce one output signal. Since a gate is a binary circuit, each input can have one of two values. In electronic circuits, these values are usually high and low voltage. We shall designate the two possible input values by 1 and 0. Likewise, the output of a gate will be either 1 or 0.

Gates and logic networks (to be defined later in this case study) are usually depicted by **block diagrams.** In the following discussion, we shall introduce the standard block diagram symbols.

There are three basic types of gates: OR, AND, and NOT gates.

1. OR Gates

The block diagram for the OR gate is shown in Figure 7.31(a). The table defining the output z for all possible combinations of inputs is shown in Figure 7.31(b).

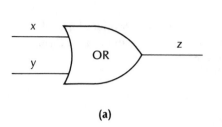

x	y	Output z
0	0	0
0	1	1
1	0	1
1	1	1

(a)   (b)

Figure 7.31

The OR gate is a realization of Boolean addition. Therefore, we denote the output of an OR gate having inputs x and y by $x + y$.

It is possible to have more than two inputs to an OR gate, as shown in Figure 7.32.

Since the OR gate of Figure 7.32 can be interpreted in terms of Boolean addition, we make the following note (Note 7.20).

Note 7.20

The output of an OR gate is 1 if one or more input values is 1.

There is a relation between an OR gate and a basic switching circuit called a **parallel circuit,** depicted in Figure 7.33.

The switches x and y are shown in the open position (value 0 for the corresponding Boolean variables). Current will flow from the source to the device if either switch x or switch y is closed (value 1 for the corresponding Boolean variables).

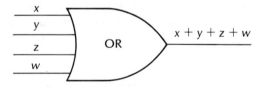

Figure 7.32

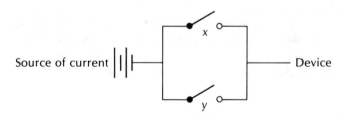

Figure 7.33

2. AND Gates

The block diagram and table for the AND gate are shown in Figure 7.34. The output of an AND gate is 1 only if both inputs are 1. Thus, an AND gate realizes Boolean multiplication. Therefore, we denote the output of an AND gate having inputs x and y by xy.

Like the OR gate, an AND gate can have more than two inputs, as shown in Figure 7.35. The output is the Boolean product of the input values (see Note 7.21).

Note 7.21

The output of an AND gate is 1 only if all input values are 1.

Also like the OR gate, the AND gate is related to a switching circuit, the **series circuit,** shown in Figure 7.36. Current can flow from the source to the device only if both switches are closed (that is, if both Boolean variables have value 1).

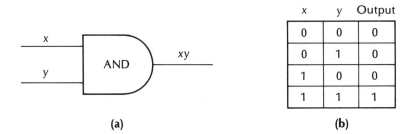

x	y	Output
0	0	0
0	1	0
1	0	0
1	1	1

(a)   (b)

Figure 7.34

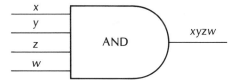

Figure 7.35

Figure 7.36

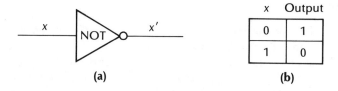

Figure 7.37

3. NOT Gates

The block diagram and table for a NOT gate are shown in Figure 7.37. The NOT gate is a realization of Boolean complementation. The output of a NOT gate having input x is, therefore, denoted by x'.

Note 7.22

The output of a NOT gate is 1 when the input is 0, and 0 when the input is 1.

The switching circuit corresponding to a NOT gate is shown in Figure 7.38. Note that switch x' is closed when switch x is open (current will flow to device 1), and x' is open when x is closed (current will flow to device 2).

Many axioms of Boolean algebra and the laws stated in Theorem 7.1 have counterparts in terms of the switching circuits just described for OR, AND, and NOT gates. The switching circuits corresponding to a law of Boolean algebra are equivalent circuits. For example, the left side of distributive law (i), $a + (bc) = (a + b)(a + c)$, is represented by the switching circuit of Figure 7.39.

The right side of distributive law (i) is represented by the switching circuit of Figure 7.40.

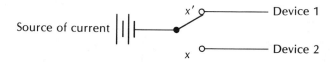

Figure 7.38

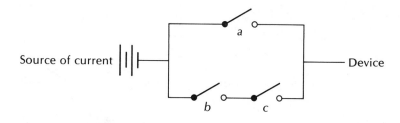

Figure 7.39

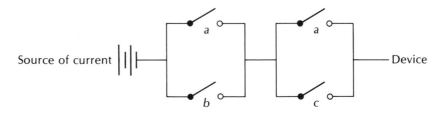

Figure 7.40

Distributive law (i) implies that the circuits of Figures 7.39 and 7.40 are equivalent. Note that the first circuit requires fewer switches.

Logic Networks

OR, AND, and NOT gates can be interconnected to form **logic networks** (also called *gating, switching,* or *combinational networks*). To simplify the block diagrams of logic networks, we shall always assume that the values of an input *and* its complement are available for input to the network. Therefore, it will be unnecessary to use NOT gates on inputs. This is not unrealistic because most input signals come from devices called **flip-flops,** which provide both an output and its complement.

Figure 7.41 shows two examples of logic networks. In each network we show the output of each gate as a Boolean expression in terms of the gate's input. The output

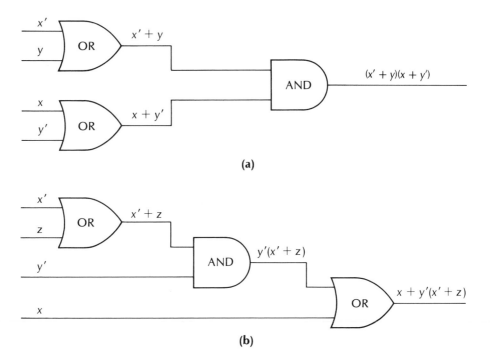

Figure 7.41

from each network can, therefore, be represented by a single Boolean expression. We can also perform the inverse operation. That is, given a Boolean expression, we can construct the logic network it represents.

EXAMPLE 7.21

Draw the block diagram of the logic network represented by the Boolean expression $(x + y)'z + z'$.

SOLUTION

The block diagram is shown in Figure 7.42.

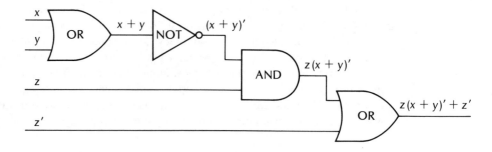

Figure 7.42

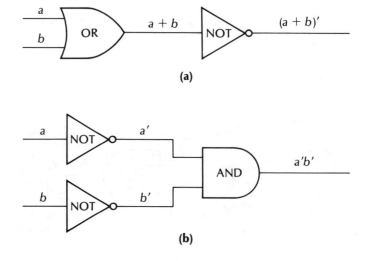

(a)

(b)

Figure 7.43

The axioms and laws of Boolean algebra have counterparts in terms of logic networks. For example, the left and right sides of De Morgan's law (i), $(a + b)' = a'b'$, are represented by the logic networks of Figure 7.43(a) and (b), respectively. De Morgan's law tells us these logic networks are equivalent.

In addition to being a tool to describe logic networks, Boolean algebra can also be used to design logic networks. A binary combinational device can be viewed as a "black box" or function. A certain combination of inputs is required to produce a particular output. The job of the engineer is to (1) design a logic network that will produce the desired effect and (2) design the network as efficiently (that is, as cheaply) as possible.

To accomplish this, we think of a logic network as a realization of a Boolean function. The techniques of this chapter can then be used to construct and minimize the Boolean function for the network. Finally, we construct the logic network corresponding to the Boolean expression. Several examples illustrate this process.

EXAMPLE 7.22

Design a logic network that has three inputs x, y, and z and produces an output of 1 if and only if at least two inputs are 1.

SOLUTION

The first step is to construct a Boolean function satisfying the description of the logic network. The table of such a function is shown in Figure 7.44.

x	y	z	f	Minterms
0	0	0	0	
0	0	1	0	
0	1	0	0	
0	1	1	1	$x'yz$
1	0	0	0	
1	0	1	1	$xy'z$
1	1	0	1	xyz'
1	1	1	1	xyz

Figure 7.44

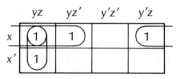

Figure 7.45

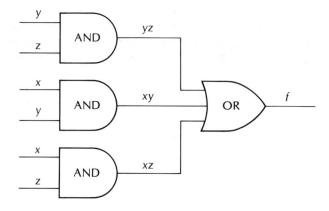

Figure 7.46

Next, we construct the complete sum-of-products expression for the function f. The minterms are shown in the minterm column of Figure 7.44. Thus,

$$f = x'yz + xy'z + xyz' + xyz$$

We next simplify this expression for f by constructing its Karnaugh map. The Karnaugh map with the minimum number (3) of basic rectangles enclosed in ovals is shown in Figure 7.45.

Thus, the minimal-sum form of f is

$$f = yz + xy + xz$$

The logic network represented by this minimal-sum Boolean expression is shown in Figure 7.46.

The type of logic network derived by the procedure used in Example 7.22 is called a **2-level AND-OR network** because each input signal must pass through two gates, the first of which is an AND, the second an OR. Frequently, the criterion for efficient construction of a logic network is the number of gates used. We call a **minimal AND/OR network** a network containing the smallest number of AND and OR gates possible to achieve the desired outputs.

Note 7.23

To find a minimal AND/OR network:

1. Construct the logic network represented by the minimal-sum form of the function.
2. Construct the logic network represented by the minimal-product form of the function.
3. Compare the networks constructed in steps 1 and 2 and choose the one that uses the fewest number of gates.

EXAMPLE 7.23 _____

Find the minimal AND/OR network for the Boolean function of Figure 7.47.

x	y	z	f
0	0	0	0
0	0	1	0
0	1	0	0
0	1	1	0
1	0	0	0
1	0	1	1
1	1	0	1
1	1	1	1

Figure 7.47

SOLUTION

From the table of the function, the complete sum-of-products form of the function is

$$f = xy'z + xyz' + xyz$$

The Karnaugh map of this expression is shown in Figure 7.48.

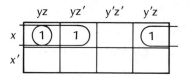

Figure 7.48

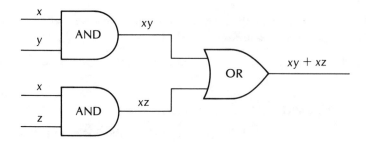

Figure 7.49

The maximal rectangles are enclosed in ovals. Thus, the minimal-sum form of f is

$$f = xy + xz$$

The logic network represented by this function is shown in Figure 7.49.

To find the minimal-product form of the function, we first construct the complementary Karnaugh map and identify the maximal rectangles. See Figure 7.50.

The expression for the complement is, therefore,

$$f' = x' + y'z'$$

Using De Morgan's laws we obtain for f:

$$f = (f')' = (x' + y'z')'$$
$$= (x')'(y'z')'$$
$$= x(y + z)$$

which is the minimal-product form of the function. The network corresponding to this expression is shown in Figure 7.51. Note that this is not a 2-level network because the input signal x passes through only one gate.

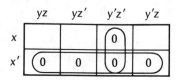

Figure 7.50

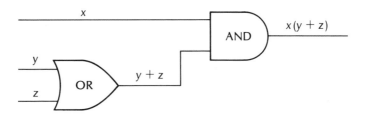

Figure 7.51

The network of Figure 7.51 is called an **OR/AND network** because a typical input signal passes through an OR gate and then an AND gate. Since the OR/AND network of Figure 7.51 requires one less gate than the AND-OR network of Figure 7.49, the OR/AND network is the minimal AND/OR network.

EXAMPLE 7.24

Design a logic network that allows a light to be independently controlled from two wall switches.

SOLUTION

The circuit has two inputs (the two wall switches) and one output (the light). Let 0 correspond to switch up and 1 to switch down. Also, let 1 mean light on and 0 mean light off. If both switches are down, the light should be off. If one switch is down, the light should be on. If two switches are up, one cancels the effect of the other and the light is off. The table of the function corresponding to the circuit is given in Figure 7.52(a).

The sum-of-products form of f is $f = x'y + xy'$, which is in minimal-sum form. The corresponding circuit is shown in Figure 7.52(b).

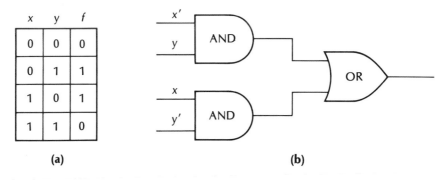

x	y	f
0	0	0
0	1	1
1	0	1
1	1	0

(a)   (b)

Figure 7.52

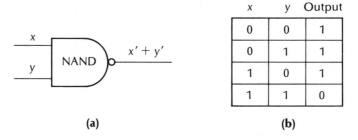

x	y	Output
0	0	1
0	1	1
1	0	1
1	1	0

(a)   (b)

Figure 7.53

NAND and NOR Gates

We conclude this case study with a few remarks about two other important types of gates: NAND and NOR gates. The block diagram and defining table for the **NAND** gate is shown in Figure 7.53.

The effect of a NAND gate is equivalent to (1) passing the inputs x and y through an AND gate, giving xy, and then (2) passing xy through a NOT gate, giving $x' + y'$ as final output. [Recall $(xy)' = x' + y'$ by De Morgan's law (ii)]. Like AND and OR gates, a NAND gate can have more than two inputs.

Note 7.24

The output of a NAND gate is 0 only when all inputs are 1.

The block diagram for a **NOR** gate and the defining table are shown in Figure 7.54.

The effect of a NOR gate is equivalent to (1) passing the inputs x and y through an OR gate, giving $x + y$, and then (2) passing $x + y$ through a NOT gate, giving $x'y'$ as final output. [Recall $(x + y)' = x'y'$ by De Morgan's law (i)]. A NOR gate can have more than two inputs.

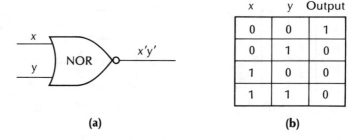

x	y	Output
0	0	1
0	1	0
1	0	0
1	1	0

(a)   (b)

Figure 7.54

Note 7.25

The output of a NOR gate is 1 only when all inputs are 0.

The utility and importance of NAND and NOR gates is that the AND, OR, and NOT operations can be implemented using only NANDs or only NORs. Figure 7.55 shows how NOT, AND, and OR can be implemented using only NANDs. We leave the corresponding circuits using NORs as an exercise.

In fact, for 2-level networks we have the following theorems.

Theorem 7.5

Given a 2-level AND-OR network, an equivalent network can be constructed by replacing all AND and OR gates with NAND gates, obtaining what is called a **NAND-only network.**

Theorem 7.6

Given a 2-level OR-AND network, and equivalent network can be constructed by replacing all OR and AND gates by NOR gates, obtaining what is called a **NOR-only network.**

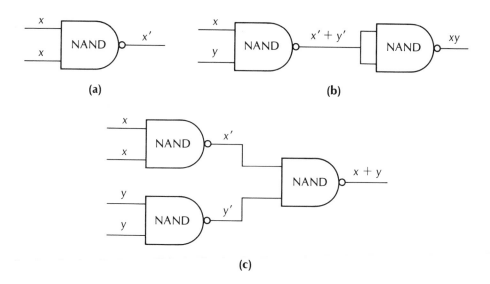

(a)

(b)

(c)

Figure 7.55

These theorems are the reason why NAND and NOR gates are very popular in the manufacture of integrated circuits. A manufacturer needs to make only one kind of gate, say NAND gates, to construct any kind of switching circuit. Moreover, because all gates are of the same type, they can be shared among several circuits. This reduces the number of gates and the number of physical connections between components of the integrated circuit. The size of the integrated circuit and its cost can, therefore, be reduced.

EXAMPLE 7.25 _____

Construct a NAND-only network for the function $f = xy + xz$ of Example 7.23 and verify algebraically that it is equivalent to the original network.

SOLUTION

In Example 7.23, we constructed the 2-level AND-OR network of this function. In Figure 7.56, we show the 2-level AND-OR network and the equivalent NAND-only network.

To show the networks are equivalent, note that the outputs from the first two NAND gates are $x' + y'$ and $x' + z'$, as shown in Figure 7.56(b). The output from the third NAND gate is

$$(x' + y')' + (x' + z')' = xy + xz$$

by De Morgan's law and the involution law.

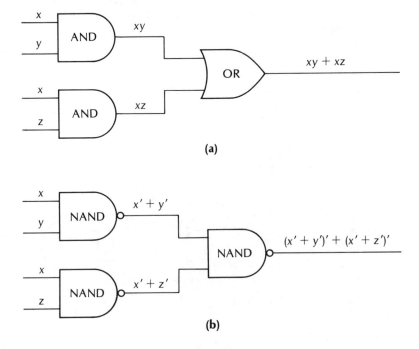

(a)

(b)

Figure 7.56

It is important to realize that Theorem 7.6 does not apply to the OR/AND network constructed in Example 7.23, shown in Figure 7.51, because it is not a 2-level network (the input signal x passes through only one gate). Theorems 7.5 and 7.6 apply only to 2-level networks.

EXERCISES: CASE STUDY 7A

1. For each of the logic networks in Figure 7.57, write the Boolean expression that gives the output in terms of the input variables.

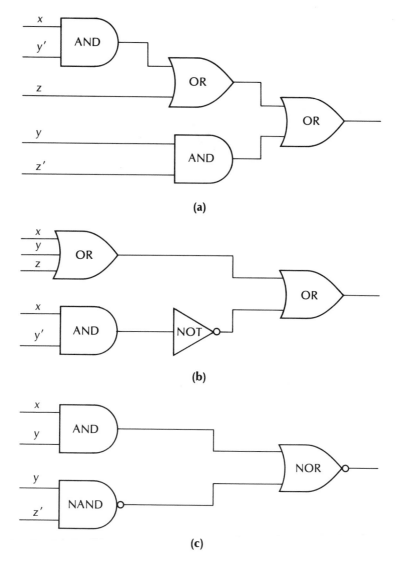

(a)

(b)

(c)

Figure 7.57

2. Draw the block diagrams of the logic networks corresponding to each of the following Boolean expressions using AND and OR gates only:

   a. $x'y' + xy$ **b.** $x(y' + z')$

   c. $x(yz' + y'z)$ **d.** $x'y + y'x(yz + y'z)$

3. Design a minimal AND/OR network for each of the functions f_1, f_2, and f_3 whose outputs are given in Figure 7.58

x	y	z	f_1	f_2	f_3
0	0	0	1	0	0
0	0	1	1	1	0
0	1	0	0	1	1
0	1	1	1	1	1
1	0	0	1	0	1
1	0	1	0	0	0
1	1	0	1	0	0
1	1	1	1	1	0

Figure 7.58

4. Design a minimal AND/OR network whose output is 1 only if exactly two of the four inputs x, y, z, and w are 1.

5. Convert any 2-level OR-AND networks derived in exercise 3 to NOR-only networks.

6. Convert any 2-level AND-OR networks derived in exercise 3 to NAND-only networks.

7. Construct switching circuits for each side of the following:

   a. Associative law (i)   **b.** Associative law (ii)

   c. Distributive law (ii)   **d.** De Morgan's law (ii)

   e. Complement law (i)   **f.** Complement law (ii)

   g. Absorption law (i)   **h.** Absorption law (ii)

8. Construct logic networks using AND, OR, and NOT gates for each side of the laws of exercise 7.

9. Construct a NAND-only network for distributive law (ii).

10. Construct a NOR-only network for distributive law (i).

11. Show how AND, OR, and NOT gates can be implemented using only NOR gates.

Problems and Projects

12. It was noted in Example 7.25 that Theorem 7.6 does not apply to the OR/AND network constructed in Example 7.23. Construct a NOR-only network equivalent to the OR/AND network of Figure 7.51 by adjusting one of the inputs of the network.

13. Design a logic network that allows a light to be independently controlled by three wall switches.

14. Four switches are set to correspond to the bits in a 4-bit binary number. Design a logic network that will turn a light on if and only if the binary number represented by the four bits is a perfect square.

REFERENCES

Bartee, T. C. *Digital Computer Fundamentals.* 4th ed. New York: McGraw-Hill, 1977.

Gilbert, W. J. *Modern Algebra with Applications.* New York: Wiley and Sons, 1976.

Mendelson, E. *Boolean Algebra and Switching Circuits.* New York: McGraw-Hill, 1970.

Reeves, C. M. *An Introduction to Logical Design and Digital Circuits.* Cambridge: Cambridge University Press, 1972.

Scott, J. R. *Basic Computer Logic.* New York: D. C. Heath, 1981.

CHAPTER 8

Graph theory is an area of mathematics that has applications in many subject areas: computer science, biology, chemistry, environmental conservation, social science, transportation, and telecommunications. The graphs of graph theory are not the familiar ones that use an (x, y)-coordinate system. An example of a graph from graph theory is given in Figure 8.1.

In this chapter we will discuss graphs, their properties, and several of their applications. Properties examined include embedding the graph on a surface (**planarity**), structural bonding (**connectivity**), and traveling around the graph (**Eulerian** and **Hamiltonian** graphs). We conclude the chapter with two case studies: a sorting algorithm called *heap sort* that uses graphs, and a discussion of the critical path method.

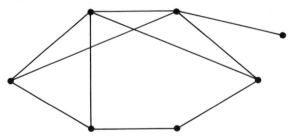

Figure 8.1

8.1 GRAPHS AND DIGRAPHS

A **graph** consists of a finite set of points, called **vertices,** and a set of line segments, called **edges,** which join distinct pairs of vertices. We denote the graph by G (and H if two graphs are being discussed), the set of vertices of G by **V(G),** and the set of edges of G by **E(G).** A single element from the set of vertices is called a **vertex** (*vertices* is the plural form of *vertex*). To facilitate discussion of graphs, we label the vertices by lowercase letters. An edge is then denoted by listing the vertices that are its endpoints (see Figure 8.2).

GRAPH THEORY

When naming an edge, the order of the vertices is unimportant. Thus, edge *bd* is the same as edge *db*. If there is an edge joining a pair of vertices, those vertices are said to be **adjacent.** Otherwise, they are **nonadjacent.** In Figure 8.2, *a* and *b* are adjacent while *a* and *c* are nonadjacent vertices. An edge is **incident** with a vertex if the edge is joined to the vertex. Thus, an edge is incident with its endpoints.

The **degree** of a vertex *v* is the number of edges incident with *v* and is denoted $d(v)$. The **degree sequence** of a graph is the list of the degrees of its vertices in non-increasing order.

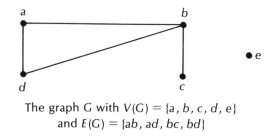

The graph *G* with $V(G) = \{a, b, c, d, e\}$
and $E(G) = \{ab, ad, bc, bd\}$

Figure 8.2

EXAMPLE 8.1

a. Find the degree of each vertex in graph *G* in Figure 8.2.

b. What is the degree sequence of *G*?

c. What is the degree sequence of the graph in Figure 8.1?

SOLUTION

a. We count how many edges emanate from each vertex. Thus, $d(a) = 2$, $d(b) = 3$, $d(c) = 1$, $d(d) = 2$, and $d(e) = 0$.

b. The degree sequence of G is 3, 2, 2, 1, 0.

c. The degree sequence of the graph in Figure 8.1 is 4, 4, 3, 3, 3, 2, 1.

Since an edge contributes 1 to the degree of each of its two endpoints, we have the following theorem, called the "first theorem of graph theory."

Theorem 8.1

In a graph G, the sum of the degrees of the vertices equals twice the number of edges.

As an immediate consequence, we have Corollary 8.1.

Corollary 8.1

The sum of the degrees of the vertices of a graph is an even number.

Theorem 8.1 and Corollary 8.1 are very useful, as we shall see at several points in this chapter. For now, note that 4, 4, 4, 3, 3, 3, 2, 2, 1, 1 cannot be the degree sequence of a graph, because the sum of these numbers is odd, not even.

Graphs are used to model problems in various fields. If a problem can be modeled using graphs, the investigator can sometimes use the theorems of graph theory to produce a solution to the original real-world problem.

In *social science*, for example, graphs can be used to model social interaction within a group. The **friendship graph** of a group of people can be constructed by letting each vertex represent a person in the group. Two vertices are adjacent if the corresponding people are friends. If the graph of Figure 8.2 is interpreted as a friendship graph, b and d are friends, but c and d are not. Person e has no friends in the group. A problem that a researcher might examine is the effect on the friendship graph if one of the people moves to a different neighborhood.

In *chemistry*, graphs are used to model the interaction of atoms in a molecule. In such a graph, each vertex represents an atom. Two vertices are adjacent if there is a bond between the corresponding atoms in the molecule. If the graph of Figure 8.2 is interpreted as a graph of a molecule, atoms a and d are bonded, but a and c are not. Graphs have been used to examine the freezing, burning, and melting properties of various chemicals and to examine the possibility of combining certain atoms to produce new types of molecules.

One of the many uses of graphs in *computer science* is to model computer networks. Each vertex represents a computer, terminal, printer, or other component of the network. Two vertices are adjacent if there is a direct communication link between the corresponding components. If the graph of Figure 8.2 is interpreted as the graph of a computer network, b is linked to all other components except e.

The degree of a vertex has different interpretations depending on how the graph is interpreted. If the graph G is a friendship graph, $d(v)$ is the number of friends person v has. If G is the graph of a molecule, $d(v)$ is the number of atoms to which v is bonded. If G is the graph of a computer network, $d(v)$ is the number of components v can communicate with.

Multigraphs and Pseudographs

There are structures that generalize the concept of a graph. A **multigraph** allows more than one edge between a pair of vertices. Such edges are called **multiple edges** [see Figure 8.3(a)]. In a computer network, there may be multiple communication lines between components. In chemistry, there may be double bonds between atoms. In such cases it may be appropriate to use a multigraph to model the problem.

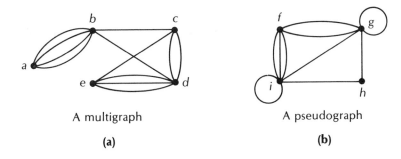

A multigraph

(a)

A pseudograph

(b)

Figure 8.3

A second generalization of a graph is the pseudograph. A **loop** is an edge that connects a vertex to itself. A **pseudograph** allows loops and multiple edges [see Figure 8.3(b)]. In a friendship graph, it may be convenient to consider a person a friend of himself. If such is the case, a pseudograph is an appropriate model.

Let $\mathcal{G}$, $\mathcal{M}$, and $\mathcal{P}$ be the set of all graphs, multigraphs, and pseudographs, respectively. Then $\mathcal{G} \subset \mathcal{M} \subset \mathcal{P}$; that is, every graph is a multigraph, and every multigraph is a pseudograph.

In most applications where they are appropriate, loops and multiple edges are usually suppressed and the problem is solved using graphs. Thus, we will not discuss pseudographs and multigraphs further in this text.

Digraphs

A **directed graph** or **digraph** (for short) consists of a set of vertices and a set of directed edges called **arcs,** which join pairs of distinct vertices. In a digraph, **parallel arcs** are a pair of arcs, one directed from vertex a to vertex b and the other directed from *b* to *a*. These are distinct arcs and are denoted by *ab* and *ba*, respectively (see Figure 8.4). Thus, the order in which we list the vertices of an arc *is* important. The direction of an arc is indicated by the arrow on the edge.

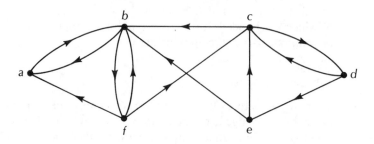

Figure 8.4

Remark   A digraph is a generalization of a graph. In a graph, if we replace each edge by a pair of parallel arcs, we obtain a digraph that is equivalent to the original graph.

EXAMPLE 8.2

a. List the arcs in the digraph of Figure 8.4.

b. List the parallel arcs.

SOLUTION

a. The arcs are *ab*, *ba*, *bf*, *cb*, *cd*, *dc*, *de*, *eb*, *ec*, *fa*, *fb*, and *fc*.

b. The parallel arcs are *ab* and *ba*, *cd* and *dc*, and *bf* and *fb*.

The **indegree** of a vertex *v* is the number of arcs directed toward *v* and is denoted *id(v)*. The **outdegree** of *v* is the number of arcs directed away from *v* and is denoted *od(v)*.

EXAMPLE 8.3

List the indegree and the outdegree of each vertex in Figure 8.4.

SOLUTION

$id(a) = 2$, $od(a) = 1$, $id(b) = 4$, $od(b) = 2$, $id(c) = 3$, $od(c) = 2$, $id(d) = 1$, $od(d) = 2$, $id(e) = 1$, $od(e) = 2$, $id(f) = 1$, and $od(f) = 3$.

Each arc contributes 1 to the indegree of the vertex it points to and 1 to the outdegree of the vertex it points away from. This gives the following result.

Theorem 8.2

In a digraph, the sum of the indegrees equals the sum of the outdegrees.

In analyzing *traffic flow* using graphs, streets correspond to arcs and street intersections correspond to vertices. In this case, a digraph would be a more appropriate model than a graph because of the existence of one-way streets. The direction of an arc would indicate the direction of the one-way street it represents. A pair of parallel arcs would correspond to a two-way street.

In *ecology*, digraphs are used to analyze predator–prey relationships. A vertex in such a digraph represents a species. Since it is important to keep track of who eats whom, a digraph is an appropriate model, with an arc from species u to species v if individuals of u prey on individuals of v.

EXERCISES 8.1

1. Each of the following items corresponds to a graph. For each of them describe (i) what the vertices correspond to, and (ii) when two vertices are adjacent.
   a. A road map
   b. A family tree
   c. The countries on a map
2. For each graph in Figure 8.5, list $V(G)$, $E(G)$, the degree of each vertex, and the degree sequence.

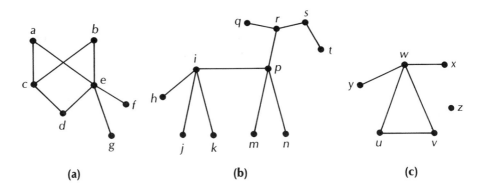

(a)   (b)   (c)

Figure 8.5

3. Construct graphs with the following vertex and edge sets:
   a. $V(G) = \{a, b, c, d\}$, $E(G) = \{ab, cd, bd, bc\}$
   b. $V(H) = \{e, f, g, h, i, j\}$, $E(H) = \{eh, ej, hi, ij, fh, hj\}$.
   c. $V(M) = \{k, m, n, p, q, r\}$, $E(M) = \{km, mp, kp, qr\}$

4. a. Three graphs have degree sequence 3, 2, 2, 1, 1, 1. Find two of them.
 b. Why is 5, 3, 3, 2, 1, 1, 1, 1 not the degree sequence of a graph?
 c. Draw the graph having degree sequence 2, 1, 1, 0.
 d. Although $5 + 2 + 1 + 1 + 1 = 10$ (which is even), 5, 2, 1, 1, 1 is not the degree sequence of a graph. Why?

5. The following items can be modeled using digraphs. For each of them, describe (i) what the vertices correspond to, and (ii) when an arc points from vertex a to vertex b.
 a. The scheduling of jobs at a manufacturing plant
 b. The divisors of 400
 c. The directions of streets in a city

6. For each digraph in Figure 8.6, (a) list the indegree and the outdegree for each vertex, and (b) list the parallel arcs.

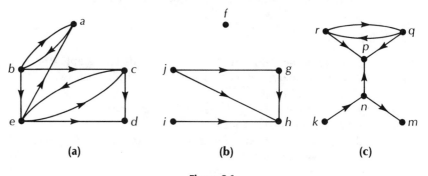

(a)   **(b)**   **(c)**

Figure 8.6

7. In a digraph, a vertex having indegree 0 is called a **transmitter,** and a vertex having outdegree 0 is called a **receiver.**
 a. List all transmitters from exercise 6.
 b. List all receivers from exercise 6.
 c. Explain why it is easy to spot a vertex that is both a transmitter and a receiver.

8. a. How would you define a multidigraph?
 b. How would you define a pseudodigraph?
 c. Draw a multidigraph that is not a digraph.
 d. Draw a pseudodigraph that is not a multidigraph.

Problems and Projects

9. Use the techniques of Chapter 3 to show the following:
 a. The maximum number of edges in a graph having n vertices is $\binom{n}{2}$.
 b. The maximum number of arcs in a digraph with n vertices is $n(n-1)$.

8.2 BASIC DEFINITIONS

A **path** in a graph G is a sequence of distinct vertices $v_0, v_1, v_2, \ldots, v_k$ such that $v_i v_{i+1}$ is an edge of G, $0 \le i \le k - 1$. The **length** of a path is the number of edges in it.

In Figure 8.7, c,e,b is a path of length 2 between c and b, while c,d,e,a,b is a path of length 4 between c and b (see Note 8.1). The edges in the first path are ce and eb, while the edges in the second path are cd, de, ea, and ab.

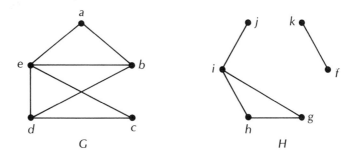

Figure 8.7

Note 8.1

A path using k distinct vertices has length $k - 1$.

Notice that graph G of Figure 8.7 is in one piece, whereas graph H is composed of two separate pieces. G is said to be connected, while H is disconnected. Formally, a graph is **connected** if there is a path between every pair of vertices; otherwise, it is **disconnected**. Graph G of Figure 8.7 is connected; however, H is disconnected since there is no path between f and i.

Subgraphs

A **subgraph** A of graph B has as its vertex set a subset of the vertices of B, and as its edge set a subset of the edges of B. We use $A \prec B$ to denote A *is a subgraph of B*. Using set notation, we have Note 8.2.

Note 8.2

For graphs A and B, $A \prec B$ if and only if $V(A) \subset V(B)$ and $E(A) \subset E(B)$.

If graph A is not a subgraph of B, we write $A \not\prec B$.

EXAMPLE 8.4

Determine which of the graphs in Figure 8.8 are subgraphs of B.

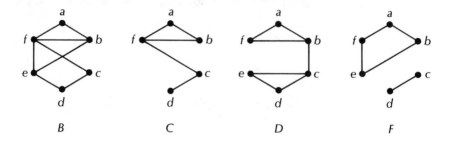

Figure 8.8

SOLUTION

We have $B \prec B$ (a graph is a subgraph of itself), $C \prec B$, and $F \prec B$. Note that $D \not\prec B$ because $bc \in E(D)$, but $bc \notin E(B)$.

If $A \prec B$, A may or may not contain all the vertices of B. In Figure 8.8, $C \prec B$, and C contains only five of the six vertices of B. Graph F contains all six vertices of B and $F \prec B$. A subgraph sometimes contains as many edges as possible for its vertex set while still remaining a subgraph. Graph A is an **induced subgraph** of B if whenever an edge appears between a pair of vertices in B, it also appears between those vertices in A. In Figure 8.8, C is an induced subgraph of B. Although $F \prec B$, F is not an induced subgraph of B, because c and f are vertices in F, and $cf \in E(B)$ but $cf \notin E(F)$.

Earlier, we saw that a connected graph is in one piece, while a disconnected graph has two or more pieces. These pieces are subgraphs and are called **components** of the graph. To define components formally, we require the concept of subgraphs maximal with respect to a property. We will use maximal subgraphs in several other contexts in this chapter.

A is a **maximal subgraph of B with respect to property P** if whenever C is a subgraph of B having property P and $E(A) \subset E(C) \subset E(B)$ and $V(A) \subset V(C) \subset V(B)$, we have $A = C$. Thus, if A is a maximal subgraph of B with respect to P, A is a subgraph of B with property P and is a "largest" subgraph of B with property P.

If P is the property of being connected, then a component of a graph G is a maximal connected subgraph of G. Note that a connected graph has only one component, namely itself. In Figure 8.8, *B*, *C*, and *D* have one component, while *F* has two components.

Isomorphic Graphs

If we describe the vertex set and edge set of a graph, the graph is completely determined. However, two people may *draw* that graph differently. For example, suppose we are asked to draw the graph *G* with vertex set $V(G) = \{a, b, c, d, e\}$ and edge set $E(G) = \{ab, ac, bc, bd, ce\}$. One person might draw *G* as in Figure 8.9(a), while another might draw it as in Figure 8.9(b). The two graphs are the same even though they appear to be different. In most problems where we are trying to determine whether two graphs are the same, the vertices have not yet been labeled to help us in our decision.

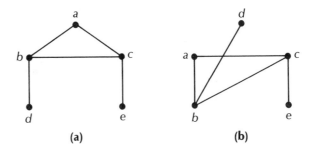

(a)   (b)

Figure 8.9

Graphs *G* and *H* are **isomorphic** if they can be labeled so that *u* and *v* are adjacent in *G* if and only if the corresponding vertices are adjacent in *H*. We can also describe isomorphic graphs in terms of the function concepts of Chapter 5. Graphs *G* and *H* are **isomorphic** if there exists a 1-1, onto function $f : V(G) \to V(H)$ such that $uv \in E(G)$ if and only if $f(u)f(v) \in E(H)$. The function *f* is called an **isomorphism.** A more compact way of stating the function description is as follows: *G* and *H* are isomorphic if there is a 1-1, onto function between their vertices that preserves adjacency.

Note 8.3 will be useful in the next example.

Note 8.3

Let $f : V(G) \to V(H)$ be an isomorphism between graphs *G* and *H*. Then for any vertex $a \in V(G)$, $d(a) = d(f(a))$. In other words, if two graphs are isomorphic, corresponding vertices have the same degree.

EXAMPLE 8.5 _____

For each pair of graphs in Figure 8.10, describe a function that shows the graphs are isomorphic.

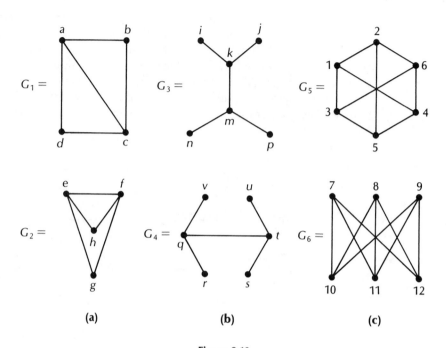

(a)   (b)   (c)

Figure 8.10

SOLUTION

a. To show the graphs are isomorphic, we must give a correspondence between the vertices of G_1 and G_2 that preserves adjacency. By Note 8.3, corresponding vertices must have the same degree. Thus, vertex a (having degree 3) must correspond to e or f; vertex b (having degree 2) must correspond to h or g. This condition on the degrees of the vertices is sufficient to yield the required correspondence between the vertices. Let $f(a) = e$, $f(b) = h$, $f(c) = f$, and $f(d) = g$. The function f induces a correspondence from the edges of G_1 to the edges of G_2 as follows: $ab \rightarrow eh$, $bc \rightarrow hf$, $ad \rightarrow eg$, $cd \rightarrow fg$, and $ac \rightarrow ef$. Because $\{eh, hf, eg, fg, ef\}$ is precisely the edge set of G_2, G_1 is isomorphic to G_2.

b. By checking degrees as in part a, we find that k must correspond to q or t; i must correspond to r, s, u, or v; and so on. In this case, we must be careful. If, for example, $f(k) = q$, then since i is adjacent to k and has degree 1, $f(i) = v$ or r. We can check that the following assignment preserves adjacencies: $f(k) = q$, $f(m) = t$, $f(i) = v$, $f(j) = r$, $f(n) = u$, and $f(p) = s$. Thus G_3 is isomorphic to G_4.

c. It does not help to check degrees in this case because all vertices in each graph have degree 3. Here we must focus closely on adjacencies. In G_6, three vertices (7, 8, and 9) are not adjacent to one another, but each is adjacent to the same set of vertices: 10, 11, and 12. In G_5, vertices 2, 3, and 4 have this property. Thus, let $f(2) = 7$, $f(3) = 8$, $f(4) = 9$, $f(1) = 10$, $f(5) = 11$, and $f(6) = 12$.

Before discussing the next example, we state Note 8.4.

Note 8.4

Let G and H be isomorphic with isomorphism $f: V(G) \to V(H)$. If $v_1, v_2, \ldots, v_k$ is a shortest path between vertices v_1 and v_k in G, then $f(v_1), f(v_2), \ldots, f(v_k)$ is a shortest path between vertices $f(v_1)$ and $f(v_k)$ in H.

EXAMPLE 8.6 _____

There are five nonisomorphic graphs with degree sequence 3, 3, 2, 2, 1, 1. Draw them and explain why they are not isomorphic to one another.

SOLUTION

Four of the graphs are connected. The other has two components. The graphs are shown in Figure 8.11. The graph (e) is distinct from the others (why?). To show (a) through (d) are distinct, we focus on the length of the shortest path between

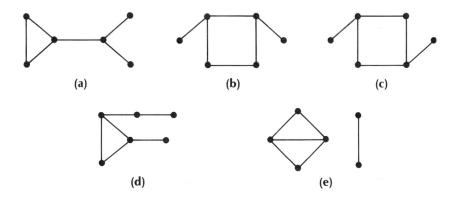

Figure 8.11

the two vertices of degree 1 (see Note 8.4). In (a) this length is 2, in (b) it is 3, and in (c) and (d) the length is 4. Thus, this check distinguishes (a) and (b) from the others.

We now only need to show that (c) is not isomorphic to (d). Let us use the same test as before, but focus on the vertices of degree 3. In (d) the vertices of degree 3 are adjacent, but in (c) the two vertices of degree 3 are not adjacent. Thus (c) and (d) are not isomorphic.

There is no efficient algorithm for distinguishing nonisomorphic graphs. Indeed, finding better isomorphism-testing algorithms is a very active area of graph theory research. There are, however, several techniques that work well for graphs of moderate size (see Note 8.5).

Note 8.5

Some items to check when trying to show that a pair of graphs are not isomorphic are:

1. The number of vertices
2. The number of components
3. The number of edges
4. The degree sequence
5. The length of the shortest path between pairs of vertices with a given degree
6. The length of the longest path in the graph

We used part 5 of Note 8.5 to distinguish graphs (c) and (d) in Example 8.6. We could have used part (6) of Note 8.5 instead. The longest path in graph (c) has length 4, whereas the longest path in graph (d) has length 5. Thus, graphs (c) and (d) of Figure 8.11 are not isomorphic.

Several other concepts we consider in this chapter will prove useful in distinguishing nonisomorphic graphs.

EXERCISES 8.2

1. a. If a graph has six vertices and five edges, what is the largest number of components it can have?
   b. How about six vertices and three edges?
   c. How about six vertices and seven edges?

2. In graph G_1 of Figure 8.12:
   a. List all paths between a and b.
   b. What is the length of each path in part a?
   c. Draw three connected subgraphs of G_1.
   d. Draw three disconnected subgraphs of G_1.
   e. Draw the induced subgraph with vertex set $\{a, b, c, e\}$.

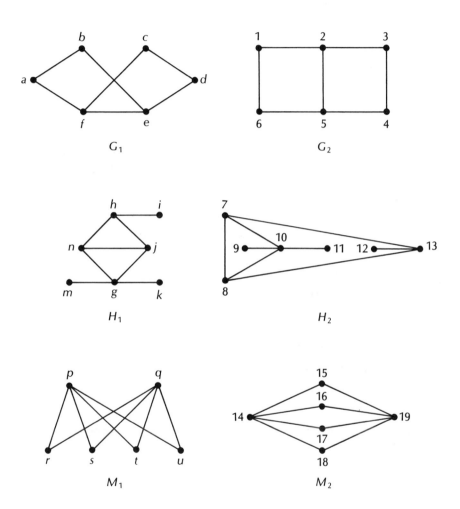

Figure 8.12

3. If G has n vertices, what is the minimum number of edges it must have to be connected? (*Hint:* Draw graphs for various values of n).
4. For each pair of graphs in Figure 8.12, give a function that shows the graphs to be isomorphic.

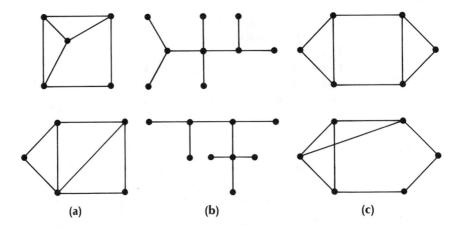

Figure 8.13

5. Explain why each pair of graphs in Figure 8.13 is not isomorphic.
6. Draw all nonisomorphic graphs having the following degree sequences:
   a. 3, 3, 2, 1, 1, 1, 1   **b.** 4, 2, 2, 1, 1, 1, 1
   c. 3, 2, 1, 1, 1, 1, 1   **d.** 5, 3, 2, 2, 2, 1, 1
7. In Note 8.4, we refer to *a* shortest path between two vertices in a graph. Give an example of a graph that has at least two shortest paths between each pair of nonadjacent vertices.

Problems and Projects

8. Prove Note 8.3.
9. Prove Note 8.4.

8.3 CLASSES OF GRAPHS

In this section, we discuss several important classes of graphs: trees, bipartite graphs, complete graphs, regular graphs, and planar graphs.

Trees

One type of graph we have already discussed is the path. A path having n vertices is denoted P_n. See Figure 8.14(a). A graph related to the path is the cycle. A **cycle**

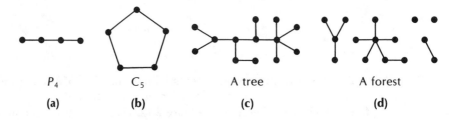

P_4 C_5 A tree   A forest
(a)   (b)   (c)   (d)

Figure 8.14

consists of a sequence of vertices $v_1, v_2, \ldots, v_k$ such that $v_i v_{i+1}$ is an edge of G ($1 \leq i \leq k - 1$) and $v_k v_1$ is an edge of G. See Figure 8.14(b). A cycle, therefore, is a path whose first and last vertices are identified. The cycle having n vertices is denoted C_n.

A **tree** is a connected graph that does not contain a cycle as a subgraph. The graph of Figure 8.14(c) is a tree. Note that we usually use T rather than G to denote a tree. A **forest** is a graph, all of whose components are trees. Thus, a forest is a graph (not necessarily connected) that does not contain a cycle. Note that a path is a tree, but not every tree is a path. Also, every tree is a forest, but not every forest is a tree.

Trees are an important class of graphs because of their simplicity and their occurrence in applications. We state some properties of trees in the following theorem.

Theorem 8.3

Suppose G is a graph having n vertices. Then the following statements are equivalent:

  i. G is a tree.

  ii. G is connected and has $n - 1$ edges.

  iii. G has $n - 1$ edges and no cycles.

  iv. Any two vertices of G are connected by a unique path.

  v. G contains no cycles, but the addition of any edge to G will produce a single cycle.

Theorem 8.3 is many theorems in one. For example, if we pair (i) and (ii), we get the theorem: G is a tree having n vertices if and only if G is connected and has $n - 1$ edges. For the proof of various parts of Theorem 8.3, see exercises 15 and 17. The characterizations in Theorem 8.3 are each useful in different situations. We shall see one such situation in our discussion of the Greedy Algorithm in Example 8.7.

In Chapter 3, we saw how trees are used for counting. In Chapter 4, we saw how trees are used to describe the sample space of a compound experiment. But trees occur in other applications as well. In the **minimum-weight spanning tree** problem, we are given a graph G that has a real number (or **weight**) associated with each of its edges. Such a graph is called a **weighted graph.** We must find a tree that is a subgraph of G containing all vertices of G such that the total of the weights on the edges in T is a minimum. A subgraph that contains all the vertices of a graph is called a **spanning subgraph.** Therefore, we are looking for a minimum-weight spanning tree.

Finding a minimum-weight spanning tree is related to creating a minimum-cost communication network. Each component of the network is represented by a vertex in a graph. A possible communication line between components is represented by an edge. The cost of constructing the communication line is, then, the weight assigned to the edge representing that communication line. A minimum-weight spanning tree for this weighted graph represents the cheapest communication network that could be built.

EXAMPLE 8.7 _____

Find a minimum-weight spanning tree for the weighted graph in Figure 8.15.

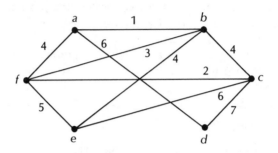

Figure 8.15

SOLUTION

One method for finding a minimum-weight spanning tree, called the **Greedy Algorithm,** is to continually select an edge having minimum weight so as to avoid forming a cycle. Stop when a spanning tree is obtained. We show the steps of the algorithm in Figure 8.16. First we select edge ab, which has weight 1; then cf, which has weight 2; and then bf, which has weight 3. There are three edges that have weight 4: af, bc, and be. Edge af would form the cycle a,b,f; so we do not select af. Edge bc would form the cycle b,c,f; so we do not select bc. We select be, which does not form a cycle. Edge ef has weight 5 but would form a cycle, so we omit it. Edge ad has weight 6. We include it. Our subgraph now has $n - 1 = 5$ edges and

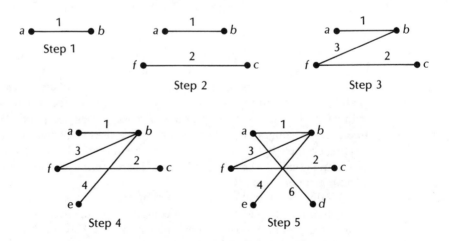

Figure 8.16

no cycles, so it is a tree. By Theorem 8.3(v), the addition of any edge to our subgraph would produce a cycle. Thus, we stop. We have obtained a spanning tree. The total weight of the spanning tree is 16.

Note that we use Theorem 8.3(iii) to tell us when to stop the Greedy Algorithm. The tree we must construct has six vertices and, therefore, must have five edges and no cycles. The graph in step 5 has five edges and no cycles, and so it must be a tree.

Bipartite Graphs

A tree is a special type of bipartite graph. A graph is **bipartite** if its vertices can be separated into two sets A and B, so that vertices within the same set are nonadjacent. A bipartite graph is usually drawn with the vertices of A on the left and vertices of B on the right [see Figure 8.17(a)], or vertices of A at the top and vertices of B at the bottom [see Figure 8.17(b)]. Note that the only adjacencies are *between* vertices of the set A and vertices of the set B. There are none *among* vertices of sets A or B.

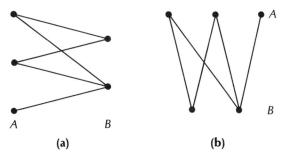

(a)   (b)

Figure 8.17

A simple labeling procedure, described in Note 8.6, will determine whether G is bipartite.

Note 8.6

G is bipartite if and only if adjacent vertices get distinct labels by the following procedure (in the procedure, we refer to the labels a and b as opposite labels): Label any vertex a. Now, label all vertices adjacent to a with the label b. Next, label by a every vertex that is adjacent to each vertex just labeled b. Continue in this manner, labeling vertices adjacent to each vertex just labeled with the opposite label. Do this for each component.

If the procedure of Note 8.6 is successful—that is, if adjacent vertices get distinct labels—the graph G is bipartite. The set A in the definition of a bipartite graph is then the set of all vertices with label a; the set B is the set of all vertices with label b.

EXAMPLE 8.8 ───

Label each graph in Figure 8.18 to determine if it is bipartite. For those that are bipartite, redraw them showing the vertex sets A and B of the definition.

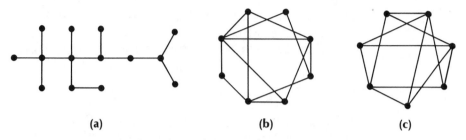

(a)   (b)   (c)

Figure 8.18

SOLUTION

We show the labeling in Figure 8.19. The subscript on each label indicates the step of the procedure of Note 8.6 during which that vertex is labeled. Graphs (a) and (c) are bipartite. The redrawings of (a) and (c) are given in Figure 8.20.

Graph (b) is not bipartite because there is a pair of adjacent vertices labeled a at step 3. The vertex a_3 on the left in Figure 8.19(b) is adjacent to the two other vertices labeled a_3.

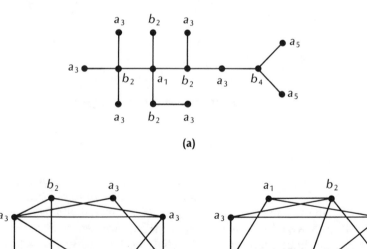

Figure 8.19

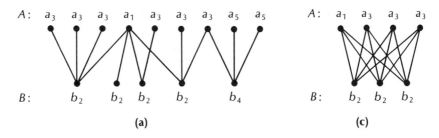

Figure 8.20

Can we tell whether a graph is bipartite without using the labeling procedure of Note 8.6? The answer is yes, as indicated in our next theorem. First, we define an **odd cycle** as a cycle containing an odd number of vertices.

Theorem 8.4

G is bipartite if and only if G does not contain an odd cycle.

The only graph in Figure 8.18 that is not bipartite is (b). It contains odd cycles. (Can you find one?) We mentioned before that a tree is a bipartite graph. We can show this using Theorem 8.4 as follows: Since a tree T does not contain any cycles, it surely does not contain any odd cycles. Thus, by Theorem 8.4, T is bipartite.

Note that there are bipartite graphs that are not trees. For example, the graph of Figure 8.18(c) is not a tree, but is bipartite.

EXAMPLE 8.9 _____

Draw all nonisomorphic bipartite graphs having four vertices.

SOLUTION

Using Theorem 8.4, we draw the graphs on four vertices having no odd cycles. The graphs are shown in Figure 8.21.

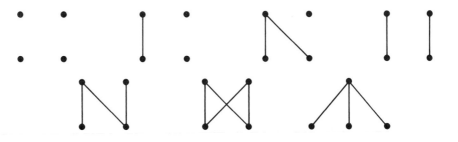

Figure 8.21

A **complete bipartite graph** is one in which each vertex in set A is adjacent to every vertex in set B. The complete bipartite graph having m vertices in A and n vertices in B is denoted $\mathbf{K_{m,n}}$. The last two graphs of Figure 8.21 are the complete bipartite graphs $K_{2,2}$ and $K_{1,3}$. The graphs $K_{2,3}$ and $K_{3,3}$ are shown in Figure 8.22.

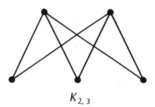 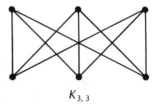

$K_{2,3}$ $K_{3,3}$

Figure 8.22

Bipartite graphs play a key role in matching problems. A **matching** M in a graph is a set of edges, no two of which share a common endpoint. The darkened edges ah, bd, and ef of Figure 8.23(a) constitute a matching. M is a **maximum matching** for G if M contains as many edges as any other matching for G. The matching in Figure 8.23(a) is not a maximum matching. The matching in Figure 8.23(b) is a maximum matching.

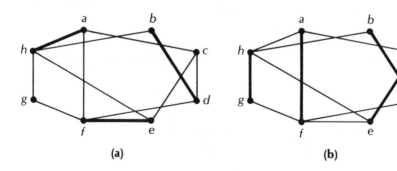

(a)   (b)

Figure 8.23

We can look for matchings in any type of graph. The graph of Figure 8.23 is not bipartite, for example. In most matching problems, however, we must find a matching for a graph that is bipartite. For example, the vertices of A could represent people, and the vertices of B could represent jobs. There is an edge ab if person a is qualified for job b. For maximum employment, we want a maximum matching. The edges in the maximum matching tell us who to hire for each job.

Complete Graphs

A graph is **complete** if its vertices are mutually adjacent—that is, if there is an edge connecting each pair of vertices. Since there are $\binom{n}{2}$ pairs of vertices in a graph with n vertices, we have Note 8.7.

Note 8.7

The complete graph on n vertices has $\binom{n}{2}$ edges. This graph is denoted K_n.

EXAMPLE 8.10 ───

Draw each K_n for $n \le 5$.

SOLUTION

See Figure 8.24.

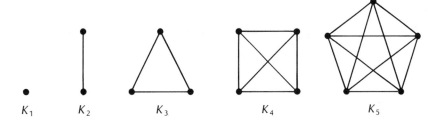

K_1 K_2 K_3 K_4 K_5

Figure 8.24

Regular Graphs

A graph is **k-regular** if all its vertices have the same degree, k. K_n is $(n - 1)$-regular, because each vertex of K_n has degree $n - 1$. C_n is 2-regular. We can show that a k-regular graph on n vertices has $nk/2$ edges (see exercise 16). This implies that n and k cannot both be odd. If either n or k is even and $0 \le k \le n - 1$, there exists a k-regular graph on n vertices (see exercise 18).

Planar Graphs

One of the uses of graphs is to model electrical networks. For example, the electrical networks in Figure 8.25(a) and (e) can be modeled by Figure 8.25(b) and

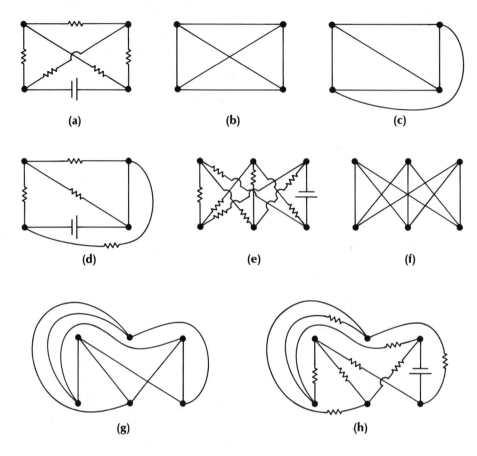

Figure 8.25

(f), respectively. The humps where wires (edges) cross in Figure 8.25(a) and (e) indicate that one wire passes above the other. Is it possible to redesign the network so there are fewer crossing wires? Usually, yes. Computers and many other electronic devices use electrical networks that are *printed* by machine on a small chip or circuit board. This can be done easily when there are no crossing edges in the graph modeling the network.

In Figure 8.25, the graph (b) can be redrawn as (c). Thus, the electrical network of (a) can be redesigned with no wires crossing as in (d). In any redrawing of graph (f), at least one pair of edges cross (we prove this in Theorem 8.5). Thus, (g) is an optimal redrawing of (f) and the network in (e) may be redesigned with one pair of crossing wires as in (h).

A **planar graph** is a graph that can be drawn in the plane (that is, on a flat surface) with no crossing edges. Thus, G is planar if there is a graph H isomorphic to G such that H has no crossing edges. A result of Wagner (1936) says that a planar graph can be drawn in the plane using *straight line segments* with no edges crossing. In Figure 8.25(c) we did not use straight line segments. Can you redraw

(c) using only straight line segments with no edges crossing? (*Hint:* Move the vertices around.) Using Wagner's result we can prove Theorem 8.5.

Theorem 8.5

$K_{3,3}$ is not planar.

Proof: Suppose the vertices of set A in $K_{3,3}$ are labeled 1, 2, 3; and those in set B are labeled 4, 5, 6. Then 1,4,2,5,3,6 is a cycle in $K_{3,3}$. No matter how the vertices are situated, in any drawing of $K_{3,3}$ using straight line segments, some pair of edges from $(1,5)$, $(2,6)$, or $(3,4)$ must cross either inside or outside the cycle (see Figure 8.26).

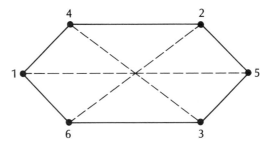

Figure 8.26

Planar graphs were characterized by the Polish mathematician Kuratowski. His characterization uses the graphs K_5 and $K_{3,3}$ and a concept called homeomorphism (see exercise 21).

A **plane graph** is a graph drawn in the plane having no edges crossing. Thus, every plane graph is planar but not conversely. However, each planar graph can be redrawn as a plane graph. In a plane graph, each cycle not containing any smaller cycles encloses a region called a **face**. The region exterior to the graph is called the **infinite face**. Let f be the number of faces (including the exterior face), e be the number of edges, and n be the number of vertices in a plane graph. The following relationship between these numbers is called **Euler's formula.**

Theorem 8.6

In a connected plane graph with f faces, e edges, and n vertices, we have the relationship: $n - e + f = 2$.

EXAMPLE 8.11 _____

Redraw the graphs in Figure 8.27 with as few edge crossings as possible. For any plane graphs you draw, verify Euler's formula.

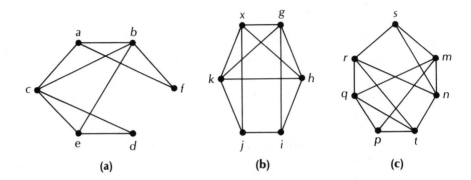

Figure 8.27

SOLUTION

The graphs are redrawn in Figure 8.28. Graphs (a) and (c) are plane graphs. In Figure 8.28(a), there are 6 vertices, 9 edges, and 5 faces. Thus $n - e + f = 6 - 9 + 5 = 2$. In Figure 8.28(c), there are 7 vertices, 13 edges, and 8 faces. Thus, $n - e + f = 7 - 13 + 8 = 2$.

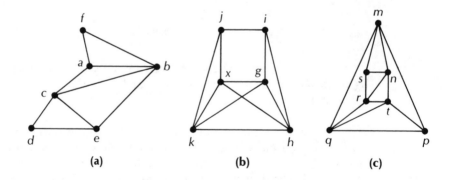

Figure 8.28

EXERCISES 8.3

1. Draw each of the following:
   a. P_6 **b.** C_4 **c.** C_3 **d.** P_3
2. Draw:
   a. All trees having five or fewer vertices

    b. All 11 trees having seven vertices

    c. All six trees having six vertices

3. There are ten forests having five vertices. Draw them.

4. a. Which complete graphs are trees?

    b. Which cycles are complete graphs?

    c. Which trees are regular?

    d. When is $K_{m,n}$ regular?

5. Find minimum-weight spanning trees for the graphs in Figure 8.29.

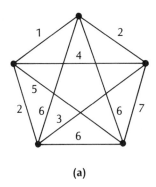

(a)

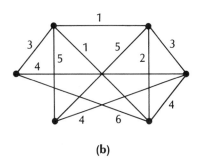

(b)

Figure 8.29

6. Draw all connected bipartite graphs having five vertices.

7. Find two maximum matchings for each of the graphs in Figure 8.29 (ignore the weights).

8. Ann, Dan, Fan, Nan, and Van are applying for jobs a, b, c, d, and e. Ann is qualified for a, b, and c; Dan for b, d, and e; Fan for b and c; Nan for c and d; and Van for b.

    a. Draw the associated bipartite graph.

    b. Find a maximum matching to create maximum employment.

9. Draw each of the following:

    a. K_4 **b.** $K_{3,4}$ **c.** K_6 **d.** $K_{2,5}$

10. Show that $K_{m,n}$ has mn edges.

11. Draw:

    a. A 3-regular graph on eight vertices

    b. Two 3-regular graphs on six vertices (be sure they are not isomorphic)

    c. A 4-regular graph on seven vertices

12. How many edges are in the following:

    a. K_{10} **b.** C_8 **c.** P_{12} **d.** $K_{3,7}$

13. A **wheel** W_n is formed from C_{n-1} $(n \geq 4)$ by adding an additional vertex v and edges from v to each vertex of C_{n-1}.

    a. Draw W_4, W_5, W_6, and W_7.

    b. How many edges are in W_n?

14. Draw the graphs in Figure 8.30 using as few crossing edges as possible. Which graphs are planar? Verify Euler's formula for the graphs that are planar.

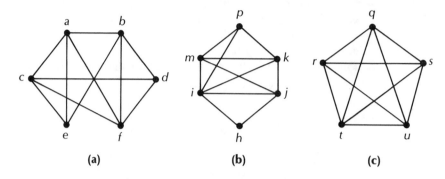

(a)   (b)   (c)

Figure 8.30

Problems and Projects

15. Prove: T is a tree if and only if any two vertices in T are connected by a unique path.

16. Use Theorem 8.1 to prove a k-regular graph on n vertices has $nk/2$ edges.

17. Prove: If T is a tree on at least two vertices, then T contains at least two endpoints (vertices of degree 1).

18. If either n or k is even, $0 \le k \le n - 1$, construct a k-regular graph on n vertices.

19. Use exercise 17 and induction to show that conditions (i), (ii), and (iii) of Theorem 8.3 are equivalent.

20. The **crossing number** of a graph G is the minimum number of crossing edges among all drawings of G. Find the crossing number of:
     a. K_4 **b.** K_5 **c.** $K_{3,3}$ **d.** W_6 **e.** $K_{3,4}$

21. To **subdivide an edge** uv, we remove uv, add an additional vertex w and edges uw and vw. Graph G is **homeomorphic** to H if G can be obtained from H by successively subdividing edges in H and resulting graphs. **Kuratowski's Theorem** states: G is planar if and only if G contains no subgraph homeomorphic to K_5 or $K_{3,3}$.
     a. Draw two graphs homeomorphic to K_5.
     b. The graph in Figure 8.23(a) is not planar. It contains a subgraph homeomorphic to $K_{3,3}$. Find and draw that subgraph.

8.4 MATRICES ASSOCIATED WITH GRAPHS

Many computer algorithms use graphs. How can we represent a graph in computer memory? The usual procedure is to represent, store, and manipulate the graph using a matrix. In this section, we discuss several matrices associated with graphs. We call these **graphical matrices.** Some matrices are used to represent graphs, while others are used to describe properties of graphs.

The Adjacency Matrix

A **(0, 1)-matrix** is a matrix each of whose entries is 0 or 1. The identity matrix and the zero matrix are examples of (0, 1)-matrices. The most widely used graphical matrix is the adjacency matrix, which tells whether a pair of vertices is adjacent. Let the vertices of G be labeled $v_1, v_2, \ldots, v_n$. The **adjacency matrix** $A(G)$ is the $n \times n$ (0, 1)-matrix, where

$$a_{ij} = \begin{cases} 1 & \text{if } v_iv_j \text{ is an edge of } G \\ 0 & \text{otherwise} \end{cases}$$

Since a vertex is never adjacent to itself, $A(G)$ has 0s on the diagonal. A graph and its adjacency matrix are displayed in Figure 8.31.

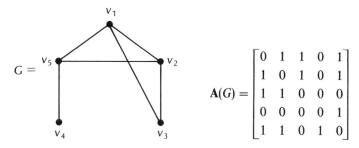

Figure 8.31

A graph G does not always come conveniently labeled. G may have labels other than $v_1, v_2, \ldots, v_n$, or G may have no labels. In either case, we arbitrarily assign labels $v_1, v_2, \ldots, v_n$ to the n vertices of G. Different label assignments will produce different matrices. Do not be concerned about this; all pertinent information is contained in the matrix no matter how we label the graph. The different label assignments correspond to isomorphic drawings of the same graph.

EXAMPLE 8.12 _____

a. From the graph G of Figure 8.32(a), construct $A(G)$.

b. From the adjacency matrix $A(H)$ of Figure 8.32(b), construct the graph H.

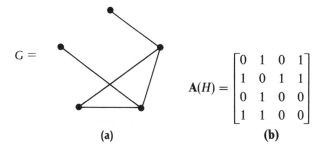

Figure 8.32

SOLUTION

a. Let the vertices of G be labeled clockwise from the top as shown in Figure 8.33(a). Then $\mathbf{A}(G)$ is displayed in Figure 8.33(b).

b. Labeling the rows and columns of the matrix 1, 2, 3, 4, as shown in Figure 8.33(c), we obtain the graph of Figure 8.33(d).

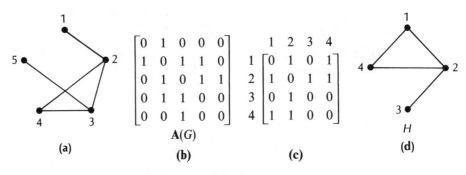

Figure 8.33

To represent a graph G in computer storage, we can store its adjacency matrix $\mathbf{A}(G)$. All information on the adjacency of vertices of G is stored in $\mathbf{A}(G)$. There is, however, a great deal of redundant information in the adjacency matrix. In a graph G, $v_i v_j$ is the same edge as $v_j v_i$. This means $a_{ij} = a_{ji}$ in $\mathbf{A}(G)$; that is, $\mathbf{A}(G)$ is symmetric. Since the value of a_{ij} is the same as that for a_{ji}, we need only the values above the diagonal of $\mathbf{A}(G)$ to determine G. If $a_{24} = 1$ then $a_{42} = 1$. We describe more efficient computer representations of graphs in exercises 13 and 14.

We now consider an interesting property of $\mathbf{A}(G)$. We first generalize the concept of a path. A **walk** of length k consists of a sequence of vertices $v_0, v_1, v_2, \ldots, v_k$ where $v_i v_{i+1} \in E(G)$ for each i, $0 \le i \le k - 1$. Thus a path is a walk that uses distinct vertices. We can use matrix multiplication to count the number of walks of a given length.

Theorem 8.7

The (i, j) entry of $\mathbf{A}^k$ is the number of walks of length k from v_i to v_j in G.

EXAMPLE 8.13 ───────────────────────────────────

Determine $\mathbf{A}$, $\mathbf{A}^2$, and $\mathbf{A}^3$ for the graph H of Figure 8.33(d).

SOLUTION

See Figure 8.34.

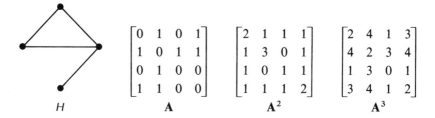

Figure 8.34

The entry 4 in $[A^3]_{1,2}$ means there are four walks of length 3 from v_1 to v_2. These walks are v_1, v_2, v_4, v_2; v_1, v_4, v_1, v_2; v_1, v_2, v_3, v_2; and v_1, v_2, v_1, v_2. The diagonal entry $[A^2]_{ii}$ counts how many sequences v_i, v_j, v_i there are. However, this is precisely the number of vertices v_j adjacent to v_i, which is the degree of v_i. The diagonal entry $[A^3]_{ii}$ counts twice the number of triangles that begin and end at v_i. The double count is caused by traversal in each direction—for example, v_1, v_2, v_4 and v_1, v_4, v_2. Each triangle in G is counted twice for each of its three vertices. Thus, it is counted six times, and we have the following.

Corollary 8.2

a. $[A^2]_{ii} =$ The degree of $v_i = d(v_i)$

b. $\left(\dfrac{1}{6}\right)\displaystyle\sum_{i=1}^{n}[A^3]_{ii} =$ The number of triangles in G, where n is the number of vertices in G.

For certain graphs G, $A(G)$ has an easily recognizable form. For example, $A(K_n)$ has 0s on the diagonal and 1s elsewhere. Bipartite graphs have the nice property described in Note 8.8.

Note 8.8

G is bipartite if and only if G can be labeled so that $A(G)$ has the form

$$\begin{bmatrix} 0 & C \\ C^t & 0 \end{bmatrix}$$

Thus, if we see two diagonal square blocks of 0s in $A(G)$, we immediately know G is bipartite.

The Incidence Matrix

Suppose G has vertex set $V(G) = \{v_1, v_2, \ldots, v_n\}$ and edge set $E(G) = \{e_1, e_2, \ldots, e_q\}$. The incidence matrix $\mathbf{B}(G)$ is the $n \times q$ $(0, 1)$-matrix defined by:

$$b_{ij} = \begin{cases} 1 & \text{if } v_i \text{ is an endpoint of } e_j \\ 0 & \text{otherwise} \end{cases}$$

Some comments are in order. Since an edge has two endpoints, there are two 1s in each column of $\mathbf{B}(G)$. Each 1 in row i of $\mathbf{B}(G)$ corresponds to an edge incident with v_i. Thus, the number of 1s in row i is the degree of v_i. If $q \neq n$, $\mathbf{B}(G)$ is not a square matrix. However, $\mathbf{BB}^t$ is square and has the same dimension $(n \times n)$ as $\mathbf{A}(G)$. There is an even stronger relationship, as shown in Note 8.9.

Note 8.9

If $\mathbf{A}$ and $\mathbf{B}$ are the adjacency and incidence matrices of G, then

$$[\mathbf{B} \cdot \mathbf{B}^t]_{ij} = \begin{cases} a_{ij} & \text{if } i \neq j \\ \text{degree of } v_i & \text{if } i = j \end{cases}$$

Should we use $\mathbf{B}(G)$ rather than $\mathbf{A}(G)$ to represent a graph in computer memory? We assume each entry of a matrix requires 1 byte of storage. If $q < n$ (so G is either a tree or disconnected), $\mathbf{B}(G)$ uses less storage (qn bytes) than $\mathbf{A}(G)$ (n^2 bytes). For most graphs, however, $q > n$ and $\mathbf{B}(G)$ wastes far more storage than $\mathbf{A}(G)$.

The Distance Matrix

The **distance between v_i and v_j**, denoted d_{ij}, is the length of the shortest path connecting v_i and v_j. If G is connected, d_{ij} is finite for every pair v_i, v_j. When G is disconnected, $d_{ij} = \infty$ for vertices in distinct components and is finite otherwise. The distance matrix $\mathbf{D}(G)$ is the $n \times n$ matrix where $[\mathbf{D}(G)]_{ij} = d_{ij}$.

EXAMPLE 8.14 _____

Find the distance matrix for the graphs in Figure 8.35.

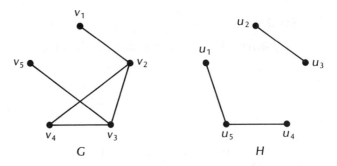

Figure 8.35

SOLUTION

$$\mathbf{D}(G) = \begin{bmatrix} 0 & 1 & 2 & 2 & 3 \\ 1 & 0 & 1 & 1 & 2 \\ 2 & 1 & 0 & 1 & 1 \\ 2 & 1 & 1 & 0 & 2 \\ 3 & 2 & 1 & 2 & 0 \end{bmatrix}, \quad \text{and} \quad \mathbf{D}(H) = \begin{bmatrix} 0 & \infty & \infty & 2 & 1 \\ \infty & 0 & 1 & \infty & \infty \\ \infty & 1 & 0 & \infty & \infty \\ 2 & \infty & \infty & 0 & 1 \\ 1 & \infty & \infty & 1 & 0 \end{bmatrix}$$

As with the adjacency matrix, the distance matrix is symmetric. Thus, if we place it in computer memory, we need only store the entries above the diagonal. Note also that $d_{ii} = 0$. That is, the diagonal entries of $\mathbf{D}(G)$ are zero.

Digraphs

Each of the matrices discussed has a counterpart for digraphs. The **adjacency matrix A(F) of a digraph** F has $a_{ij} = 1$ if the arc $v_i v_j$ is in F and $a_{ij} = 0$ otherwise. $\mathbf{A}(F)$ need not be symmetric since arc $v_i v_j$ in F does not imply $v_j v_i$ is in F. The **incidence matrix B(F) of a digraph** F is a $(0, 1, -1)$-matrix with $b_{ij} = 1$ if arc e_j points toward v_i, $b_{ij} = -1$ if arc e_j points away from v_i, and $b_{ij} = 0$ otherwise. The **distance matrix D(F) of a digraph** F has d_{ij} equal to the length of the shortest directed path from v_i to v_j (in a **directed path,** one must follow the direction of the arrows on the arcs). Note that $d_{ij} = \infty$ if there is no *directed* path from v_i to v_j.

EXAMPLE 8.15

List $\mathbf{A}(F)$, $\mathbf{B}(F)$, and $\mathbf{D}(F)$ for the digraph in Figure 8.36.

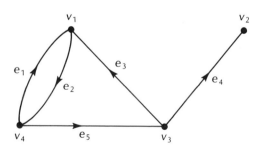

Figure 8.36

SOLUTION

$$\mathbf{A}(F) = \begin{bmatrix} 0 & 0 & 0 & 1 \\ 0 & 0 & 0 & 0 \\ 1 & 1 & 0 & 0 \\ 1 & 0 & 1 & 0 \end{bmatrix}, \quad \mathbf{B}(F) = \begin{bmatrix} 1 & -1 & 1 & 0 & 0 \\ 0 & 0 & 0 & 1 & 0 \\ 0 & 0 & -1 & -1 & 1 \\ -1 & 1 & 0 & 0 & -1 \end{bmatrix},$$

and

$$\mathbf{D}(F) = \begin{bmatrix} 0 & 3 & 2 & 1 \\ \infty & 0 & \infty & \infty \\ 1 & 1 & 0 & 2 \\ 1 & 2 & 1 & 0 \end{bmatrix}$$

EXERCISES 8.4

For exercises 1–6, use the graphs and digraphs of Figure 8.37.

1. **a.** Find $\mathbf{A}(G)$ and $\mathbf{A}(H)$.
   b. Find $[\mathbf{A}(G)]^2$ and $[\mathbf{A}(G)]^3$.
   c. How many walks of length 3 from x_1 to x_5 are there?
2. Find $\mathbf{B}(G)$ and $\mathbf{B}(H)$.
3. Find $\mathbf{D}(G)$ and $\mathbf{D}(H)$.
4. **a.** Find $\mathbf{A}(L)$ and $\mathbf{A}(M)$.
   b. What property must a digraph have for its adjacency matrix to be symmetric?
   c. A graph may be considered a special type of digraph. Consider $\mathbf{A}(H)$ from exercise 1(a). Draw the digraph to which $\mathbf{A}(H)$ corresponds.
   d. For a digraph F, what do the row sums and column sums of $\mathbf{A}(F)$ represent?
5. **a.** Find $\mathbf{B}(L)$ and $\mathbf{B}(M)$.
   b. State a property of the incidence matrix of a digraph.
6. Find $\mathbf{D}(L)$ and $\mathbf{D}(M)$.

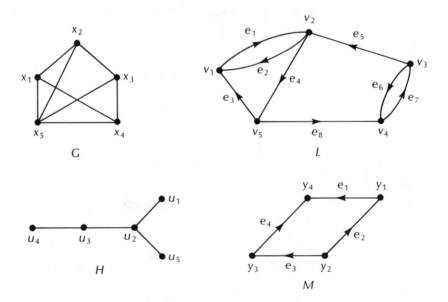

Figure 8.37

7. In Example 8.14, graph H has components of size 3 and 2. $\mathbf{D}(H)$ has 12 ∞'s.
   a. Try several examples to test the conjecture that if G has components of size n_1 and n_2, then $\mathbf{D}(G)$ has $2n_1n_2$ ∞'s.
   b. Suppose a graph G has three components whose sizes are n_1, n_2, and n_3. Find a formula for the number of ∞'s in $\mathbf{D}(G)$.

Problems and Projects

8. Prove Note 8.8.
9. Determine which graphs G satisfy $\mathbf{A}(G) = \mathbf{B}(G)$.
10. Determine which graphs G satisfy $\mathbf{A}(G) = \mathbf{D}(G)$.
11. Write a program that will input the adjacency matrix of a graph and calculate the degree of each vertex of the graph.
12. Prove: If G is bipartite, $[\mathbf{A}^3]_{ii} = 0$ for all i.
13. Let G be a graph with vertex set $V = \{v_1, v_2, \ldots, v_n\}$ and maximum degree k. The **connection matrix** $\mathbf{C}(G)$ is an $n \times k$ matrix with row i corresponding to v_i. In row i, we list the subscript of each vertex adjacent to v_i and fill the rest of the row with zeros. Determine $\mathbf{C}(G)$ and $\mathbf{C}(H)$ for the graphs G and H in Figure 8.37. [Note that $\mathbf{C}(G)$ is not a $(0, 1)$-matrix.]
14. When G has a vertex of high degree, but most vertices have low degree, a standard method is to store G as a vector. Suppose the vertices of G are labeled $v_1, v_2, \ldots, v_n$. The **connection vector** lists successively for each i ($1 \le i \le n - 1$) the subscript of vertices adjacent to v_i that are greater than i. The list for each vertex is followed by 0, which indicates the end of the adjacencies for that vertex.
   a. Find the connection vector for each of G and H in Figure 8.37.
   b. Suppose a graph has degree sequence 6, 3, 3, 3, 2, 1, 1, 1, 1, 1, 1, 1. Compare the storage requirements for the connection vector, the connection matrix, and the adjacency matrix.

8.5 CONNECTIVITY

In Chapter 4, we dealt with a computer system problem involving a power source, computer, telecommunications line, and CRT. We found that by adding redundancy to the system, the probability of system failure decreases. In this section, we examine connectivity in graphs, a concept related to redundancy of paths between vertices. Because of its applications to industrial problems, connectivity is one of the most important graph theory concepts.

If v is a vertex of G, the graph $G - v$ is formed by removing from the graph v and all edges incident with v. See Figure 8.38(a) and (b). When we remove an edge from a graph, we do not remove its endpoints. If e is an edge of G, the graph $G - e$ is formed by removing e from G. See Figure 8.38(a) and (c).

The **connectivity** $\kappa(G)$ of a graph G is the minimum number of vertices we must remove to disconnect G. The connectivity $\kappa(G)$ of graph G in Figure 8.38 is 3. If we remove v, t, and x (and their incident edges), we are left with two isolated vertices u and w. It is not possible to disconnect G by removing only two vertices.

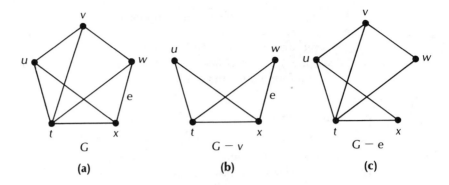

Figure 8.38

Note that $\kappa(K_n)$ is defined as $n - 1$. K_n cannot become disconnected by removing vertices and is the only such graph.

The **edge-connectivity** $\lambda(G)$ of a graph G is the minimum number of edges whose removal disconnects the graph. $\lambda(K_1)$ is defined to be 0. In Figure 8.38, $\lambda(G) = 3$ because removing edges uv, tv, and vw would disconnect G, but removing only two edges would not disconnect G.

EXAMPLE 8.16 _____

Find κ and λ for each graph in Figure 8.39.

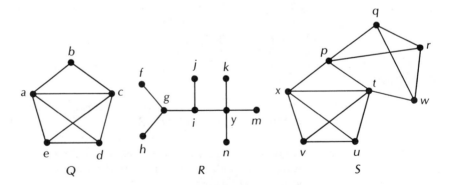

Figure 8.39

SOLUTION

$\kappa(Q) = 2$; remove a and c. $\lambda(Q) = 2$; remove ab and bc. $\kappa(R) = 1$; remove any vertex of degree greater than 1. $\lambda(R) = 1$; remove any edge. $\kappa(S) = 2$; remove p and w. $\lambda(S) = 3$; remove xp, pt, and tw.

Graphs Q and R of Figure 8.39 satisfy $\kappa = \lambda$. For graph S, we have $\kappa < \lambda$. Let $\delta(G)$ denote the **minimum degree** of vertices in G. We have the following relationship called **Whitney's Inequality.**

Theorem 8.8

For any graph G, $\kappa(G) \leq \lambda(G) \leq \delta(G)$.

We now prove the last inequality. By removing the edges adjacent to a vertex of minimum degree, that vertex becomes isolated and the graph becomes disconnected. Thus, we can disconnect the graph by removing δ edges (and possibly fewer). So $\lambda(G) \leq \delta(G)$. For the inequality $\kappa(G) \leq \lambda(G)$, see exercise 10.

Some graphs (for example, trees) are easy to disconnect. A **cutpoint** is a vertex whose removal increases the number of components. A **bridge** is an edge whose removal increases the number of components. A graph can have a cutpoint without having a bridge. (Can you give an example of such a graph?) However, we have Note 8.10.

Note 8.10

If $n \geq 3$, any component having n vertices that contains a bridge also contains a cutpoint.

EXAMPLE 8.17 _____

Identify all cutpoints and bridges in the graph of Figure 8.40.

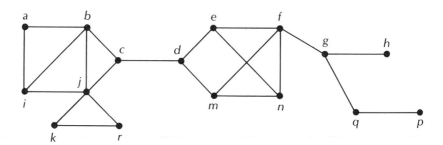

Figure 8.40

SOLUTION

The cutpoints are c, d, f, g, j, and q. Edges cd, fg, gh, gq, and qp are bridges.

The following properties of cutpoints and bridges are useful.

Theorem 8.9

a. Vertex v is a cutpoint if and only if there exist vertices x and y not equal to v such that v is on every path connecting x and y.

b. Edge e is a bridge if and only if there exist vertices a and b such that e is on every path connecting a and b.

c. Edge e is a bridge if and only if e is not contained in a cycle.

Trees can be characterized using bridges, thereby extending the list of equivalent conditions from Theorem 8.3.

Theorem 8.10

G is a tree if and only if G is connected and each edge of G is a bridge.

Remark   Connectivity is an example of a **global** property. Determining whether G is still connected after removing several vertices involves checking for paths between all remaining vertices. This is in contrast to the *degree* of a vertex, which is a **local** property. In Chapter 5, we mentioned NP-Complete problems—that is, problems for which there is no polynomial time algorithm. Determining the value of $\kappa(G)$ is an NP-Complete problem. There is, however, an efficient way to determine a good lower bound for $\kappa(G)$. G is **k-connected** if $\kappa(G) \geq k$. Thus, if $\kappa(G) = 3$, G is 3-connected. G is also 2-connected, 1-connected, and 0-connected. A theorem of Bondy uses the local property of vertex degree to obtain a lower bound on $\kappa(G)$. For details see Behzad, Chartrand, and Lesniak-Foster (1979).

Digraphs

The **underlying graph** of a digraph D is the (multi)graph obtained by removing the directions from the arcs of D. The underlying graph may be a multigraph rather than a graph, because any parallel arcs of D become multiple edges in G (see Figure 8.41).

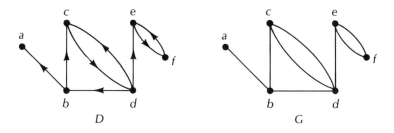

Figure 8.41.   A Digraph and Its Underlying Graph

A digraph is **connected** (also called **weakly connected**) if its underlying graph is connected (that is, if the digraph is in one piece). Recall that we use directed paths in digraphs; that is, we must follow the direction on an arc. Thus, although there is a path from f to a in the underlying graph G of Figure 8.41, there is no directed path from f to a in D (nor is there one from a to f). A digraph D is **strongly connected** if for each vertex v, there is a directed path from v to every other vertex of D. Digraph D of Figure 8.41 is connected but not strongly connected. There is no directed path, for example, from a to b.

EXAMPLE 8.18

Show the digraphs in Figure 8.42 are strongly connected.

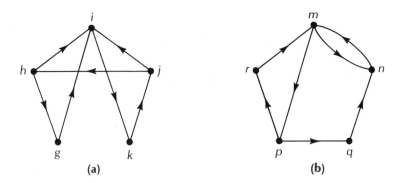

Figure 8.42

SOLUTION

a. We could list directed paths from each vertex to every other vertex. For this digraph, there is an easier way: g, i, k, j, h is a directed cycle through all the vertices (that is, a **spanning cycle**). To obtain a path between a particular pair of vertices, travel around the directed cycle: To get from k to h, use k, j, h; to get from h to k, use h, g, i, k.

b. There is no spanning cycle in this second digraph. To show the digraph is strongly connected, we first denote the directed cycle m, p, q, n by C. There is a directed path from each vertex in C to every other vertex in C. Thus, we need consider only directed paths between vertex r and vertices of C. For a path from r to a vertex of C, use arc rm and then go as far around C as necessary. For a path from a vertex in C to r, go around C (if necessary) to p, then use arc pr.

One way to check whether a digraph is strongly connected is described in Note 8.11.

Note 8.11

A digraph is strongly connected if and only if it is connected and every arc is contained in a directed cycle.

Orientable Graphs

Consider the following problem. On an island, all streets are two-way. It is, therefore, possible to travel from any intersection to any other intersection on the island. The Island Council, however, just passed a resolution declaring that all streets must be made one-way. Can one-way street signs be put up so that it will still be possible to travel from any intersection to any other intersection (without, of course, breaking the law!)?

We construct a graph model of the problem by representing each intersection by a vertex and each road connecting a pair of vertices (intersections) by an edge. Our problem can now be posed in graphical terms as follows: Given the (assumedly connected) graph G, representing the island's road system, when is it possible to assign directions to its edges so that the resulting digraph is strongly connected?

An **orientable graph** is a connected graph for which we can assign directions to its edges so that the resulting digraph is strongly connected. The assignment of directions is called an **orientation** of the graph. The following theorem solves our problem.

Theorem 8.11

A connected graph is orientable if and only if it does not contain a bridge.

EXAMPLE 8.19 _____

If the graph of Figure 8.43 is the road map of the island of our example, show how to make all streets one-way. That is, find an orientation for the graph that makes it strongly connected.

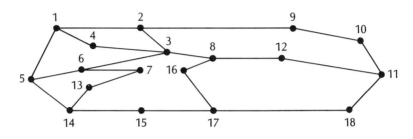

Figure 8.43

SOLUTION

Use Note 8.11 to check that the orientation in Figure 8.44 makes the graph a strongly connected digraph.

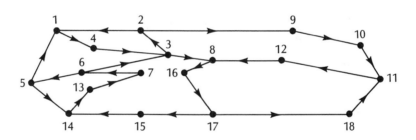

Figure 8.44

_____

EXERCISES 8.5

1. Find $\kappa(G)$ (in terms of n) for each of the following graphs:
   a. P_n **b.** C_n **c.** K_n **d.** W_n
2. Find $\lambda(G)$ (in terms of n) for each of the following graphs:
   a. P_n **b.** C_n **c.** K_n **d.** W_n
3. Find formulas for $\kappa(K_{m,n})$ and $\lambda(K_{m,n})$ in terms of m and n.
4. List all cutpoints and bridges in the graphs of Figure 8.45.
5. **a.** For each cutpoint in Figure 8.45, draw the graph obtained by deleting that cutpoint.

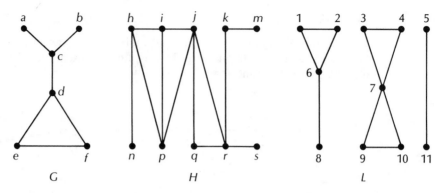

Figure 8.45

b. For each bridge in Figure 8.45, draw the graph obtained by deleting that bridge.

6. A **block** is a maximal connected subgraph having no cutpoints.
   a. Draw the blocks for each graph in Figure 8.45.
   b. Construct a graph G satisfying $\kappa(G) < \lambda(G) < \delta(G)$. (*Hint:* This can be done with a graph having two blocks, one cutpoint of degree 4, no bridges, and at least five vertices in each block.)

7. In Figure 8.44, we found a strongly connected orientation of the graph in Figure 8.43. Find an orientation that is not strongly connected.

8. Find an orientation of the graph in Figure 8.46 so that the resulting digraph is strongly connected.

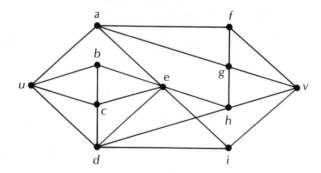

Figure 8.46

9. A digraph is **unilateral** if for each pair of vertices u and v, there is a directed path from u to v or a directed path from v to u.
   a. Draw a unilateral digraph on four vertices that is not strongly connected.
   b. Draw a connected digraph on four vertices that is not unilateral.

Problems and Projects

10. Show $\kappa(G) \leq \lambda(G)$ for all graphs G.

11. a. Prove Theorem 8.9(a).
   b. Prove Theorem 8.9(b).

12. Prove Theorem 8.10.

13. Let S be a subset of $V(G)$ and let $u, v \notin S$. S **separates** u and v if every path connecting vertices u and v passes through a vertex of S. Two paths connecting u and v are **vertex-disjoint paths** if they have no vertices in common other than u and v. **Menger's Theorem** is as follows.

Theorem 8.12

Let u and v be distinct nonadjacent vertices in G. Then the maximum number of vertex-disjoint paths connecting u and v equals the minimum number of vertices in a set that separates u and v.

Find a minimum sized separating set and a maximum set of vertex-disjoint paths connecting u and v in the graph of Figure 8.46, thereby illustrating Menger's Theorem.

14. Whitney characterized k-connected graphs as follows:

Theorem 8.13

A graph G on $n \geq 2$ vertices is k-connected if and only if for each pair of distinct vertices, there are k vertex-disjoint paths connecting them.

For each of the 15 pairs of vertices using a through f in Figure 8.46, find three vertex-disjoint paths connecting them.

15. Prove that a 3-regular (also called **cubic**) graph has a cutpoint if and only if it has a bridge.

16. For a digraph D, the (i, j)-entry of $[\mathbf{A}(D)]^k$ is the number of (*directed*) walks of length k from v_i to v_j. If D is strongly connected and has n vertices, then all entries of $[\mathbf{A}(D)]^{n-1}$ are nonzero. Write a program that inputs the adjacency matrix of a digraph and determines whether the digraph is strongly connected.

8.6 TRAVERSING GRAPHS

A familiar puzzle is to try to draw a given figure (usually an envelope) without taking your pencil off the paper. Can you draw each of the graphs in Figure 8.47 without taking your pencil off the paper? Before you start, we warn you that only

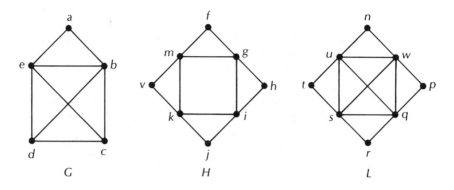

Figure 8.47

two of them can be drawn in this manner, and it sometimes matters where you start.

Eulerian Graphs

In this section, we will discuss a theorem that tells us exactly which figures we can draw without taking a pencil off the paper. We also state an algorithm that shows how to make the drawing. First, we need a concept intermediate to those of walks and paths.

Recall that a walk is a sequence of vertices $v_0, v_1, v_2, \ldots, v_k$ where $v_i v_{i+1}$ is an edge of G. In a walk, we may reuse vertices and/or edges already used. A path, however, requires that vertices (and therefore edges) be distinct. A **trail** is a walk in which the edges (but not necessarily the vertices) are distinct. Thus, a path is a special type of trail, and a trail is a special type of walk.

An **Eulerian trail** is a trail that includes each edge of the graph. (The *Eu* in Eulerian is pronounced like *oi* in oil.) The drawing puzzle asks whether a graph contains an Eulerian trial. If the trail begins and ends at the same vertex, it is said to be **closed.** A graph is **Eulerian** if it contains a closed Eulerian trail. (Some authors call a closed Eulerian trail an *Eulerian circuit*.) Thus, a graph is Eulerian if it contains a walk that includes each edge exactly once and ends at the original vertex. G is **semi-Eulerian** if it contains an Eulerian trail (the trail need not be closed).

EXAMPLE 8.20 ——————————————————————————————

Find an Eulerian trail in each of the graphs of Figure 8.47, if possible. Which graph is Eulerian?

SOLUTION

For G we have the Eulerian trail d,e,a,b,e,c,b,d,c. Other trails are possible. Each trail, however, must start at c or d. If it starts at c, it will end at d; and if it starts at d, it will end at c. G is semi-Eulerian.

For H, we have the Eulerian trial $f,g,h,i,j,k,v,m,g,i,k,m,f$. Other trails are possible. H is Eulerian.

L does not contain an Eulerian trail.

Remark   Eulerian graphs are named in honor of the famous Swiss mathematician Leonhard Euler (pronounced "oiler") who solved the Königsberg Bridge Problem in the eighteenth century. The problem asked whether it was possible to tour the town of Königsberg by crossing each of the town's seven bridges exactly once. [For more details, see Biggs, Lloyd, and Wilson (1976).] The Königsberg Bridge Problem corresponds to asking whether the multigraph in Figure 8.48 has a closed Eulerian trail. Euler proved that it does not have a closed Eulerian trail. Thus, the desired tour of the town is not possible. Euler is often called the father of graph theory, and the first theorem of graph theory (our Theorem 8.1) was proved by Euler.

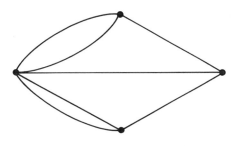

Figure 8.48

How can we tell whether a graph is Eulerian? The answer is contained in the following theorem of Euler.

Theorem 8.14

Let G be a connected graph.

a. G is Eulerian if and only if each vertex of G has even degree.

b. G is semi-Eulerian but not Eulerian if and only if G contains precisely two vertices having odd degree. Furthermore, an Eulerian trail in G must begin at one of the odd vertices and end at the other.

In Figure 8.47, G has two odd vertices (d and c), so G is semi-Eulerian; H has all even vertices, so H is Eulerian; L has four odd vertices (u, q, s, and w), so L does not have an Eulerian trail.

The following algorithm will find a closed Eulerian trail in an Eulerian graph.

Fleury's Algorithm

Suppose G is Eulerian. To find an Eulerian trail, begin at any vertex. Record and erase each edge as it is used, subject to the following condition: Never use a bridge unless there is no alternative.

For a semi-Eulerian graph, we alter the algorithm by requiring one to begin at an odd vertex.

For example, applying Fleury's Algorithm to graph H of Figure 8.47, we begin at f. Record and erase the edges in the trail f,g,m. At this point fm is a bridge and there is an alternative. We use the alternative—thus, f,g,m,v (now vk is a bridge but there is no alternative, so we use it). Thus, from the beginning $f,g,m,v,k,i,g,h,i,j,k,m,f$. Note that we use bridges only when there is no alternative.

Hamiltonian Graphs

Hamiltonian graphs are named after the famous Irish mathematician Sir William Rowan Hamilton, who invented the Around the World Game in the mid-1800s. The game involves a solid dodecahedron (commonly used nowadays for desk-calendar paperweights), a solid that has 20 vertices, 30 edges, and 12 faces (shaped like pentagons). Each vertex is labeled with the name of a well-known city. The object of the game is to find a round-the-world tour that visits each city exactly once.

A path that contains every vertex of a graph is called a **Hamiltonian path.** If a graph contains a Hamiltonian path it is called **semi-Hamiltonian. A Hamiltonian cycle** is a cycle that includes every vertex. Thus, a Hamiltonian path is a spanning path, and a Hamiltonian cycle is a spanning cycle. If G contains a Hamiltonian cycle, G is **Hamiltonian.** In Hamiltonian problems, we must pass through each vertex exactly once, whereas in Eulerian problems we must pass through each edge exactly once. It is common to confuse these two concepts. As a memory technique, remember Eulerian begins with "e" as does "edge."

EXAMPLE 8.21 _____

Draw a connected graph having four vertices that is

a. Hamiltonian and Eulerian

b. Hamiltonian and semi-Eulerian (but not Eulerian)

c. Eulerian and semi-Hamiltonian (but not Hamiltonian)

d. Eulerian but not semi-Hamiltonian

e. Hamiltonian but not semi-Eulerian

f. Neither semi-Eulerian nor semi-Hamiltonian

g. Semi-Eulerian and semi-Hamiltonian

SOLUTION

See Figure 8.49. Graphs (c) and (d) do not exist.

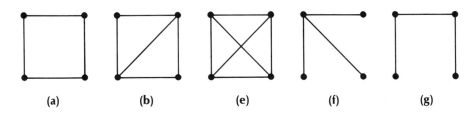

(a)   (b)   (e)   (f)   (g)

Figure 8.49

EXAMPLE 8.22

Find a Hamiltonian cycle for each graph in Figure 8.50.

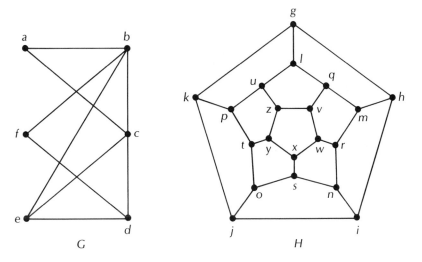

Figure 8.50

SOLUTION

A Hamiltonian cycle for G is a,b,f,d,e,c,a. Graph H is the graph of the dodecahedron. Thus, we are to find a tour in the Around the World Game. A Hamiltonian cycle in H is $g,h,i,j,k,p,t,o,s,n,r,m,q,v,w,x,y,z,u,g$.

The characterization of Eulerian graphs was given in Theorem 8.14: We need only check the degrees of a connected graph to determine whether it is Eulerian. It may be surprising, therefore, that there is no useful characterization of Hamiltonian graphs. Indeed, finding a useful characterization of Hamiltonian graphs is one of the major unsolved problems in graph theory. There are, however, quite a few sufficient conditions for Hamiltonicity—that is, theorems of the form: If G has property P then G is Hamiltonian. See exercise 11.

The Traveling Salesman Problem

Hamiltonian graphs are related to the Traveling Salesman Problem. In this problem, a salesman plans to visit various cities to show his merchandise. He would like to stop in each city once and return to the home office while minimizing his travel time. Thus, we want to find a minimum-weight Hamiltonian cycle in a weighted graph. The weights correspond to travel times.

EXAMPLE 8.23 _____

Find a minimum-weight Hamiltonian cycle for the weighted graph in Figure 8.51.

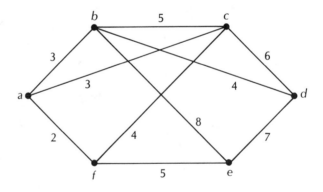

Figure 8.51

SOLUTION

We check all Hamiltonian cycles: a,b,c,d,e,f,a has weight 28; a,b,d,e,f,c,a has weight 26; a,b,e,d,c,f,a has weight 30; a,c,b,d,e,f,a has weight 26; and a,c,d,b,e,f,a has weight 28. Thus, we choose either a,b,d,e,f,c,a or a,c,b,d,e,f,a as our minimum-weight Hamiltonian cycle.

_____

Example 8.23 illustrates the difficulty of the Traveling Salesman Problem. For the small graph in Figure 8.51, we find five Hamiltonian cycles whose weights are determined. In applications, the problems are often much larger and more overwhelming. Unfortunately, there is no efficient method for solving the Traveling Salesman Problem. This is another example of an NP-Complete problem. (See Case Study 5B.)

There are many generalizations and extensions of Hamiltonicity. They involve either graphs having many Hamiltonian cycles or graphs that are "almost" Hamiltonian. Both Hamiltonian and Eulerian graphs have (follow the arrows) digraph counterparts. For discussion of these topics, see Behzad, Chartrand, and Lesniak-Foster (1979).

EXERCISES 8.6

1. Use Theorem 8.14 to determine which graphs in Figure 8.52 are Eulerian or semi-Eulerian.

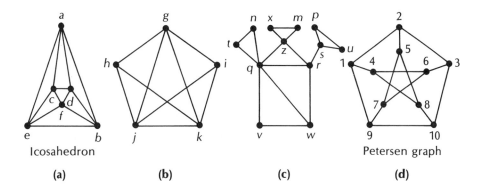

Figure 8.52

2. List an Eulerian trail for each Eulerian graph and each semi-Eulerian graph of Figure 8.52.
3. Which graphs in Figure 8.52 are Hamiltonian? semi-Hamiltonian?
4. List a Hamiltonian cycle for each Hamiltonian graph in Figure 8.52, and list a Hamiltonian path for each semi-Hamiltonian graph.
5. For which values of n are each of the following Eulerian:
   a. K_n **b.** P_n **c.** C_n **d.** W_n
6. For which values of n are each of the following Hamiltonian?
   a. K_n **b.** P_n **c.** C_n **d.** W_n
7. For which values of m and n is $K_{m,n}$
   a. Eulerian?   **b.** Hamiltonian?
8. Draw a connected graph having five vertices that is
   a. Hamiltonian and Eulerian
   b. Hamiltonian and semi-Eulerian (but not Eulerian)
   c. Eulerian and semi-Hamiltonian (but not Hamiltonian)
   d. Eulerian but not semi-Hamiltonian

   e. Hamiltonian but not semi-Eulerian

   f. neither semi-Hamiltonian nor semi-Eulerian

   g. semi-Hamiltonian and semi-Eulerian

9. Solve the Traveling Salesman Problem for each graph in Figure 8.53.

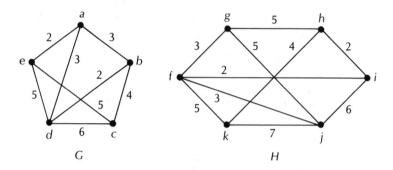

Figure 8.53

Problems and Projects

10. Prove: If G has a cutpoint, G is not Hamiltonian.

11. The following theorem of Chvátal is one of the strongest sufficient conditions for Hamiltonicity.

Theorem 8.15

Suppose G has n vertices whose degrees d_i satisfy $d_1 \le d_2 \le \cdots \le d_n$. If $d_j \le j < n/2$ implies $d_{n-j} \ge n - j$, then G is Hamiltonian.

Use Theorem 8.15 to determine whether the graphs having the following degrees are Hamiltonian:

   a. 2, 2, 2, 3, 3, 4, 6, 7, 7   **b.** 3, 3, 3, 4, 6, 6, 6, 8, 8, 9

12. Prove: If G has n vertices such that for each pair of distinct nonadjacent vertices u, v we have $d(u) + d(v) \ge n$, then G is Hamiltonian.

CASE STUDY 8A HEAP SORT

We examined and compared sorting algorithms in Case Study 6B. In this case study, we discuss the heap sort, an efficient sorting algorithm that uses trees. A **(rooted) binary tree** is a tree that has a special vertex, the **root,** whose degree is at most 2 and has all other vertices of degree at most 3. The root is drawn at the top with the tree extending downward. The distance from the root r to a vertex v is called the **level** of v and is denoted by $L(v)$. If v has level k, then the vertex adjacent to v at level $k-1$ is called the **parent** of v, and a vertex adjacent to v at level $k+1$ is called a **child** of v. The tree is called a binary tree because each vertex can have at most two children.

EXAMPLE 8.24

In the binary tree of Figure 8.54,

a. What is the root?

b. At what level is a? j? p?

c. Who is the parent of c?

d. Who are the children of g?

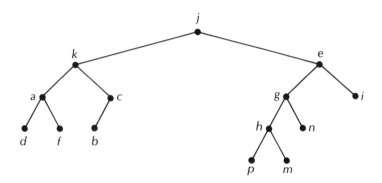

Figure 8.54

SOLUTION

a. The root is j.

b. $L(a) = 2$, $L(j) = 0$, and $L(p) = 4$.

c. k is the parent of c.

d. g has two children. They are h and n.

___

The root is the only vertex without a parent. The only vertices without children are (nonroot) vertices of degree 1 or the root if it has degree 0. A vertex without a child is called an **endpoint** or **leaf** of the tree. A **branch** at vertex v in a binary tree is a maximal subtree with root v and only one child of v. For example, the left branch at k in Figure 8.54 is the subtree induced by k, a, d, f.

The **heap sort** is a recursive algorithm that sorts a list using a binary tree. The numbers on the list become labels of the vertices. These labels are switched according to given rules until a particular type of labeling called a *heap* is obtained. The algorithm continues by converting smaller and smaller subtrees to heaps and ends with a sorted list. A **heap** is a labeling of a binary tree in which each child's label is less than or equal to the parent's label.

We begin with a binary tree we call the base tree. A **base tree** is a binary tree having as few levels as possible, which is constructed as follows: (1) Fill each level from left to right, and (2) fill each level completely before beginning a new one. We display base trees for sorting lists of n elements, $1 \le n \le 10$, in Figure 8.55. The **index** of a vertex is the smallest value of n for which the vertex appears in a base tree. Thus, the index of vertices a and b in Figure 8.55 are 5 and 3, since they first appear in base trees for $n = 5$ and $n = 3$, respectively.

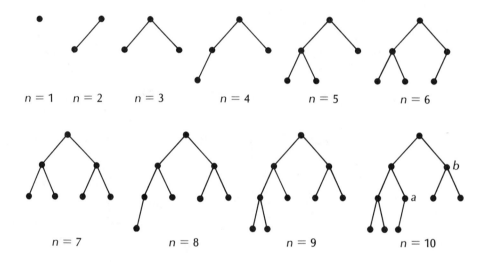

Figure 8.55   Base Trees

Creating a Heap

To create a heap for a k-element list, we begin by labeling the successive vertices of the base tree for $n = k$ with the list elements. To distinguish list elements from indices, we place list elements inside the vertices and indices outside the vertices. See Figure 8.56(a). For a number x, the greatest integer less than or equal to x is denoted $[x]$. Thus, $[4.13] = 4$, $[7] = 7$, and $[-1.3] = -2$. Note that if the tree has k vertices, the parents have indices $1, 2, \ldots, [k/2]$. Now, compare the label of the parent having the highest index to the labels of its children, and use the following process:

Step 1: If a parent's label is less than its child's label, switch the parent's label with its highest labeled child.

Step 2: Each time a switch is made, compare an element moved down in the tree to its new children (if any) using step 1.

Step 3: Repeat steps 1 and 2 successively for the parent having the next smaller index until the root has been processed.

When we reach the root (vertex 1) and no further switches are required, we have a heap.

EXAMPLE 8.25

Create the initial heap for the list $-1, 5, 2, 3, 1, 7, 4$.

SOLUTION

The successive binary trees for the procedure are shown in Figure 8.56. The indices are underlined at the right of each vertex; the number in the circle is the label for the related list element of the vertex. To get tree (b), 2 is the label of the highest indexed parent and is compared to its children (with labels 7 and 4) and switched with 7. The next highest indexed parent 5 (with index 2) is compared to its children (3 and 1) and no switch is required. Now, -1 (with index 1) is compared to its children in tree (b) and is switched with 7. Since -1 is moved down in the tree, we must perform step 2. Thus, -1 is compared to its new children (2 and 4) in tree (c) and switched with 4, giving tree (d). Since the roots has been processed, we have a heap.

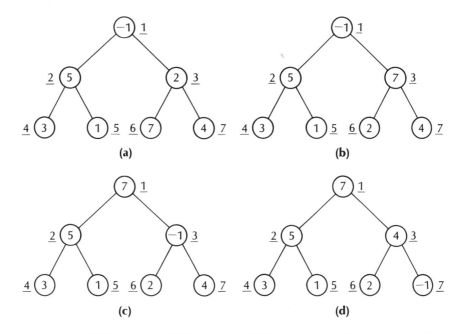

Figure 8.56

EXAMPLE 8.26

Write procedures to create a heap.

SOLUTION

We use two procedures, one of which is recursive. The main program will call the procedure CREATE to create a heap. The procedure CREATE calls the recursive procedure FIX each time we want to compare a new, lower indexed parent to its children. The procedure FIX calls itself after each switch is made to compare an element moved down in the tree to its new children (if any). The inputs to CREATE are the length of the list M and the list itself B. The inputs to FIX are the parent's number P, the length of the list K, and the list itself B.

```
procedure create (m: integer; var b: passlist);
var  i,j: integer;
begin
    j:=trunc(m/2);
    for i:=j downto 1 do
        fix(i,m,b)
end;
```

In the procedure FIX, LC and RC are the vertex indices of the left child and right child, respectively.

```
procedure fix(p,k:integer; var b: passlist);
var v: real;
begin
    v:=b[p];
    lc:=2*p;
    rc:=2*p+1;
    if lc<=k then
        if v<b[lc] then
            if rc<=k then
                if b[lc]<b[rc] then
                    begin
                        b[p]:=b[rc];
                        b[rc]:=v;
                        fix (rc,k,b)
                    end
                else
                    begin
                        b[p]:=b[lc];
                        b[lc]:=v;
                        fix(lc,k,b)
                    end
            else
                begin
                    b[p]:=b[lc];
                    b[lc]:=v;
                    fix(lc,k,b)
                end
        else
            if (rc<=k) and (v<b[rc]) then
                begin
                    b[p]:=b[rc];
                    b[rc]:=v;
                    fix(rc,k,b)
                end
end;
```

Sorting the Heap

To complete the heap sort algorithm:

1. Switch the root with the highest indexed vertex (remaining).
2. Suppress the new highest indexed vertex as if it were eliminated from the tree.
3. Create a heap from the smaller tree.
4. Repeat steps 1, 2, and 3 until there is only one vertex remaining.

   Step 3 is required because after steps 1 and 2, the smaller tree is usually not a heap.

EXAMPLE 8.27

Begin with the heap in Figure 8.56(d) and show the heap sort for −1, 5, 2, 3, 1, 7, 4.

SOLUTION

We illustrate the suppressed vertices with dotted circles in Figure 8.57. We switch 7 and −1, and suppress 7 to get (b). Switch −1 and 5, then 3 and −1 to get the heap (c). Switch 5 and 2, and suppress 5 to get (d). Switch 2 and 4 to get a heap; switch 4 and 1, and suppress 4 to get (e). Switch 1 and 3 to get a heap; switch 3 and −1, and suppress 3 to get (f). Switch −1 and 2, switch 2 and −1, and suppress 2 to get (g). Switch −1 and 1, switch 1 and −1, and suppress 1 to get (h). Since there is only one

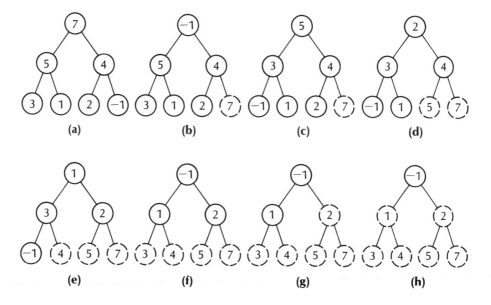

Figure 8.57

vertex remaining, the list is sorted. Read the successive vertex labels in (h) in order of the indices.

EXAMPLE 8.28

Write a program to sort a list A using heap sort.

SOLUTION

We write a program that calls the procedure CREATE, which in turn calls the procedure FIX, both from Example 8.26.

```
program heap(input,output);
type passlist = array[1..10] of real;
var p,k,n,j,i,m,lc,rc: integer;
    big,v :real;
    a: passlist;


          (*  insert procedure fix here   *)

          (*  insert procedure create here   *)


begin
    writeln('input a positive integer');
    read(n);
    if  n>10 then
        writeln('array is too large')
    else
       begin
          writeln('input',n,' real numbers in any order');
          for i:=1 to n do
              read (a[i]);
          m:=n;
          while m>=2 do
             begin
                create (m,a);
                big:=a[1];
                a[1]:=a[m];
                a[m]:=big;
                m:=m-1;
             end;
          writeln('the sorted array is as follows');
          for i:=1 to n do
              write(a[i])
       end
end.
```

Efficiency Comparison

In Case Study 6B, we compared the order of efficiency for several sorting algorithms: the method of successive minima, insertion sort, bubble sort, and quicksort. We can now include heap sort in the comparison. In Table 8.1, we list the order of efficiency for the average case (that is, for a randomly generated array) and the worst case for

	ORDER OF EFFICIENCY	
Sorting Algorithm	Average Case	Worst Case
Successive Minima	n^2	n^2
Insertion Sort	n^2	n^2
Bubble Sort	n^2	n^2
Quicksort	$n \cdot \log_2 n$	n^2
Heap Sort	$n \cdot \log_2 n$	$n \cdot \log_2 n$

Table 8.1

the five sorting methods we have discussed. [For a derivation of the orders of efficiency, see, for example, Scheid (1982).]

We expended a great deal of effort to sort the list in Example 8.27. Although less efficient, the bubble sort and selection sort (see Case Study 6B) seem easier to apply than heap sort. How can we explain this apparent contradiction? From Table 8.1, we see that the average order of efficiency for bubble sort is n^2, while that for heap sort is $n \cdot \log_2 n$. Recall from Chapter 5 that constants are eliminated when stating orders of efficiency. The *actual* average number of comparisons for bubble sort is $(1/4)n^2$. When $n = 7$, $(1/4)n^2 \approx 12$, whereas $n \cdot \log_2 n \approx 20$. In the bubble sort of Example 6.45, we used 18 comparisons, and in the heap sort of Example 8.27 we used (including forming the heap) 28 comparisons. (We used a slightly different list in Example 8.25 and Example 8.27 than that used in Example 6.45 in order to illustrate the case where a switch is made at step 2 in the create-a-heap process. A heap sort for the list of Example 6.45 would use 28 comparisons as well.) Again we ask: How can heap sort be the more efficient method? The answer lies in size. The usual sorting problem involves thousands of items, not seven. For $n = 1000$, $(1/4)n^2 = 250,000$ for bubble sort, whereas $n \cdot \log_2 n \approx 9966$ for heap sort!

EXERCISES: CASE STUDY 8A

1. Draw the base tree for:
   a. $n = 12$ **b.** $n = 15$ **c.** $n = 19$
2. Show the steps in creating the initial heap for:
   a. 5, 7, −2, 8, 4, 1   **b.** 20, 13, 14, 17, 12, 16, 11, 18
3. Show the steps in performing a heap sort for the lists in exercise 2 using a process similar to that we used in Example 8.27.
4. **a.** If $n = 100$, how many comparisons do you expect with bubble sort? With heap sort?
   b. If $n = 8000$, how many comparisons do you expect with bubble sort? With heap sort?

Problems and Projects

5. a. Run the program from Example 8.28 with the lists in exercise 2.

    b. Run your program again with the list 5, −2, 1, 7, 18, 11, 3, 4, 7, 9, 21, 18, 15, 13, 2, 6. (You will have to make a small adjustment in two lines of the program.)

6. A **ternary tree** has a root with degree at most 3 and each other vertex with degree at most 4.

    a. Draw base ternary trees for $n = 1, 2, \ldots, 15$.

    b. How would the programs of this case study differ if heap sort were based on ternary trees?

CASE STUDY 8B   THE CRITICAL PATH METHOD

Every large project consists of many activities. The manager of such a project must make time estimates for these activities and must define the precedence relations among them. For example, on a construction project the wallboard cannot be installed before the electrical wiring. The walls cannot be painted before the wallboard is installed, and so on.

A digraph called an activity digraph can be used to model the times and precedence relations among activities. To consider an example, suppose you are opening a new diner. The activities you must complete before opening are listed in Table 8.2.

Activity	Description	Immediate Predecessors	Duration in Days
START	—	—	0
A	Meet lawyer	START	3
B	Meet accountant	A	2
C	Sign lease	A	4
D	Hook up electricity	B, C	1
E	Buy stock	B, C	3
F	Hook up natural gas	B, C	1
G	Get state tax number	B	1
H	Get license	G	1
I	Set up equipment	D, E, F	2
J	City health inspection	I	1
FINISH	—	H, J	0

Table 8.2

The **activity digraph** is constructed as follows. Each vertex represents an activity. If activity i is an immediate **predecessor** of activity j, then a directed edge is drawn from vertex *i* to vertex *j*. A vertex *k* is a **successor** of vertex *i* if there is an edge in the activity digraph directed from vertex *i* to vertex *k*. One vertex in the activity digraph, the **START vertex**, represents the start of the project and, thus, has no immediate predecessors. One vertex in the activity digraph, the **FINISH vertex,** represents the finish of the project and, thus, has no successors. The duration of each activity is written beside its vertex. The activity digraph for opening the diner is shown in Figure 8.58.

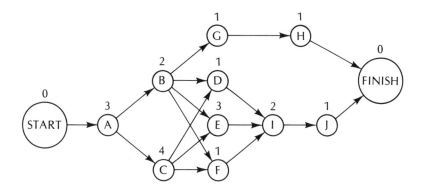

Figure 8.58

The questions we wish to answer are:

1. What is the earliest we can finish this project? This time is called the **project completion time.**
2. What is the earliest time each activity can start and finish?
3. What is the latest time each activity can start without jeopardizing the project completion time?
4. Which activities are critical in the sense that any delay in their completion will cause a delay in the entire project?

The technique we shall use to answer these questions is called the **CPM,** or **critical path method,** and was developed in 1957 by E. I. du Pont de Nemours and Company to control construction projects. The first step in CPM is to identify the activities involved in the project, their duration and each activity's immediate predecessors. The second step is to construct an activity digraph for the project. We have completed both of these steps.

The next step is to use the activity digraph to determine the earliest start and earliest finish times. The **earliest start time (EST)** of an activity is the earliest time an activity can begin, assuming all preceding activities are completed as soon as possible.

The **earliest finish time (EFT)** of an activity is the EST for the activity plus the activity's duration. To find the EST and EFT for each activity we make a **forward pass** through the activity digraph and use the rules in Note 8.12.

Note 8.12 Rules for Computing EST and EFT

1. EST of start $= 0$
2. EFT of a vertex $=$ EST of the vertex $+$ duration of the activity
3. EST of a vertex $=$ Maximum of the EFTs of all the vertex's immediate
   predecessors

Applying Note 8.12 to our activity digraph, the EST of the start is 0. The EFT of the start is, therefore, 0. The EST of A is the maximum of the EFTs of all its predecessors. Vertex A has only one predecessor, the start, so its EST is 0. The EFT of A is $0 + 3 = 3$. The ESTs for B and C are 3 because each has only one predecessor, namely A. See Table 8.3, columns 3 and 4. The EFTs for B and C are 5 and 7, respectively. Vertex D has two immediate predecessors, namely B and C, with EFTs of 5 and 7. By rule 3 of Note 8.12, the EST of D is 7, the maximum of 5 and 7. Continuing in this manner, we complete columns 3 and 4 of the table. Note that the EST of the finish, 13, equals the EFT of the finish. This is the project completion time—the earliest the project can possibly be completed (Note 8.13).

(1) Activity	(2) Duration	(3) EST	(4) EFT	(5) LST	(6) LFT	(7) Slack Time
START	0	0	0	0	0	0
A	3	0	3	0	3	0
B	2	3	5	5	7	2
C	4	3	7	3	7	0
D	1	7	8	9	10	2
E	3	7	10	7	10	0
F	1	7	8	9	10	2
G	1	5	6	11	12	6
H	1	6	7	12	13	6
I	2	10	12	10	12	0
J	1	12	13	12	13	0
FINISH	0	13	13	13	13	0

Table 8.3

Note 8.13

The EFT of the finish vertex is the project completion time.

The next step in the CPM is to determine the latest start time and the latest finish time for each activity. The **latest start time (LST)** of an activity is the latest time the activity can start without causing delay in the project completion time. The **latest finish time (LFT)** of an activity is the latest time an activity can be completed without causing delay in the project completion time. To compute the LST and LFT of each activity, we make a **backward pass** through the activity digraph and use the rules in Note 8.14.

Note 8.14   Rules for Computing LST and LFT

1. LFT of finish = Project completion time
2. LST of a vertex = LFT of the vertex — duration of the activity
3. LFT of a vertex = Minimum of the LSTs of all the successors of the vertex

---

Applying Note 8.14 to our activity digraph, the LFT and LST of the finish vertex are 13. Since vertices H and J have only one successor, the finish, their LFTs are the LST of the finish, namely 13. Continuing backward in the activity digraph, we fill in columns 5 and 6 of Table 8.3. Note, in particular, that vertex B has four successors, D, E, F and G, having LSTs of 9, 7, 9, and 11, respectively. By rule 3 of Note 8.14, the LFT of B is 7.

The final step in the CPM is to determine the slack time for each activity. The **slack time** of an activity is the time we can delay the start of the activity beyond its EST without delaying the project completion time. The slack time of an activity can be computed by either formula in Note 8.15.

Note 8.15   Computing Slack Time

The slack time of an activity can be computed by either of the following:

1. Slack time = LST — EST
2. Slack time = LFT — EFT

Column 7 of Table 8.3 contains the slack times for each activity in our example. The slack time for activity F, for example, is 2. This means that we can start F as early as day 7 or as late as day 9 without jeopardizing the project completion time. That is, we have a 2-day leeway or slack in starting the activity.

Of particular interest are those activities having a slack time of 0. These activities cannot be delayed beyond their EST without jeopardizing the project completion time. An activity is called a **critical activity** if its slack time is 0. Note that, by definition, the start and finish activities are always critical. The critical activities in our example are START, A, C, E, I, J, FINISH. A directed path from start to finish consisting of all critical activities is called a **critical path** (there may be several critical paths). If any activity on a critical path is delayed, the project will be delayed beyond its earliest completion time.

Since the slack time for every critical activity is zero, we have Note 8.16.

Note 8.16

The project completion time equals the sum of the durations of the activities on a critical path.

In our example, the sum of the durations on the critical path is

$$0 + 3 + 4 + 3 + 2 + 1 + 0 = 13.$$

We summarize the steps in the CPM in Note 8.17.

Note 8.17   Outline of the CPM

1. Identify the activities, their durations, and their immediate predecessors.
2. Construct the activity digraph.
3. Using the activity digraph, compute the EST and EFT for each activity (See Note 8.12).
4. Using the activity digraph, compute the LST and LFT for each activity (See Note 8.14).
5. Compute the slack time for each activity (See Note 8.15).
6. Identify all critical activities and the critical paths.

Remark   A technique related to CPM is **PERT** (Program Evaluation and Review Technique). CPM assumes the durations of a project's activities can be estimated with relative certainty. This is the case in training, construction, maintenance, and advertising projects. Other types of projects, such as research projects and computer software development, consist of activities whose durations cannot be estimated with a great degree of certainty. PERT considers each activity duration time as a random variable (see Chapter 4). Estimates are made of each activity's duration and an activity digraph is constructed. The

EST, EFT, LST, LFT, slack times, and critical paths are computed as described in this case study. Probability calculations are then made to estimate the project completion time. The interested reader is referred to Anderson (1982).

EXERCISES: CASE STUDY 8B

1. For the data in Table 8.4.
   a. Construct the activity digraph.
   b. For each activity, compute EST, EFT, LST, LFT, and slack time.
   c. Find a critical path and the project completion time.

Activity	Immediate Predecessors	Duration
START	—	0
A	—	2
B	—	3
C	A	5
D	A, B	4
E	C, D	6
F	B, C	2
FINISH	E, F	0

Table 8.4

2. For the activity digraph in Figure 8.59,
   a. Compute EST, EFT, LST, LFT, and the slack time for each activity.
   b. Find a critical path and the project completion time.

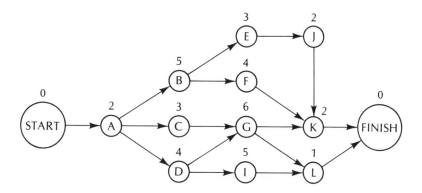

Figure 8.59

3. For the activity digraph in Figure 8.60,
   a. Compute EST, EFT, LST, LFT, and the slack time for each activity.
   b. Find a critical path and the project completion time.

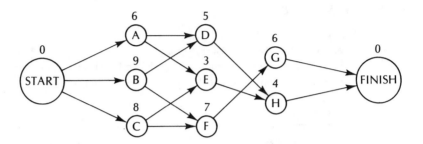

Figure 8.60

4. An investment company prepares an annual report for its customers. This report will be mailed to all customers, stockholders, and selected individuals and organizations. The activities to be performed, their immediate predecessors, and durations are shown in Table 8.5.
   a. What is the fewest number of weeks in which the report can be prepared?
   b. What are the critical activities? Find a critical path.

Activity	Description	Immediate Predecessors	Duration in Weeks
START	—	—	0
A	Decide on a theme for the report	START	1
B	Have articles written for the report	A	4
C	Get art work for the report	A	2
D	Lay out the report	B, C	2
E	Prepare mailing list	D	1
F	Proofread and send first draft to the printer	D	3
G	Make final changes and send to printer	F	2
H	Get final report from printer and send to mailroom	E, G	1
I	Send out report	H	2
FINISH	—	I	0

Table 8.5

5. Not satisfied with the quality of new housing, you decide to be your own contractor and build your new home yourself by subcontracting the various phases of the project. The activities you must perform, their immediate predecessors and durations are contained in Table 8.6.
   a. Find the minimum number of weeks for the completion of the project.
   b. Find the critical activities and a critical path.
   c. Develop a schedule to tell when contractors should arrive at the site to do their work.

Activity	Description	Immediate Predecessors	Duration in Weeks
START	—	—	0
A	Purchase land	START	2
B	Clear land	A	1
C	Build foundation	B	1
D	Build frame and exterior walls	C	2
E	Rough electrical wiring	D	1
F	Rough plumbing	D	2
G	Finish interior walls	E, F	1
H	Finish exterior	D	2
I	Finish electric and plumbing	G	1
J	Paint interior	I	2
K	Landscape	H	1
L	Carpet and furnish	J	2
FINISH	—	K, L	0

Table 8.6

Problems and Projects

6. In an activity digraph, the **time-length** of a directed path from start to finish is defined as the sum of the durations of the activities on the path.
   a. Prove Theorem 8.16.

Theorem 8.16

The time-length of a critical path in an activity digraph is the maximum time-length of all the directed paths from START to FINISH.

---

b. Theorem 8.16 gives another method for finding the critical path in an activity digraph. List all directed paths from START to FINISH and find their time-lengths. A critical path is a path of maximum time-length. Use this method to find a critical path in the activity digraph of Figure 8.61.

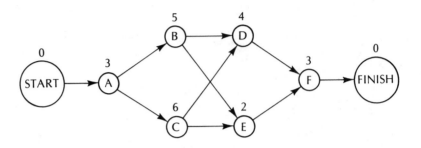

Figure 8.61

c. Explain why this technique for finding a critical path is not suitable for large digraphs.

7. a. Explain why an activity digraph cannot contain a directed cycle.

b. Explain why an activity digraph cannot contain the digraph of Figure 8.62.

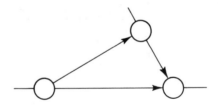

Figure 8.62

REFERENCES

Anderson, M. Q. *Quantitative Management Decision Making: With Models and Applications.* Monterey, Calif.: Brooks/Cole, 1982.

Behzad, M., Chartrand, G., and Lesniak-Foster, L. *Graphs and Digraphs.* Belmont, Calif.: Wadsworth, 1979.

Belford, G. G., and Liu, C. L. *Pascal.* New York: McGraw-Hill, 1984.

Biggs, N. L., Lloyd, E. K., and Wilson, R. J. *Graph Theory 1736–1936.* Oxford: Clarendon, 1976.

Chartrand, G. *Graphs as Mathematical Models.* Boston: Prindle, Weber & Schmidt, 1977.

Harary, F. *Graph Theory.* Reading, Mass.: Addison-Wesley, 1969.

Scheid, F. *Computers and Programming.* New York: McGraw-Hill, 1982.

Wagner, K. Bemerkungen zum vierfarben problem. *Jber. Deutsch. Math. Verein.* 46 (1936): 21–22.

Wilson, R. J. *Introduction to Graph Theory.* 2d ed. New York: Academic Press, 1979.

APPENDIX ON PASCAL

Recently, Pascal has become one of the most popular languages for teaching programming. It is also being used widely in business and industry. We discuss several of the basic features and keywords of Pascal in this appendix.

OPERATIONS AND EXPRESSIONS

Addition, subtraction, multiplication, and division are denoted by $+$, $-$, $*$, and $/$, respectively, in Pascal. The rules of precedence from algebra, which are used to decide which operations to perform first in an algebraic expression, are also used in Pascal. Thus, we first multiply or divide (from left to right), then add or subtract (from left to right). In the expression

$$A + B * C$$

the multiplication $B * C$ is performed first. If A, B, and C have values 6, 10, and 2, the value of $A + B * C$ is 26. Just as in algebra, parentheses are used to override the rules of precedence. If we wanted to perform the addition first, we would write $(A + B) * C$.

DECLARING VARIABLES

In Pascal, the type (integer, real, Boolean, or character) of all variables must be declared at the beginning of the program. The declaration is begun with the reserved word VAR, followed by a list of the names of all variables of one type, followed by a colon, the type, and a semicolon. All variables of a second type (if any) are then listed, followed by a colon, the type, and a semicolon. Here is an example.

VAR   AVER,AREA,BASE,ALTITUDE: REAL;

          NUMBER,INDEX,K,J: INTEGER;

          CODE: CHAR;

Note that the reserved word VAR is only used once. The variable CODE will be used to store a character. The type CHAR denotes a character variable and can store one character.

In many problems we need a list of numbers (or characters). Such a list is called an **array.** Suppose, for example, we store the integer test grades for a class of 30 students in an array called GRADES. We declare the array as follows:

VAR GRADES: ARRAY [1 .. 30] OF INTEGER;

If other variables were already declared, we would omit the word VAR and simply include the line

GRADES: ARRAY [1 .. 30] OF INTEGER;

following the description of the other variables. Note that *square brackets,* not parentheses, follow the word ARRAY. The numbers 1 and 30 inside the brackets are separated by two consecutive periods. These numbers describe the range of subscripts for the array variable GRADES. We refer to a particular element of the array by following the array name with the element number enclosed in square brackets. Thus, the twelfth element is GRADES[12].

Two-dimensional arrays are described similarly. Thus

VAR Q:ARRAY [1 .. 5,1 .. 12] OF INTEGER;

is the declaration for a two-dimensional array (a matrix) with 5 rows and 12 columns. The element in row 2, column 7 is Q[2, 7].

ASSIGNMENT STATEMENTS

The assignment symbol in Pascal consists of a colon followed by an equal sign. To assign 7 to the variable K we write

$$K := 7;$$

Note that K is a variable whose storage space has been allocated by the compiler because we declared K in the VAR section of the program. If K had not been declared, we would receive an error message from the compiler.

On the right side of the assignment symbol, we place the value or expression whose calculated value will be assigned to the variable to the left of the assignment symbol. Thus, the sequence of statements

BASE := 10;

ALTITUDE := 4.2;

AREA := 0.5 * BASE * ALTITUDE

will cause 10 to be stored in BASE, 4.2 to be stored in ALTITUDE, and 21 to be stored in AREA. The last statement is, of course, the familiar formula for the area of a triangle. When an assignment statement is used, the value of each variable to the right of := must already be known to the computer. Thus, the value of the variables named to the right of := must have either been assigned, calculated, or read in earlier.

You may have noticed that some lines end with a semicolon and some do not. There are certain rules in Pascal that govern when a semicolon is used. For example, a statement immediately preceding an END statement does not require a semicolon.

INPUT AND OUTPUT

To read in the value of a variable ALTITUDE use the statement

READ(ALTITUDE)

Of course, ALTITUDE must have been declared earlier in the VAR section of the program. Many versions of Pascal are interactive, so we input data directly from the terminal while the program is running. This is in contrast to other versions of Pascal, where the data is read from punched cards or from previously stored files. If the program runs interactively, we should prompt before executing a read by having the computer type out a message telling us to input the data. To print something on the terminal, we use WRITELN to print on a new line and WRITE to print on the same line. Thus, to read in the value of ALTITUDE directly from the terminal, we prompt first and then read in the value of ALTITUDE.

WRITELN('INPUT THE VALUE OF ALTITUDE');

READ(ALTITUDE)

To print a message, we must enclose the message in quotes. Note that *both* the opening and closing quote marks are *single* quotes *curving forward*. If a double quote or a backward curving quote mark (some keyboards have these) is used, an error message results. To print the *value* of ALTITUDE we use

WRITELN(ALTITUDE)

We can identify the value by including an appropriate quoted message in the statement.

WRITELN('ALTITUDE=', ALTITUDE)

Note that the quoted message is separated from the variable name by a comma. We can print out the values of the variables BASE, ALTITUDE, and AREA by using the statement

WRITELN(BASE,ALTITUDE,AREA)

Analogous to the situation with the assignment statement, the value of each variable must have already been determined before it is used in a WRITELN or WRITE statement.

BEGIN BLOCKS

A sequence of statements to be performed together as a group is enclosed in a **BEGIN block** (also called a **compound statement**), which starts with the reserved word BEGIN, is followed by the sequence of statements, and is completed by the word END. The word BEGIN never has a semicolon after it, while END follows the usual rules for semicolons except that the last END in the program is followed by a period.

THE PROGRAM STATEMENT

The **PROGRAM statement** consists of the reserved word PROGRAM followed by the program name as in

PROGRAM TRIANGLE;

If the program will print out information using WRITE or WRITELN, we must tell this to the computer immediately by attaching the word OUTPUT in parentheses following the program name. Thus,

PROGRAM TRIANGLE(OUTPUT);

If data is read in, we also include the word INPUT in the parentheses as in

PROGRAM TRIANGLE(INPUT,OUTPUT);

We are now ready for a simple program. The following program reads in the values of BASE and ALTITUDE and calculates the area of a triangle. Study this program carefully. Notice the declarations, the prompt for the READ statement, the assignment statement, the BEGIN block, and the WRITELN statements.

```
program triangle(input,output);
var base,altitude,area: real;
begin
   writeln('input the base and then the altitude');
   read(base, altitude);
   area:=0.5*base*altitude;
   writeln('base=',base,'  altitude=',altitude);
   writeln('the area of the triangle is', area);
end.
```

The word BEGIN and the final WRITELN statement are not followed by a semicolon; the (final) word END is followed by a period.

IF-THEN-ELSE

Sometimes we want to perform certain actions only when a certain condition is true. For example, in the sample program just given, the area would be calculated even if we input negative numbers for BASE and ALTITUDE. We can build into the program a test to check that the data values are positive numbers by using an IF statement. There are two general types of IF statements: the one-way IF (sometimes called IF-THEN), and the two-way IF (sometimes called IF-THEN-ELSE). The format for the one-way IF is

IF *condition* THEN

statement or BEGIN block

The condition usually has the form

XXXXX   R   YYYYY

where XXXXX and YYYYY denote a variable, number, or expression, and R denotes a relational operator. The relational operators are =, <, >, <>, <=, >=. These are read equals, less than, greater than, not equal, less than or equal, and greater than or equal, respectively. The statement following THEN is executed if the condition is true; otherwise, the statement is skipped. If we want a whole sequence of statements to be performed when the condition is true, we must enclose those statements within a separate BEGIN block. An example using a single statement is

IF INDEX <= 10 THEN

NUMBER := NUMBER + 1;

An example using several statements is

IF K <> (J-INDEX) THEN

BEGIN

K := K + 1;

WRITELN('INPUT ANOTHER NUMBER');

READ(NUMBER)

END;

Note that there is no punctuation following THEN. There is a semicolon following END in the one-way IF, provided the statement following it is not an END statement. The format for the two-way IF is

IF *condition* THEN

statement or BEGIN block

ELSE

statement or BEGIN block

If the condition is true, the statement (or BEGIN block) following THEN is executed and the ELSE part is skipped. If the condition is false, the statement (or BEGIN block) following ELSE is executed and the THEN part is skipped.

We can form **compound conditions** by enclosing individual conditions in parentheses and connecting the conditions by AND or OR. To **negate** a condition, we enclose it in parentheses and precede the parentheses with the word NOT. With minor adjustments to our TRIANGLE program, we can check the input data so we calculate the area only when BASE and ALTITUDE are positive numbers:

```
program triangle(input,output);
var base,altitude,area: real;
begin
   writeln('input the base and then the altitude');
   read(base, altitude);
   if (base>0) and (altitude>0) then
      begin
         area:=0.5*base*altitude;
         writeln('base=',base,'  altitude=',altitude);
         writeln('the area of the triangle is', area)
      end
   else
      writeln('bad data')
end.
```

Note that the END of the inner BEGIN block is not followed by a semicolon. This is because the next word is ELSE. Here is the second instance where a semicolon is omitted. Thus, even when THEN is followed by a single statement in a two-way IF, that statement does not end with a semicolon because it is followed by ELSE.

ITERATIVE DO LOOPS—THE FOR STATEMENT

There are many problems in which certain calculations or other procedures are performed repeatedly. As a simple example, suppose we want to add the integers between 5 and 500 inclusive. This can be accomplished with the FOR statement in Pascal, as follows:

NUMBER := 0;

FOR J := 5 TO 500 DO

NUMBER := NUMBER + J;

In the FOR statement, the value of J is initially set to 5 and the statement NUMBER := NUMBER + J is executed, making NUMBER equal 5. Then J is incremented by 1 making J equal to 6 and the statement NUMBER := NUMBER + J is executed again, making NUMBER equal 11. Then J becomes 7 and NUMBER becomes 18, J becomes 8 and NUMBER becomes 26, and so on until J becomes 500 and NUMBER becomes 125,240. The program would then proceed with the statement beyond NUMBER := NUMBER + J;.

In this example, the number 5 is called the **initial value** and 500 is the **final value** of the variable J. The variable J is called the **control variable** (also the **counter** or **index**).

We mentioned arrays briefly when we discussed declaring variables. The FOR statement is used most often to process arrays. Recall that GRADES was an array of 30 integers corresponding to the grades for a class of students. We can input the grades as follows:

WRITELN('INPUT THE 30 GRADES FOR THE CLASS');

FOR J := 1 TO 30 DO

READ(GRADES[J]);

Note the square brackets used for the array. Also, we prompt for all 30 grades before the FOR statement. If we want to calculate the class average we can use the statements

NUMBER := 0;

FOR J := 1 TO 30 DO

NUMBER := NUMBER + GRADES[J];

AVER := NUMBER/30;

In this example, we *initialize* NUMBER to 0 because NUMBER is the variable where we will accumulate the sum of the grades. The statement NUMBER := NUMBER + GRADES[J]; is performed 30 times because J begins at 1 and runs through each integer up to and including 30. Thus, the grades of the first student, second student, . . . , and thirtieth student will be accumulated in NUMBER. The computer will then move on to process the statement AVER := NUMBER/30;, thereby calculating the class average.

To process a sequence of statements rather than a single statement we use a BEGIN block. This was done for the one-way and two-way IF statements. The technique is the same with the FOR statement. Thus we can input the grades, count the number of grades below 60 as they are read in, and accumulate the sum of all grades by using a single FOR statement and a BEGIN block.

```
INDEX := 0;
NUMBER := 0;
WRITELN('INPUT THE 30 GRADES');
FOR J := 1 TO 30 DO
    BEGIN
        READ(GRADES[J]);
        IF GRADES[J] < 60 THEN
            INDEX := INDEX + 1;
        NUMBER := NUMBER + GRADES[J]
END;
```

In some problems it is useful to have the control variable decrease in value from its highest to its lowest limiting values. This is accomplished easily by listing the largest value after :=, changing TO to DOWNTO, and then listing the lowest value. Thus, the initial value becomes the final value and vice versa. Instead of

```
FOR J := 5 TO 500 DO
    NUMBER := NUMBER + J;
```

we write

```
FOR J := 500 DOWNTO 5 DO
    NUMBER := NUMBER + J;
```

These two FOR statements have the same effect. Note that DOWNTO is *one word*. The initial value and final value of the control variable could be either variables or expressions, as long as the values of the variables or expressions are determined before the FOR statement is reached.

THE WHILE STATEMENT

The **WHILE statement** is used to execute a certain statement or BEGIN block repeatedly as long as some given condition is true. The format for the WHILE statement is

WHILE *condition* DO

statement or *BEGIN block*

When execution reaches the WHILE statement, the condition is tested. If the condition is true, the statement or BEGIN block will be executed and the condition tested again. If the condition is still true, statement* or the BEGIN block will be executed again. The process continues until the condition is tested and found to be false. Execution then proceeds to the statement following statement∗ or the BEGIN block.

The WHILE statement is often used when reading and processing an unknown number of data items, as in the following example:

```
WRITELN('INPUT A NUMBER');
READ(X);
WHILE X <> 9999 DO
    BEGIN
        .
        .
        .

        WRITELN('INPUT A NUMBER');
        READ(X)
    END;
```

Note that X is read *before* we enter the loop; otherwise, when the condition is tested X would have no value and an error message would be received. The number 9999 is used as a sentinel to tell when the data is finished. The second READ is the last statement in the BEGIN block, so that when 9999 is finally input, it will not be processed because the condition $X <> 9999$, which is tested next, is false and the program exits the loop.

To see if you understand how the WHILE statement works, try to determine the output of the following Pascal program fragment where X, Y, and W denote integer variables.

```
X := 6;
WHILE X < 15 DO
    BEGIN
        Y := 4 ∗ X;
        X := X + 6;
        W := 3 ∗ X;
        X := X - 2;
        WRITE('Y =',Y,'W =',W)
    END;
WRITE('X =',X);
```

The program fragment would have the following output:

Y = 24   W = 36   Y = 40   W = 48   Y = 56   W = 60   X = 18

If you trace the program fragment, you will notice that X becomes greater than 15 several times but the looping continues because the condition is only tested when we first enter the loop and when we reach the *end* of the BEGIN block and loop back.

As with the one-way and two-way IF statements, the condition in a WHILE statement could be a compound condition formed by connecting two or more conditions using AND, OR, or NOT.

BUILT-IN FUNCTIONS

There are several standard functions available in Pascal. It may be useful to know the following built-in functions for exercises in this text: ABS, DIV, MOD, MAXINT, EXP, LN, ROUND, TRUNC, SQR, SQRT. All these functions except DIV, MOD, and MAXINT have a single argument, which could be a variable, a constant, or an expression. The functions are used on the right side of an assignment statement as in

$$Y := ABS(X);$$
$$W := EXP(A - B);$$
$$K := ROUND(X);$$
$$Z := SQRT(SQR(B) - 4 * A * C)$$

The functions DIV and MOD each have two integer arguments and the result is an integer. $Y := MAXINT$ has no argument since it is a built-in constant (function). MAXINT is machine- and compiler-dependent and the result is an integer. DIV and MOD are integer operators (functions) that require two operands. The expression

$$INDEX := J\ DIV\ K;$$

stores the quotient (the whole number part) of J divided by K in INDEX. The expression

$$NUMBER := J\ MOD\ K;$$

stores the remainder in NUMBER when J is divided by K. Thus, if J is 31 and K is 4, then INDEX is 7 while NUMBER is 3.

The role of each of the functions is described in the following table:

Function	Role
ABS(X)	Finds the absolute value of X
J DIV K	Finds the quotient of J divided by K
J MOD K	Finds the remainder when J is divided by K
EXP(X)	Finds e raised to the power X, that is, e^x
MAXINT	Finds the largest representable integer
LN(X)	Finds the logarithm, base e, of X—that is, ln (X)
ROUND(X)	Rounds X to the nearest integer
TRUNC(X)	Finds the whole number part of X
SQR(X)	Finds the square of X, that is, X^2
SQRT(X)	Finds the square root of X, that is, $\sqrt{X}$

See Case Study 5A for a further discussion of functions.

SETS AND SET OPERATIONS

Pascal is one of the few computer languages that implements sets and operations on sets. Before a set is used or manipulated, the elements must be specified. To specify a set variable, first use the **TYPE statement** to designate the universal set. Then describe the set variable as being a set of that type (that is, a subset of the universal set). Thus, we use the TYPE statement to specify a power set and the VAR statement to state that a particular variable is from that power set. For example, suppose we want to use WEEKEND to denote the set {SATURDAY, SUNDAY}, WEEKDAY the other days of the week, and NDAYS to denote the days containing an "N." To do so, we use

```
TYPE
     DAY = (SUNDAY,MONDAY,TUESDAY,WEDNESDAY,
              THURSDAY,FRIDAY,SATURDAY);
     DAYSET = SET OF DAY;
VAR NDAYS, WEEKDAY, WEEKEND: DAYSET;
BEGIN
     WEEKEND := [SATURDAY,SUNDAY];
     WEEKDAY := [MONDAY .. FRIDAY];
     NDAYS := [SUNDAY,MONDAY,WEDNESDAY]
```

Note that the elements of a set are enclosed in *square brackets* in Pascal. The empty set is denoted [ ].

The operations of set **union** and **intersection** are denoted by + and * in Pascal. Thus, NDAYS + WEEKEND is the set [SATURDAY,SUNDAY,MONDAY,WEDNESDAY], while NDAYS * WEEKEND is the set [SUNDAY]. **Complements** can be obtained using the difference operator, −. In the following fragment, LARGE becomes the set $[10, 11, 12, \ldots, 50]$.

```
TYPE
     NUMBER = 1 .. 50;
     A = SET OF NUMBER;
VAR UNIVERSE,LARGE,SMALL: A;
     NEXT: NUMBER;
BEGIN
     SMALL := [1 .. 9];
     UNIVERSE := [1 .. 50];
     LARGE := UNIVERSE − SMALL;
```

The difference operator can also be used outside the context of complementation. Thus $[3, 5, 8] - [4, 5, 9]$ is the set $[3, 8]$.

The set relational operators $<$, $>$, $=$, $<=$, $>=$, and $<>$ are used to compare sets. The symbol $<$ for sets in Pascal corresponds to the relation "is a proper subset of" in mathematics, while $< \, >$ means not equal ($\neq$), and $< \, =$ means $\subset$. Thus,"

$$[1,5] < [1,2,5,9] \qquad \text{is TRUE}$$
$$[1,5] < [1,5] \qquad \text{is FALSE}$$
$$[1,5] < \, = [1,2,5,9] \qquad \text{is TRUE}$$
$$[1,5] < \, > [\,] \qquad \text{is TRUE}$$

Set membership ($\in$) is denoted by IN in Pascal. Therefore, 5 IN $[1,5]$ is TRUE, whereas 2 IN $[1,5]$ is FALSE.

The only troublesome part of dealing with sets in Pascal is printing them. In fact, with most compilers, sets of only certain restricted types can be printed. Almost all versions of Pascal allow printing a set whose universe is a small (less than 512 elements) subrange of the positive integers. Some also allow printing a set whose universe is a small (less than 50) set of single characters. Even when a set can be printed, however, there is still work to be done because a set cannot be used as an operand in a WRITE statement. Thus, to print a set, we must print each element individually. The printing can be handled by running through the elements of the universal set using the built-in function **SUCC**, which gives the successor, that is, the next element in the set. We test to see whether each element is in the set to be printed. If it is, we print the element and remove it from the set. We continue testing, printing, and removing until the empty set is obtained. We illustrate this process in the following program, which prints the set large = $\{x \,|\, x \in N, \, 10 \leq x \leq 50\}$.

```
program sets2(input,output);
type
   number = 1..50;
   a = set of number;
var universe,large,small: a;
   next: number;
begin
   small:= [1..9];
   universe:= [1..50];
   large:= universe - small;
   next:= 1;
   while large <> [] do
      begin
         if next in large then
            begin
               write(next);
               large:= large - [next]
            end;
         if next <> 50 then next:= succ(next);
      end
end.
```

PROCEDURES

A **procedure** is a group of statements that will perform a particular subtask in our program. A procedure is placed at the beginning of a program following the VAR statements and begins with a PROCEDURE statement having the form

PROCEDURE name;

A procedure may also have a *formal* parameter list containing value parameter names and/or variable parameter names. A **value parameter** is a parameter whose *value* will be passed to the called procedure. A **variable parameter** is a parameter whose *storage location* is passed to the called procedure so the called procedure can retrieve or alter the value of the variable in the calling procedure. Some examples of procedure statements with formal parameter lists are as follows:

PROCEDURE COUNTING (INDEX,NUMBER: INTEGER);
PROCEDURE TRIG (SIDE1,SIDE2,SIDE3: INTEGER; VAR AREA:REAL;
                   VAR RIGHT: INTEGER);
PROCEDURE TEST (VAR A,B:REAL);

The procedure COUNTING has only value parameters. The procedure TRIG has three value parameters: SIDE1, SIDE2, and SIDE3; and two variable (VAR) parameters. The procedure TEST has two variable (VAR) parameters: A and B.

To use a procedure that has no parameter list, we **call** (sometimes referred to as **invoke** or **activate**) the procedure by stating its name, such as

HEADING;

if HEADING is the name of the procedure. When a procedure has a parameter list, we call the procedure by stating the name of the procedure followed by the *actual parameter list*. The actual parameter for a variable parameter can only be a variable name. The actual parameter for a value parameter can be an expression (including constants, variables, and arithmetic operators and function names). For example,

TRIG(X,2.3,Z + 9,A,B);

or

TEST(U,W);

are valid while

TEST(62.5,W);

is invalid because 62.5 is not a variable name.

As an example of using procedures, we present a program with two procedures: HEADING and TRIG. The first procedure will be called once to print the heading for a table of output values. The procedure TRIG will be called as long as there is still data to process. The input to the procedure TRIG (which has three value parameters and two variable parameters) is passed by way of the three value parameters, representing the three sides of a triangle. The output of TRIG is passed back to the main procedure by way of the variable parameters AREA and RIGHT. If the values input do not represent a triangle, RIGHT is set to -1. Otherwise, the value of RIGHT is set to 1 if the triangle is a right triangle and set to 0 when the triangle has no right angle.

```
program sides (input,output);
var small1,small2,number,i,j,signal: integer;
    s,area: real;
    a: array[1..10,1..3] of integer;
procedure trig(side1,side2,side3:integer;
                 var area:real;var right:integer);
begin
    if (side1+side2)<=side3 then
      right:=-1
    else
      begin
         s:=(side1+side2+side3)/2;
            area:=sqrt(s*(s-side1)*(s-side2)*(s-side3));
            if (sqr(side1)+sqr(side2))=sqr(side3) then
               right:=1
            else
               right:=0
      end
end;

procedure heading;
begin
    writeln ('our program reads in 3 numbers and determines');
    writeln ('if they are the sides of a triangle. if so');
    writeln ('our program determines the area and determines');
    writeln ('whether we have a right triangle');
    writeln;
    writeln ('side 1','side 2','side 3','area','right triangle')
end;

begin    (*main program*)
    writeln('input the number of triangles');
    read (number);
    if number>10 then
      writeln ('too many triangles')
    else
      begin
         writeln ('input', 3*number,'positive integers');
         writeln ('each group of 3 are the sides of a triangle');
         for i:=1 to number do
            for j:=1 to 3 do
               read(a[i,j]);
         heading;
         for i:=1 to number do
           begin
              writeln;
              if (a[i,1]<=0) or (a[i,2]<=0) or (a[i,3]<=0) then
                 writeln (a[i,1],a[i,2],a[i,3],' not a triangle')
              else
                begin
                   if (a[i,1]>=a[i,2]) and (a[i,1]>=a[i,3]) then
                      trig(a[i,2],a[i,3],a[i,1],area, signal)
                   else
                      if (a[i,2]>=a[i,1]) and (a[i,2]>=a[i,3]) then
                         trig(a[i,1],a[i,3],a[i,2],area,signal)
                      else
                         trig(a[i,1],a[i,2],a[i,3],area,signal);
                   if signal =-1 then
                      writeln (a[i,1],a[i,2],a[i,3],' not a triangle')
                   else
                      if signal=0 then
                         writeln(a[i,1],a[i,2],a[i,3],area,'  no')
                      else
                         writeln(a[i,1],a[i,2],a[i,3],area,'  yes')
                end
           end
      end
end.
```

RECURSIVE PROCEDURES

A **recursive procedure** is a procedure that calls itself. Such procedures are available in Pascal and do not need to be declared in any special way. At the appropriate places within the procedure, simply call the procedure in the usual way; that is, state the name of the procedure and the values of the value parameters and list the variable parameters. Following is a recursive procedure to calculate the exponential N^X, where N and X are positive integers.

```
procedure powers(n,x:integer; var result:integer);
begin
            if x=1 then
              result:= n*result
            else
               begin
                  result:= result*n;
                  powers(n,x-1,result)
               end
end;
```

Answers to Selected Odd-Numbered Problems

CHAPTER 1

Exercises 1.1

1.

17	10001	25	11001
18	10010	26	11010
19	10011	27	11011
20	10100	28	11100
21	10101	29	11101
22	10110	30	11110
23	10111	31	11111
24	11000	32	100000

3. a. 100110 **c.** 10001110 **5. a.** 45 **c.** 461 **7. a.** 10001 **c.** 1001
9. a. 1011 **c.** 1011
11. a. i. 0.171875 **ii.** 1.65625 **b. i.** 100111.01101 **ii.** 11.101

Exercises 1.2

1. a. 201_8 or 81_{16} **c.** 641_8 or $1A1_{16}$ **3. a.** 555_8
5. a. 110001_2 **c.** 110001111_2 **e.** 1101000001_2 **7. a.** 6253_8
9. a. 2652_8 **c.** $ED28_{16}$ **11. a.** 112_8
13. a. i. 7.46875 **b. i.** 41.44_8 23.9_{16}

Exercises 1.3

1. 1, 2, 3, 4, 6, 9, 12, 18, 36; 1, 2, 4, 31, 62, 124
3. 8 and 3; 4 and 21; 18 and 9
7. a. False **c.** True **9. a.** 28 **c.** 15

Exercises 1.4

1. a. 0.375 **c.** $0.\overline{6}$ **e.** 0.35 **3.** There is none!
5. a. The additive inverses are -4, $+7$, -2, $+5$, $+6/7$.
   b. The multiplicative inverses are $1/8$, $-1/2$, $5/4$, $-9/4$.
7. No

Exercises: Case Study 1A

1. 64;   4096;   68719476736
7. When adding numbers of opposite signs, we subtract the smaller absolute value from the larger. Therefore, the sum must be smaller in absolute value than the absolute values of the numbers being added.

Exercises: Case Study 1B

1. **a.** 0.51237×10^2 **c.** -0.893×10^{-7}
3. **a.** $+0.239 \times 10^2 = 23.9$ **c.** $-0.269 \times 10^4 = -2690$
5. The two smallest numbers are 0.100×10^{-5} and 0.101×10^{-5}. They differ by $0.001 \times 10^{-5} = 0.00000001$.
7. $ab000$, where a is 0 or 1 and b is any digit will suffice to represent 0.
9. **a.** -50 to 49   **c.** 0.9999×10^{49} **e. i.** 0.3795×10^{-24} **iii.** 0.3141×10^1
   g. With truncation: 0.533745; With rounding: 0533746

Exercises: Case Study 1C

1. **a.** 0.138×10^3 **b.** 0.406×10^{-1} **3. a.** Overflow   **b.** -0.383×10^{-1}
5. **a.** $yz = 0.164 \times 10^0$, $x(yz) = 0.437 \times 10^{-1}$; $xy = 0.251 \times 10^0$, $(xy)z = 0.439 \times 10^{-1}$
7. **a.** -0.100×10^{-5} **b.** 0.200×10^{-5} **c.** absolute error $= 0.3 \times 10^{-5}$
                                                                         relative error $= 1.5$
                                                                         percent error $= 150\%$
9. **a.** 0.000999   **b.** 0.0133   **c.** It is 13 times greater.

Exercises: Case Study 1D

1.
Character		Hex	Binary
a.	C	C3	11000011
b.	c	83	10000011
c.	#	7B	01111011
d.	4	F4	11110100
e.	,	6B	01101011

3. **a.** +   **c.** D   5. The first line of the dump is: I hope that you

CHAPTER 2

Exercises 2.1

1. The collection of tall people, the collection of difficult subjects, and the collection of expensive cars
3. **a.** $\{5, 10, 15, 20, \ldots\}$ **b.** $\{1, 2, 3, 4, 5, 6\}$ **c.** $\{21, 22, 23, \ldots, 75\}$
5. Sets a, b, c, and f are finite.

Exercises 2.2

1. Statements a, c, and d are true.
3. $\mathcal{P}(C)$ contains $2^4 = 16$ sets. They are ϕ, $\{1\}$, $\{2\}$, $\{3\}$, $\{4\}$, $\{1, 2\}$, $\{1, 3\}$, $\{1, 4\}$, $\{2, 3\}$, $\{2, 4\}$, $\{3, 4\}$, $\{1, 2, 3\}$, $\{1, 2, 4\}$, $\{2, 3, 4\}$, $\{1, 3, 4\}$, $\{1, 2, 3, 4\}$.
5. The first set contains the element 6, but the second does not.
7. Yes. If $A \subset B$ and $C \subset A$, then $C \subset B$. Thus each such C is a set in $\mathcal{P}(B)$, so $\mathcal{P}(A) \subset \mathcal{P}(B)$.

Exercises 2.3

1.

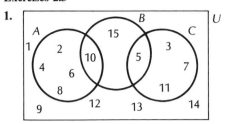

3. **a.** 2, 3, and 5
  b. 1 and 8
  c. −1, −2, and −5
  d. d and f are empty.

e.

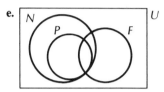

Exercises 2.4

1. **a.** $\{2, 6, 8, 12, 15\}$ **b.** $\{6, 12, 15\}$ **c.** $\{4, 6, 8, 12\}$ **d.** $\{10\}$
3. **a.** $\{a: a = 2b \ \text{ or } \ a = 3b, \ b \in N\}$ **b.** $\{c: c = 6d, \ d \in N\}$
  c. $\{2, 3, 4, 6, 8, 9, 10, 12\}$; $\{6, 12, 18, 24, 30, 36, 42, 48\}$
  d. $\{a: a \equiv 0 \bmod 2 \ \text{ or } \ a \equiv 0 \bmod 3, \ a \in N\}$; $\{c: c \equiv 0 \bmod 6, \ c \in N\}$
  e. $\{f: f \equiv 1 \text{ or } 2 \bmod 3, f \in N\}$
5. **a.**

A' $\qquad\qquad$ $(A')'$

7. **a.** $A \cap (B \cup C)$ **b.** $C \cap A'$ **c.** $C \cup (A \cap B)$ **d.** $(A \cup C)'$
11. **a.** $B \in A$, $\ B \subset A$, $\ C \notin A$, $\ C \not\subset A$
  b. $\mathscr{P}(A) = \{\phi, 1, \{1\}, \{1, 2\}, \{1, \{1\}\}, \{1, \{1, 2\}\}, \{\{1\}, \{1, 2\}\}, \{1, \{1\}, \{1, 2\}\}\}$
  c. $W = \{\phi, \{\phi\}, \{\{\phi\}\}, \{\{\{\phi\}\}\}, \dots\}$

Exercises 2.5

1. c and e are true statements; d is a false statement.
3. **a.** $p \wedge \sim q$ **b.** $\sim(\sim p \vee \sim q)$ **c.** $p \vee q$
  d. Either I don't drive a Chevy or I smoke.   **e.** Either I smoke or I don't smoke.
  f. I do not smoke and drive.
5. **a.**

p	q	s	$\sim q$	$p \wedge \sim q$	s	$(p \wedge \sim q) \vee s$
T	T	T	F	F	T	T
T	T	F	F	F	F	F
T	F	T	T	T	T	T
T	F	F	T	T	F	T
F	T	T	F	F	T	T
F	T	F	F	F	F	F
F	F	T	T	F	T	T
F	F	F	T	F	F	F

c.

p	q	s	t	$p \vee q \vee s$	t	$(p \vee q \vee s) \wedge t$
T	T	T	T	T	T	T
T	T	T	F	T	F	F
T	T	F	T	T	T	T
T	T	F	F	T	F	F
T	F	T	T	T	T	T
T	F	T	F	T	F	F
T	F	F	T	T	T	T
T	F	F	F	T	F	F
F	T	T	T·	T	T	T
F	T	T	F	T	F	F
F	T	F	T	T	T	T
F	T	F	F	T	F	F
F	F	T	T	T	T	T
F	F	T	F	T	F	F
F	F	F	T	F	T	F
F	F	F	F	F	F	F

Exercises 2.6

1. a. $p \rightarrow \sim q$ **b.** $\sim(p \vee q) \rightarrow \sim p$ **c.** $q \leftrightarrow \sim p$

3. a.

a	b	c	$a \vee b$	$\sim c$	$a \vee b \leftrightarrow \sim c$
T	T	T	T	F	F
T	T	F	T	T	T
T	F	T	T	F	F
T	F	F	T	T	T
F	T	T	T	F	F
F	T	F	T	T	T
F	F	T	F	F	T
F	F	F	F	T	F

c.

a	b	$\sim b$	$a \wedge \sim b$	$\sim a$	b	$\sim a \wedge b$	$a \wedge \sim b \leftrightarrow \sim a \wedge b$
T	T	F	F	F	T	F	T
T	F	T	T	F	F	F	F
F	T	F	F	T	T	T	F
F	F	T	F	T	F	F	T

f.

a	b	$\sim a$	b	$\sim a \wedge b$	$\sim(\sim a \wedge b)$	$b \vee \sim a$	$\sim(\sim a \wedge b) \rightarrow b \vee \sim a$
T	T	F	T	F	T	T	T
T	F	F	F	F	T	F	F
F	T	T	T	T	F	T	T
F	F	T	F	F	T	T	T

Exercises 2.7

1. Every entry in the last column of each truth table should be T.

a.

p	$\sim p$	$p \vee \sim p$
T	F	T
F	T	T

c.

p	q	$p \leftrightarrow q$	$\sim p$	$(p \leftrightarrow q) \wedge \sim p$	$\sim q$	$((p \leftrightarrow q) \wedge \sim p) \to \sim q$
T	T	T	F	F	F	T
T	F	F	F	F	T	T
F	T	F	T	F	F	T
F	F	T	T	T	T	T

e.

p	q	$\sim q$	$p \vee \sim q$	$\sim p$	q	$\sim p \wedge q$	$\sim(\sim p \wedge q)$	$p \vee \sim q \leftrightarrow \sim(\sim p \wedge q)$
T	T	F	T	F	T	F	T	T
T	F	T	T	F	F	F	T	T
F	T	F	F	T	T	T	F	T
F	F	T	T	T	F	F	T	T

3. Part a will be shown. The others are done analogously.

a. This is equivalent to showing that

$$a \vee (b \wedge c) \leftrightarrow (a \vee b) \wedge (a \vee c)$$

is a tautology.

a	b	c	$b \wedge c$	$a \vee (b \wedge c)$	$a \vee b$	$a \vee c$	$(a \vee b) \wedge (a \vee c)$
T	T	T	T	T	T	T	T
T	T	F	F	T	T	T	T
T	F	T	F	T	T	T	T
T	F	F	F	T	T	T	T
F	T	T	T	T	T	T	T
F	T	F	F	F	T	F	F
F	F	T	F	F	F	T	F
F	F	F	F	F	F	F	F

Since columns 5 and 8 are identical,

$$a \vee (b \wedge c) \leftrightarrow (a \vee b) \wedge (a \vee c)$$

is a tautology.

5. Show that $\sim(p \wedge q) \leftrightarrow \sim p \vee \sim q$ is a tautology.

p	q	$p \wedge q$	$\sim(p \wedge q)$	$\sim p$	$\sim q$	$\sim p \vee \sim q$	$\sim(p \wedge q) \leftrightarrow \sim p \vee \sim q$
T	T	T	F	F	F	F	T
T	F	F	T	F	T	T	T
F	T	F	T	T	F	T	T
F	F	F	T	T	T	T	T

7. a.

p	q	$p \to q$	$\sim p$	q	$\sim p \vee q$	$(p \to q) \leftrightarrow (\sim p \vee q)$
T	T	T	F	T	T	T
T	F	F	F	F	F	T
F	T	T	T	T	T	T
F	F	T	T	F	T	T

c.

p	q	$p \leftrightarrow q$	$p \to q$	p	$q \to p$	$(p \to q) \wedge (q \to p)$
T	T	T	T	T	T	T
T	F	F	F	T	T	F
F	T	F	T	F	F	F
F	F	T	T	F	T	T

Since columns 3 and 7 are identical,

$$(p \leftrightarrow q) \leftrightarrow ((p \to q) \wedge (q \to p))$$

is a tautology.

9. *Inverse* **a.** If x is not odd, then x^2 is not odd.   **c.** $x > y \rightarrow f(x) > f(y)$
    Converse **a.** If x^2 is odd, then x is odd.   **c.** $f(x) \leq f(y) \rightarrow x \leq y$
    Contrapositive **a.** If x^2 is not odd, then x is not odd.   **c.** $f(x) > f(y) \rightarrow x > y$

Exercises 2.8

1. **a.** Valid   **b.** Valid   **c.** Valid   **d.** Fallacious
3. The argument may be written as $\sim e \rightarrow g, e \vee f \vdash f \rightarrow g$. Use a truth table to show it is fallacious.
5. Conditions
    and Actions

Raining	T	T	T	T	F	F	F	F
Sunny	T	T	F	F	T	T	F	F
Above 95	T	F	T	F	T	F	T	F
Beach						✓		✓
Movies	✓	✓	✓	✓	✓		✓	

Exercises: Case Study 2A

1. 4, 7, 12, 8, 5, 6, 9;   4, 5, 12, 8, 7, 6, 9;   4, 5, 6, 8, 7, 12, 9;   4, 5, 6, 7, 8, 12, 9;
    4, 5, 6, 7, 8, 9, 12;   4, 5, 6, 7, 8, 9, 12
3. 100, higher;   150, lower;   125, lower;   112, higher;   118, correct
7. **a.** $p \wedge (\sim q \wedge \sim s)$ **b.** $((p \vee q) \vee \sim s) \wedge \sim q$ **c.** $(p \wedge q) \vee (\sim q \wedge s)$
9. **a.** (NOT P) OR Q   **b.** (NOT P) AND (NOT Q)   **c.** (P AND Q) AND (S OR T)
    d. P OR ((NOT Q) AND R)

CHAPTER 3

Exercises 3.1

1. **a.** $\displaystyle\sum_{i=10}^{17} (i-8) = \sum_{j=2}^{9} j$ by Property 3.1   **c.** $\displaystyle\sum_{k=1}^{50} 9 = 50(9)$ by Property 3.5

    e. $\displaystyle\sum_{i=36}^{50} 4(i-30) = 4 \sum_{i=36}^{50} (i-30) = \sum_{j=6}^{20} j$ by Properties 3.3 and 3.1

3. **a.** $\displaystyle\sum_{i=1}^{20} 6i$ or $6 \sum_{i=1}^{20} i$ **b.** $\displaystyle\sum_{j=3}^{10} (3j+1)$ or $\sum_{k=4}^{11} (3k-2)$ **c.** $4 \displaystyle\sum_{i=1}^{15} i^2$

5. Rewrite the sum on the left as $4 \displaystyle\sum_{i=1}^{n} i^2$. Use induction as in exercise 4 to show

    $\displaystyle\sum_{i=1}^{n} i^2 = n(n-1)(2n+1)/6$. Multiplying by 4 gives $2n(n-1)(2n+1)/3$.

7. We must show $\displaystyle\sum_{i=1}^{n} [1/(i(i+1))] = n/(n+1)$. When $n = 1$, each side of the equation yields $1/2$. Assume the formula is true for $n = k$, and show it is true for $n = k + 1$:

    $$\sum_{i=1}^{k+1} [1/(i(i+1))] = 1/((k+1)(k+2)) + \sum_{i=1}^{k} [1/(i(i+1))]$$
    $$= 1/((k+1)(k+2)) + k/(k+1) = (k+1)/(k+2)$$

9. When $n = 1$, $2(1) + 1 = 3 \leq 3^1$. Assume that $2k + 1 \leq 3^k$, and show that $2(k+1) + 1 \leq 3^{k+1}$:

    $$2(k+1) + 1 = 2k + 1 + 2 \leq 3^k + 2 \leq 3^k + 3^k = 2 \cdot 3^k \leq 3 \cdot 3^k = 3^{k+1}$$

11. Write the sum as $\sum\limits_{i=1}^{n} (4i - 1)$. When $n = 1$, $\sum\limits_{i=1}^{1} (4i - 1) = 3 = 1(2(1) + 1)$. Assume that

$\sum\limits_{i=1}^{k} (4i - 1) = k(2k + 1)$, and show that $\sum\limits_{i=1}^{k+1} (4i - 1) = (k + 1)(2k + 3)$:

$$\sum\limits_{i=1}^{k+1} (4i - 1) = 4(k + 1) - 1 + \sum\limits_{i=1}^{k} (4i - 1) = 4k + 3 + k(2k + 1)$$

$$= 2k^2 + 5k + 3 = (k + 1)(2k + 3)$$

13. **a.** For $n = 2$, $(A_0 \cup A_1)' = A_0' \cap A_1'$, which is one of De Morgan's laws for two sets. Assume $(A_0 \cup A_1 \cup \cdots \cup A_{k-1})' = A_0' \cap A_1' \cap \cdots \cap A_{k-1}'$ and show $(A_0 \cup A_1 \cup \cdots \cup A_k)' = A_0' \cap A_1' \cap \cdots \cap A_k'$:

$$(A_0 \cup A_1 \cup \cdots \cup A_k)' = ((A_0 \cup A_1 \cup \cdots \cup A_{k-1}) \cup A_k)'$$

$$= (A_0 \cup A_1 \cup \cdots \cup A_{k-1})' \cap A_k'$$

$$= (A_0' \cap A_1' \cap \cdots \cap A_{k-1}') \cap A_k'$$

$$= A_0' \cap A_1' \cap \cdots \cap A_k'$$

Exercises 3.2

1. 40   **3.** 6; *abc, acb, bac, bca, cab, cba*
5. $9!/(2! \cdot 2! \cdot 2!) - 1 = 45,359$ (assuming he uses all nine letters)
7. a. $12!/8!$ **b.** $30!/28!$ **9.** 90,000   **11.** 12,960   **13.** $5! \, 3! \, 4! \, 3! = 103,680$
15. 3   **19.** 250,000

Exercises 3.3

1. a. 120   **b.** 4005   **c.** 0.3553   **3.** $\binom{16}{4} = 1820$

5. a. $\binom{26}{5} = 65,780$ **b.** $26\binom{25}{3} = 59,800$

7. a. $\binom{10}{5} = 252$ **b.** $\binom{10}{4} = 210$ **c.** $\binom{10}{5}/2 = 126$ **9.** $4\binom{13}{5} = 5148$

11. a. $\binom{6}{2}\binom{9}{2} = 540$ **b.** $\binom{6}{4} + \binom{9}{4} = 141$ **13.** $\binom{12}{10} = 66$

15. $\binom{25}{3} + 25\binom{25}{2} = 9800$

Exercises 3.4

1. $a^4 + 8a^3 + 24a^2 + 32a + 16$ **3.** $\binom{10}{6} = 210$ **5.** $\binom{5}{3}(-2)^3 = -80$

7. $\binom{5}{0}100^5 + \binom{5}{1}100^4 + \binom{5}{2}100^3 + \binom{5}{3}100^2 + \binom{5}{4}100 + \binom{5}{5} = 10,510,100,501$

11. $\binom{2n}{2} = 2n(2n - 1)/2 = 2n^2 - n = n^2 + 2n(n - 1)/2 = 2\binom{n}{2} + n^2$

13. $\binom{n}{1} + 6\binom{n}{2} + 6\binom{n}{3} = n + 6(n(n - 1)/2) + 6(n(n - 1)(n - 2)/6)$

$$= n + 3n^2 - 3n + n^3 - 3n^2 + 2n = n^3$$

15. $x^4 + y^4 + z^4 + 4x^3y + 4xy^3 + 4x^3z + 4xz^3 + 4yz^3 + 4y^3z + 12xyz^2 + 12xy^2z + 12x^2yz + 6x^2y^2 + 6x^2z^2 + 6y^2z^2$

17. $(7!/(2! \cdot 3! \cdot 2!))2^3 \cdot 3^2 = 15,120$

Exercises 3.5

1. $3^4 = 81$ **3. a.** 1   **b.** 1   **c.** $\binom{4}{2} = 6$

5. $\binom{15+3-1}{2} = \binom{17}{2} = 136$

7. a. 1260   **b.** 1,576,575   **c.** 495   **9.** $9!/(4! \cdot 2! \cdot 1! \cdot 2!) = 3780$

11. $\binom{5+3-1}{2} = \binom{7}{2} = 21$ **13.** $4^2 \binom{17+4-1}{3} \Big/ 4! = 760$

15. $\binom{10+4-1}{3} = \binom{13}{3} = 286$ **17. a.** 71   **b.** 16   **c.** 8

Exercises: Case Study 3A

1. *Initial assertion:* N is a natural number and A is a list of n real numbers. *Intermediate assertion:* Each time the value of I is incremented, the largest A_k, $1 \le k < I$, is stored in LARGE. *Correctness assertion:* The program will terminate and print the largest element in the list A.

5. a. $\{x: x \ge 130, x \in N\}$ **b.** $a - b \equiv 0 \bmod 2$

Exercises: Case Study 3B

5. a.

x	5							
e	0.01							
l	1		2			2.125	2.1875	2.21875
u	5	3		2.5	2.25			
m	3	2	2.5	2.25	2.125	2.1875	2.21875	2.23438
d	4	1	1.25	0.0625	0.48438	0.21484	0.07715	0.00757
$e \le 0$ or $x < 0$	no							
$x \ge 1$	yes							
$d > e$	yes	yes	yes	yes	yes	yes	yes	no
$m * m > x$	yes	no	yes	yes	no	no	no	

The square root of 5 with tolerance 0.01 is 2.23438.

CHAPTER 4

Exercises 4.1

1. a. $S = \{1, 2, 3, 4\}$, $P(1) = 1/3$, $P(2) = 1/6$, $P(3) = 1/3$, $P(4) = 1/6$

b. $A = \{2\}$ $B = \{4\}$

c. $A \cap B = \varnothing$: An even prime number greater than 3 comes up. There is no such number.

$A \cup B = \{2, 4\}$: An even prime or a number greater than 3 comes up.

$B' = \{1, 2, 3\}$: A number less than or equal to 3 comes up.

d. $P(A) = 1/6$, $P(B) = 1/6$, $P(A \cap B) = 0$, $P(A \cup B) = 1/3$, $P(B') = 5/6$

3. a. $S = \{0, 1, 2, 3, \ldots\}$. Record the number of cars passing through the intersection during the indicated time interval for a large number of days. The probability of simple event $\{n\}$ is the proportion of times exactly n cars passed through the intersection.
b. $F = \{0, 1, 2, \ldots, 49\} = \{x : 0 \le x \le 49, \ x \text{ an integer}\}$
$G = \{20, 21, 22, \ldots\} = \{x : x \ge 20, \ x \text{ an integer}\}$
c. $F \cup G = S$, the entire sample space. Either fewer than 50 or at least 20 cars is the entire sample space.
$F \cap G = \{x : 20 \le x \le 49\}$. At least 20 but fewer than 50 cars pass the intersection.
$F' = \{x : x \ge 50\}$. At least 50 cars pass the intersection.
$G' = \{x : x < 20\}$. Fewer than 20 cars pass the intersection.

5. a. 34/100   **b.** 79/100   **c.** 25/100   **d.** 79/100

7. a. $\binom{7}{4} \Big/ \binom{20}{4}$ **b.** $\binom{13}{4} \Big/ \binom{20}{4}$ **c.** $\binom{7}{2}\binom{13}{2} \Big/ \binom{20}{4}$ **d.** $1 - \binom{13}{4} \Big/ \binom{20}{4}$

e. $\left[\binom{13}{3}\binom{7}{1} + \binom{13}{4} \right] \Big/ \binom{20}{4}$

9. a. $(13)(48) \Big/ \binom{52}{5}$ **b.** $(1)(48) \Big/ \binom{52}{5}$ **c.** $40 \Big/ \binom{52}{5}$

Exercises 4.2

1. a. $P(A|B) = 1/2$ $P(B|A) = 2/11$ **b.** No
3. a. No   **b.** No   **c.** $P(A \cup D) = 0.02285$ $P(B \cup E) = 0.137828$
5. The probability of system failure is 0.3165.
7. If $F = \{\text{A student knows FORTRAN}\}$ and $P = \{\text{A student knows Pascal}\}$, then
   a. $P(F|P) = 0.222$ **b.** $P(P|F) = 0.4$ **c.** $P(F \cup P) = 0.6$

Exercises 4.3

1. The probability of two balls the same color is 20/72.
3. a. 3/8   **b.** 4/8   **c.** 6/8
5. a. 120/1024   **b.** 45/1024   **c.** 55/1024   **d.** 969/1024
7. The probability of at least one hit is 0.76.
9. a. 10/32   **b.** 0.367   **c.** 0.237

Exercises 4.4

1. a. 44.58   **b.** 298.5
3. a.

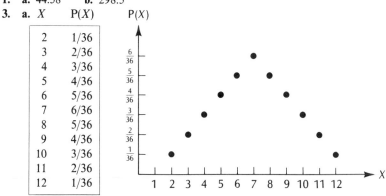

X	$P(X)$
2	1/36
3	2/36
4	3/36
5	4/36
6	5/36
7	6/36
8	5/36
9	4/36
10	3/36
11	2/36
12	1/36

b. Mean = 7, Variance = 5.83, Standard deviation = 2.41

5. a.

X	P(X)
+2	1/6
+3	1/3
−1	1/3
−4	1/6

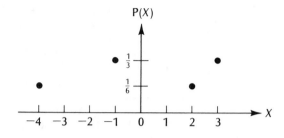

b. 1/3   **c.** $333.33

7. 100, 200, 300, 400, 500, 600, 500, 400, 300, 200, 100, respectively

Exercises: Case Study 4A

1. a. 41, 41, 41, 41, 41, 41, 41, 41, 41, 41, 41   **b.** 73, 78, 83, 88, 93, 98, 3, 8, 13, 18
   c. 61, 5551, 5141, 67831, 72621, 17511, 93501, 8591, 81781, 42071

CHAPTER 5

Exercises 5.1

1. a. $A \times B = \{(a, 1), (a, 2), (a, 3), (a, 4)\}$

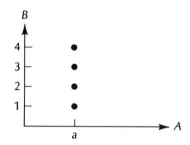

c. $B \times C = \{(1, x), (1, y), (2, x), (2, y), (3, x), (3, y), (4, x), (4, y)\}$

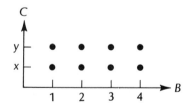

3. a. Yes   **b.** No, the transitive property does not hold.
   c. No, reflexivity and transitivity do not hold.

Exercises 5.2

1. a, b, and c are functions; d is not.
3. $f(-4) = -64, f(0) = 0, f(2) = 8; g(-3) = -20, g(0) = -2$
5. a. Domain = {All real numbers}, Range = $\{-1, 0, 1\}$
   b. SGN$(-11) = -1$,   SGN$(14) = +1$

7. INDEX("I", S) = 2, INDEX ("S", S) = 3, INDEX ("I", T) = 6, INDEX("U", S) = 0, INDEX ("U", T) = 2

9. The null string—that is, the string with no characters

11. **a.** Domain = $\{x : x \geq 1/4\}$ **b.** $f(1/2) = 1, f(3) = \sqrt{11}, f(9) = \sqrt{35}$

Exercises 5.3

1.
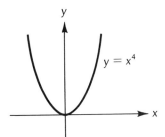

3.
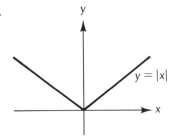

5.
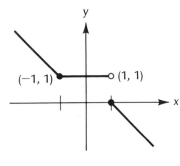

7.
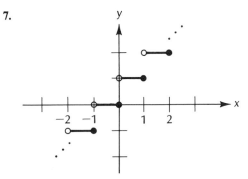

9. The function is constant in value.

11. **a.** No. If it did, the function would have two y values corresponding to $x = 0$.
     b. No. Two values of y would then correspond to one value of x.

Exercises 5.4

1. **a.**
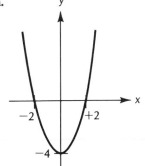

    b.
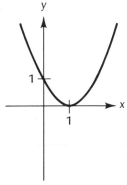

c.

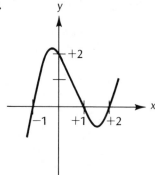

5. a. 5/7   **b.** ∞   **c.** 1   **d.** 0

7. a.

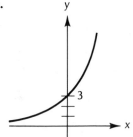

c.

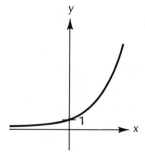

9. The limit is 1.

Exercises 5.5

1. a. $(f + g)(x) = 3x - 3$,   Domain = Range = {All real numbers}
   $(f - g)(x) = x + 3$,   Domain = Range = {All real numbers}
   $(fg)(x) = 2x^2 - 6x$,   Domain = {All real numbers}
                Range = $\{x : x \geq -9/2\}$
   $(f/g)(x) = 2x/(x - 3)$,   Domain = $\{x : x$ is a real number, $x \neq 3\}$
                Range = $\{x : x$ is a real number, $x \neq 2\}$
 c. $(f + g)(x) = x^2 + x^3$,   Domain = Range = {All real numbers}
   $(f - g)(x) = x^2 - x^3$,   Domain = Range = {All real numbers}
   $(fg)(x) = x^5$,   Domain = Range = {All real numbers}
   $(f/g)(x) = 1/x$,   Domain = Range = $\{x : x$ is a real number, $x \neq 0\}$

3. No   **5.** The constant function, $f(x) = c$

7. a. $(f \circ g)(x) = f(g(x)) = \sqrt{(x - 1)}$,   Domain = $\{x : x \geq 1\}$
 b. (ALPH ∘ LEN) (x) = ALPH(LEN(x)) maps the string x to a lowercase letter of the alphabet.

9. LEN ∘ FIRSTCHR(S) = $\begin{cases} 0 & \text{if } S = \text{the null string} \\ 1 & \text{otherwise} \end{cases}$

11. a. $f^{-1}(x) = (x + 1)/5$ **b.** $f^{-1}(x) = (x + 1)/(x - 1)$

Exercises: Case Study 5A

9. a. $A(2, 3) = 9$ **b.** 16

Exercises: Case Study 5B

5. $a < b$

Exercises: Case Study 5C

1. a. KD WYBX KD XYYX KD DGSVSQRD NSW
     WKBSK DRYE RKCD ROKBN WI RIWX
 b. QF UMND QF DMMD QF FGKLKSBF RKU
     UQNKQ FBMO BQWF BAQNR UY BYUD
3. IT WAS BEAUTY THAT KILLED THE BEAST
5. a. $x \equiv 3(y - 2) \bmod 26$
 b. WHAT LIGHT THROUGH YONDER WINDOW BREAKS

CHAPTER 6

Exercises 6.1

1. a. Scalar   **b.** Vector   **c.** Vector   **3.** $\begin{bmatrix} 10 & 33 & -17 & -57 \end{bmatrix}$

5. a. $\begin{bmatrix} -5 \\ 4 \\ 20 \end{bmatrix}$ **c.** $\begin{bmatrix} 29 \\ -48 \\ -60 \end{bmatrix}$ **e.** $\begin{bmatrix} 7 \\ 6 \\ -11 \end{bmatrix}$

7. a. $x = 3, \quad y = 10$ **c.** $x = 10, \quad y = 7$ **e.** $x = 3, \quad y = -5$

9. a. 450   **b.** 318   **c.** 961   **d.** 1069

11. a. $\begin{bmatrix} 10 & 10 & -12 \\ -2 & 2 & 16 \end{bmatrix}$ **b.** $\begin{bmatrix} -1 & -7 \\ -16 & 7 \\ -16 & 11 \end{bmatrix}$ **c.** Undefined

13. a. $\begin{bmatrix} -1 & -3 & -5 & -7 \\ 0 & -2 & -4 & -6 \end{bmatrix}$ **b.** $\begin{bmatrix} 0 & -1 & 0 \\ -1 & -6 & -9 \\ -2 & -11 & -18 \end{bmatrix}$

   c. $\begin{bmatrix} 1 & 2 \\ 2 & 2 \\ 3 & 3 \end{bmatrix}$ **d.** $\begin{bmatrix} 9 & 16 \\ 11 & 18 \\ 13 & 20 \end{bmatrix}$

Exercises 6.2

1. a. $\begin{bmatrix} -25 & -64 \\ 10 & -22 \end{bmatrix}$ **b.** $\begin{bmatrix} -25 & 166 \\ -3 & -6 \\ 23 & -62 \end{bmatrix}$ **c.** $\begin{bmatrix} 215 \end{bmatrix}$ **3.** 6×4

5. a. $\begin{bmatrix} 22 & 39 \\ -19 & 27 \end{bmatrix}$ **b.** $\begin{bmatrix} -40 & 79 & -118 & 157 \end{bmatrix}$

7. $AB = \begin{bmatrix} -2 & 13 \\ -10 & -7 \end{bmatrix}$ $BA = \begin{bmatrix} -10 & 7 \\ -22 & 1 \end{bmatrix}$

9. $\begin{bmatrix} 3 & 1 & -2 & 7 \\ 1 & -1 & 0 & 3 \\ -2 & 0 & 3 & -8 \end{bmatrix} \begin{bmatrix} x \\ y \\ z \\ w \end{bmatrix} = \begin{bmatrix} 12 \\ 18 \\ -4 \end{bmatrix}$

11. **a.** 1. $[(A^t)^t]_{ij} = [A^t]_{ji} = [A]_{ij}$ for each pair i, j
     2. $[(A + B)^t]_{ij} = [A + B]_{ji} = [A]_{ji} + [B]_{ji} = [A^t]_{ij} + [B^t]_{ij}$
     3. $[(kA)^t]_{ij} = [kA]_{ji} = k[A]_{ji} = k[A^t]_{ij}$

13. **a.** $[A + B]_{ij} = [A]_{ij} + [B]_{ij} = [A]_{ji} + [B]_{ji} = [A + B]_{ji}$
     b. $[C + C^t]_{ij} = [C]_{ij} + [C^t]_{ij} = [C]_{ij} + [C]_{ji} = [C]_{ji} + [C]_{ij}$
     $= [C]_{ji} + [C^t]_{ji} = [C + C^t]_{ji}$

15. **a.** Maximize $\mathbf{p} = \begin{bmatrix} 2 & 3 \end{bmatrix} \begin{bmatrix} x_1 \\ x_2 \end{bmatrix}$

   Subject to $\begin{bmatrix} x_1 \\ x_2 \end{bmatrix} \geq 0, \quad \begin{bmatrix} 5 & 2 \\ 7 & 6 \\ 0 & 1 \end{bmatrix} \begin{bmatrix} x_1 \\ x_2 \end{bmatrix} \leq \begin{bmatrix} 20 \\ 30 \\ 4 \end{bmatrix}$

   b. Minimize $\mathbf{c} = \begin{bmatrix} 20 & 30 & 4 \end{bmatrix} \begin{bmatrix} y_1 \\ y_2 \\ y_3 \end{bmatrix}$

   Subject to $\begin{bmatrix} y_1 \\ y_2 \\ y_3 \end{bmatrix} \geq 0, \quad \begin{bmatrix} 5 & 7 & 0 \\ 2 & 6 & 1 \end{bmatrix} \begin{bmatrix} y_1 \\ y_2 \\ y_3 \end{bmatrix} \geq \begin{bmatrix} 2 \\ 3 \end{bmatrix}$

Exercises 6.3

1. **a.** $x = 7, y = -4$ **b.** $x = -22/9, y = 26/27$ **c.** $x = -2, y = 2, z = -11$
    d. $x = 1, y = 3/2, z = -5/3$
3. **a.** Multiply to show $A^{-1}A = I$ **b.** Multiply to show $AB \neq I$
5. $AA^{-1} = (A^{-1})^{-1}A^{-1} = I$ and $A^{-1}A = A^{-1}(A^{-1})^{-1} = I$
7. **a.** $x = -17, y = -3/2, z = 89/2$ **b.** $x = -21, y = -5/2, z = 113/2$
    c. $x = 15, y = 17/2, z = 13/2$
9. **a.** 5   **b.** $-1/6$ **c.** -81 **d.** 30   **e.** -14 **f.** $123\frac{2}{3}$
11. **a.** $\begin{vmatrix} 3 & 1 & -1 \\ 2 & -1 & 2 \\ 0 & 0 & -1 \end{vmatrix}$ **b.** $(-1)^6 \begin{vmatrix} 3 & 0 & -1 \\ 1 & -1 & 0 \\ 2 & 0 & 2 \end{vmatrix} = -8$ **c.** -10 **d.** -10

    e. Row 4 and column 3, because they have the most 0s.
13. **a.** $x_1 = 3 - x_3, \quad x_2 = -1 + 2x_3, \quad x_3 = x_3$
15. **A** and **B** are square and **AB** exists, so **A** and **B** have the same size. By part 2 of Note 6.12, det (**AB**) = det (**A**) · det (**B**). Since the product is zero, one of its factors must be zero.
17. It must be either 1 or -1.
19. Yes, det (**AB**) = det (**A**) · det (**B**) = det (**B**) · det (**A**) = det (**BA**) when **A** and **B** are both $n \times n$. Otherwise, no.

Exercises: Case Study 6A

1. **a.** After three steps, one stack contains 3   8, and the other contains 1. There is no way to get 1 to precede 3 in either stack without violating the nondecreasing condition.
3. **b.** 31

5. a.

peg A	peg B	peg C
37	456	12
7	3456	12
17	3456	2
17	23456	
7	123456	
	123456	7

    b. Completing the puzzle after part a is equivalent to solving the puzzle with six hoops, which requires $2^6 - 1 = 63$ moves. Adding the 5 moves from part a, we see 68 moves are needed.

CHAPTER 7

Exercises 7.1

1. **a.** $3 + 6 = 6$ $10 + 15 = 30$ $5 + 1 = 5$
    b. $5 \cdot 15 = 5$ $2 \cdot 6 = 2$ $6 \cdot 15 = 3$
    c. $3' = 10$ $10' = 3$ $1' = 30$ $30 = 1$
    d. $2 + (3 \cdot 5) = 2 + 1 = 2$ $(15' + 6)' \cdot 3 = (2 + 6)' \cdot 3 = 6' \cdot 3 = 5 \cdot 3 = 1$
    e. Zero element $= 1$, Unit element $= 30$

3. **a.** $a + (b + c')$ **b.** $(a'b)'(a + b')$

5. **a.**

a	b	a'	$a' + b$	$a(a' + b)$	ab
0	0	1	1	0	0
0	1	1	1	0	0
1	0	0	0	0	0
1	1	0	1	1	1

    b.

a	b	c	b'	ab'	$ab' + c$	bc	$a + c$	$b'(a + c)$	$bc + b'(a + c)$
0	0	0	1	0	0	0	0	0	0
0	0	1	1	0	1	0	1	1	1
0	1	0	0	0	0	0	0	0	0
0	1	1	0	0	1	1	1	0	1
1	0	0	1	1	1	0	1	1	1
1	0	1	1	1	1	0	1	1	1
1	1	0	0	0	0	0	1	0	0
1	1	1	0	0	1	1	1	0	1

c.

a	b	$a + b$	b'	$a + b'$	$(a + b)(a + b')$
0	0	0	1	1	0
0	1	1	0	0	0
1	0	1	1	1	1
1	1	1	0	1	1

7. a. $x' + yz$ **b.** $y' + (x' + z')(x + z)$ **c.** $x'(x + y')$

Exercises 7.2

1. a. $f = x'y + xy, \quad f = (x + y)(x' + y)$
   b. $g = x'y'z' + x'yz' + x'yz + xyz'$
     $g = (x + y + z')(x' + y + z)(x' + y + z')(x' + y' + z')$
   c. $h = x'y'z' + x'y'z + xy'z' + xy'z$
     $h = (x + y' + z)(x + y' + z')(x' + y' + z)(x' + y' + z')$

3. a. They are equivalent. Both have complete sum-of-products form $x'y + xy'$.
   b. They are not equivalent. The expressions have the complete sum-of-products forms xy' and $x'y + xy'$, respectively.

Exercises 7.3

1. a. The cell with minterm $xy'z'$ has adjacent cells xyz', $x'y'z'$, and $xy'z$.
   The cell with minterm $x'y'z$ has adjacent cells $x'y'z'$, $xy'z$, and $x'yz$.
   b. The cell with minterm $xy'zw$ has adjacent cells $x'y'zw$, $xyzw$, $xy'z'w$, and $xy'zw'$.
   The cell with minterm $xy'z'w'$ has adjacent cells $x'y'z'w'$, $xyz'w'$, $xy'zw'$, and $xy'z'w$.
   The cell with minterm $x'yz'w$ has adjacent cells $xyz'w$, $x'y'z'w$, $x'yzw$, and $x'yz'w'$.

3. We show several basic rectangles in each case.

a.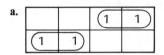

b.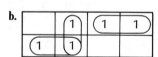

c.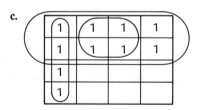

d.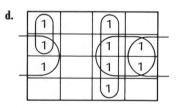

Exercises 7.4

1. a. $E = xz + x'y'$ **b.** $E = x'yz'$ **c.** $E = yz + xz$
   d. $E = xy + y'w + x'y'w' + x'y'z$

3. $f_1 = x'z' + xy'z$
   $f_2 = xyz + y'z' + x'y'$
   $f_3 = xz$

Exercises: Case Study 7A

1. **a.** $(xy' + z) + yz'$ **b.** $(x + y + z) + (xy')'$ **c.** $(xy)'(y' + z)'$
3. For f_1,

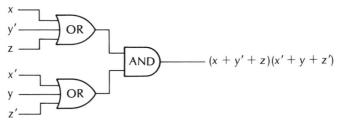

$(x + y' + z)(x' + y + z')$

For f_2,

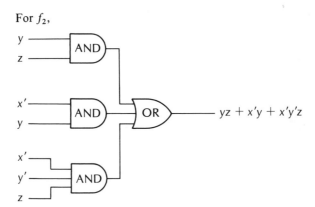

$yz + x'y + x'y'z$

For f_3,

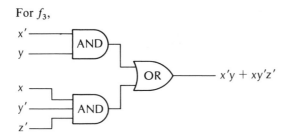

$x'y + xy'z'$

5. The minimal AND/OR network for f_1 is an OR/AND network. The equivalent NOR-only network is

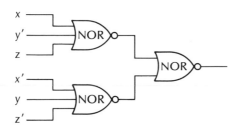

7. a.

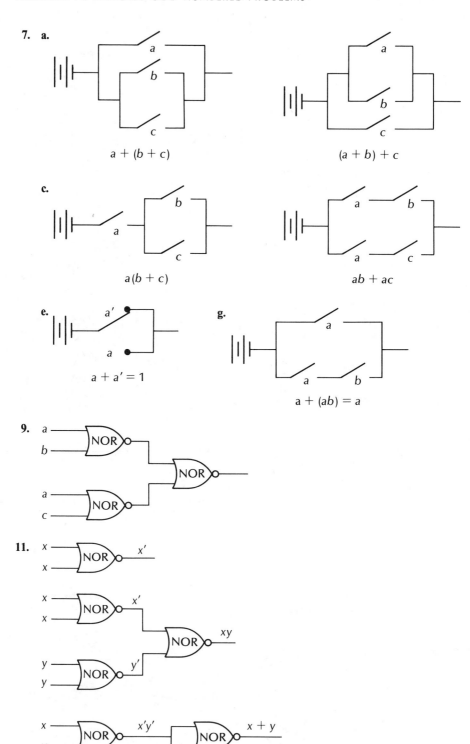

$a + (b + c)$

$(a + b) + c$

c.

$a(b + c)$

$ab + ac$

e.

$a + a' = 1$

g.

$a + (ab) = a$

9.

11.

13.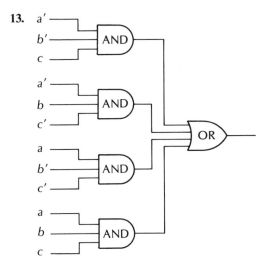

The three switches are a, b, and c. A value of 0 means the switch is off; 1 means the switch is on.

CHAPTER 8

Exercises 8.1

1. a. Each vertex corresponds to a city or town; each edge is a road joining a pair of cities.
   c. Each vertex represents a country; two countries are joined by an edge if they share a common border.

3. a.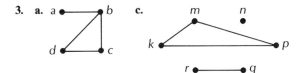

5. a. Each vertex represents a job. There is an arc directed from vertex a to vertex b if job a should be performed immediately before job b during the manufacturing process.
   c. Each vertex corresponds to an intersection. There is an arc from vertex a to vertex b if it is legal to drive directly from intersection a to intersection b without crossing another intersection.

7. a. Vertices f, j, i, and k are transmitters.
   c. Such a vertex is isolated; that is, it has degree 0.

9. a. Each edge corresponds to a pair of vertices. There are $\binom{n}{2}$ ways of selecting a pair of vertices from the set of n vertices. Thus, the maximum number of edges is $\binom{n}{2}$.

   (The result can also be proved by induction.)

Exercises 8.2

1. a. 3   **c.** 2   **3.** $n - 1$
5. a. They have different degree sequences.
   c. The vertices of degree 2 are adjacent in only one graph.

7.

Exercises 8.3

1. a.   **c.**

3.

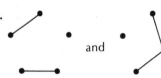

5.

7. a.

(Others are possible.)

9. a.   **d.**

11. a.

(Others are possible.)

c.

(Others are possible.)

13. a.

$W_4 =$ $\qquad W_6 =$

21. a.

 (Others are possible.)

Exercises 8.4

1. a. $A(G) = \begin{bmatrix} 0 & 1 & 0 & 1 & 1 \\ 1 & 0 & 1 & 0 & 1 \\ 0 & 1 & 0 & 1 & 1 \\ 1 & 0 & 1 & 0 & 1 \\ 1 & 1 & 1 & 1 & 0 \end{bmatrix}$ $\qquad A(H) = \begin{bmatrix} 0 & 1 & 0 & 0 & 0 \\ 1 & 0 & 1 & 0 & 1 \\ 0 & 1 & 0 & 1 & 0 \\ 0 & 0 & 1 & 0 & 0 \\ 0 & 1 & 0 & 0 & 0 \end{bmatrix}$

b. $[A(G)]^2 = \begin{bmatrix} 3 & 1 & 3 & 1 & 2 \\ 1 & 3 & 1 & 3 & 2 \\ 3 & 1 & 3 & 1 & 2 \\ 1 & 3 & 1 & 3 & 2 \\ 2 & 2 & 2 & 2 & 4 \end{bmatrix}$ $\qquad [A(G)]^3 = \begin{bmatrix} 4 & 8 & 4 & 8 & 8 \\ 8 & 4 & 8 & 4 & 8 \\ 4 & 8 & 4 & 8 & 8 \\ 8 & 4 & 8 & 4 & 8 \\ 8 & 8 & 8 & 8 & 8 \end{bmatrix}$

c. From the $(1, 5)$ entry of $[A(G)]^3$, we see that there are eight walks of length 3 from x_1 to x_5.

3. $D(G) = \begin{bmatrix} 0 & 1 & 2 & 1 & 1 \\ 1 & 0 & 1 & 2 & 1 \\ 2 & 1 & 0 & 1 & 1 \\ 1 & 2 & 1 & 0 & 1 \\ 1 & 1 & 1 & 1 & 0 \end{bmatrix}$ $\qquad D(H) = \begin{bmatrix} 0 & 1 & 2 & 3 & 2 \\ 1 & 0 & 1 & 0 & 1 \\ 2 & 1 & 0 & 1 & 2 \\ 3 & 0 & 1 & 0 & 3 \\ 2 & 1 & 2 & 3 & 0 \end{bmatrix}$

5. a. $B(L) = \begin{bmatrix} -1 & 1 & 1 & 0 & 0 & 0 & 0 & 0 \\ 1 & -1 & 0 & -1 & 1 & 0 & 0 & 0 \\ 0 & 0 & 0 & 0 & -1 & -1 & 1 & 0 \\ 0 & 0 & 0 & 0 & 0 & 1 & -1 & 1 \\ 0 & 0 & -1 & 1 & 0 & 0 & 0 & -1 \end{bmatrix}$

$B(M) = \begin{bmatrix} -1 & 1 & 0 & 0 \\ 0 & -1 & -1 & 0 \\ 0 & 0 & 1 & -1 \\ 1 & 0 & 0 & 1 \end{bmatrix}$

b. Each column has exactly one 1 and one -1.

7. b. $2(n_1 n_2 + n_1 n_3 + n_2 n_3)$ $\qquad$ **9.** The graphs for which each component is a cycle

13. $\mathbf{C}(G) = \begin{bmatrix} 2 & 4 & 5 & 0 \\ 1 & 3 & 5 & 0 \\ 2 & 4 & 5 & 0 \\ 1 & 3 & 5 & 0 \\ 1 & 2 & 3 & 4 \end{bmatrix}$ $\mathbf{C}(H) = \begin{bmatrix} 2 & 0 & 0 \\ 1 & 3 & 5 \\ 2 & 4 & 0 \\ 3 & 0 & 0 \\ 2 & 0 & 0 \end{bmatrix}$

Exercises 8.5

1. **a.** $\kappa(P_n) = 1$ **c.** $\kappa(K_n) = n - 1$
3. $\kappa(K_{m,n}) = \min\{m, n\}$ and $\lambda(K_{m,n}) = \min\{m, n\}$

5. **a.**

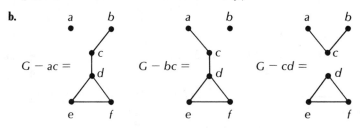

(Results for H and L are obtained similarly.)

b.

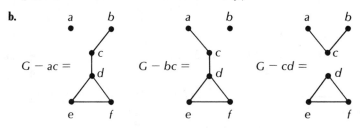

7. Choose any vertex v and direct all its incident edges toward v. No matter how the other edges are oriented, the resulting digraph will not be strongly connected.

9. **a.**   **b.**   (Others are possible.)

13. $\{a, h, i\}$ is a minimum-sized separating set. $\{uafv, uceiv, udhv\}$ is a maximum set of vertex-disjoint paths. (Other answers are possible.)

Exercises 8.6

1. **a.** Eulerian   **b.** Semi-Eulerian   **c.** Semi-Eulerian   **d.** Neither
3. **a.** Hamiltonian   **b.** Hamiltonian   **c.** Neither   **d.** Semi-Hamiltonian
5. **a.** Odd n **b.** $n = 1$ **c.** All n (≥ 3)   **d.** None
7. **a.** m and n both even   **b.** m and n both even
9. For G use $adbcea$, with total cost 16; for H use $fgjkhif$, with total cost 23.

Exercises: Case Study 8A

1. a.

3. a.

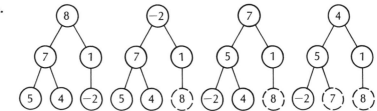

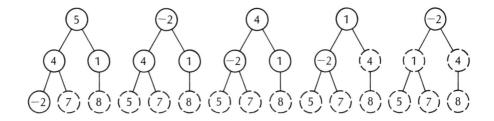

Exercises: Case Study 8B

1. a.

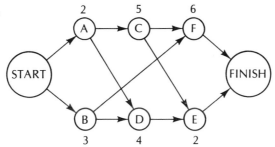

b.

Activity	EST	EFT	LST	LFT	Slack time
START	0	0	0	0	0
A	0	2	0	2	0
B	0	3	4	7	4
C	2	7	2	7	0
D	3	7	7	11	4
E	7	9	11	13	4
F	7	13	7	13	0
FINISH	13	13	13	13	0

c. Critical path is START, A, C, F, FINISH.

3. **a.**

Activity	EST	EFT	LST	LFT	Slack time
START	0	0	0	0	0
A	0	6	7	13	7
B	0	9	0	9	0
C	0	8	1	9	1
D	9	14	13	18	4
E	8	11	15	18	7
F	9	16	9	16	0
G	16	22	16	22	0
H	14	18	18	22	4
FINISH	22	22	22	22	0

b. START, B, F, G, FINISH. Completion time is 22.

7. **a.** If you follow around the cycle, the precedence relations would imply that an activity would have to precede itself, which is impossible.

b. The left-most activity does not immediately precede the right-most one if the top activity must be performed between the other two.

Index